Wireless Communications Research Trends

WIRELESS COMMUNICATIONS RESEARCH TRENDS

TONG S. LEE
EDITOR

Nova Science Publishers, Inc.
New York

For permission to use material from this book please contact us:
Telephone 631-231-7269; Fax 631-231-8175
Web Site: http://www.novapublishers.com

LIBRARY OF CONGRESS CATALOGING-IN-PUBLICATION DATA

Wireless communications research trends / Tong S. Lee (editor).
p. cm.
Includes index.
ISBN-13: 978-1-60021-674-9 (hardcover)
ISBN-10: 1-60021-674-9 (hardcover)
1. Wireless communication systems. I. Lee, Tong S.
TK5103.2.W5725 2007
621.384--dc22
2007010435

Published by Nova Science Publishers, Inc. ✢ New York

CONTENTS

Preface **vii**

Chapter 1 Topology Construction for Bootstrapping Peer-to-peer Systems over Ad-hoc Networks **1**
Wei Ding

Chapter 2 SARC: Secure Anonymous Routing for Cluster based MANET **55**
Lijun Qian, Ning Song and Xiangfang Li

Chapter 3 Trends and Challenges for Mobility Management in IP-based Next Generation Wireless Networks **83**
Christian Makaya and Samuel Pierre

Chapter 4 Energy Aware Medium Access Control Protocols **123**
Jagoba Arias, Itziar Marín and Aitzol Zuloaga

Chapter 5 Distributed Data Management In Sensor Networks **165**
Stefano Chessa, Francesco Nidito and Susanna Pelagatti

Chapter 6 On Reliability of Mobile Ad-hoc Wireless Networks **191**
Jason L. Cook and Jose Emmanuel Ramirez-Marquez

Chapter 7 MEMS Micro-Antennas for Wireless Biomedical Systems **229**
P. M. Mendes and J. H. Correia

Chapter 8 A Transport Layer Security Protocol for Hybrid Networks **267**
Nikos Komninos

Chapter 9 How can 3G and 2G Systems Cooperate? **287**
S-E. Elayoubi, L. Sartori and B. Fourestié

Chapter 10 Impact of Ultra Wide Band (UWB) on Macrocell Downlink of UMTS, CDMA-450, DCS-1800 and GSM-900 Systems **301**
Bazil Taha-Ahmed, Miguel Calvo-Ramón and Leandro Haro-Ariet

Chapter 11 Dependable Public Wireless LANs without Hardware Support **325**
Jenn-Wei Lin and Ming-Feng Yang

Chapter 12 An Efficient Mobile-Agent-Based Platform for Dynamic Service Provisioning in 3G/UMTS **349**
Yuan-Lin Ko, Kuochen Wang and Hung-Cheng Shih

Index **371**

PREFACE

The scope of this new and important book includes personal portable telephones, multimedia devices, digital assistants, and communicating palmtop computers; Registration and handoff protocols, messaging, and communications and computing requirements; Network control and management for protocols associated with routing and tracking of mobile users; Location-independent numbering plans for movable personal services; Personal profiles, personalized traffic filtering, and other database-driven aspects of personal communications; Link access technologies and protocols; Radio and infrared channel characterization and other microcell-based personal communication systems; Satellite Systems and Global Personal Communications; Traffic management and performance issues; Policy issues in spectrum allocation, industry structure, and technology evolution; Applications, case studies, and field experience; Intelligent vehicle highway systems.

Chapter 1 - As leaders of the decentralization movement, Mobile Ad-hoc Networks (MANETs) and Peer-to-Peer (P2P) systems are hot topics in networking community. Decentralization model puts ordinary network users on the driver's seat, gives them much more control, and stimulates their enthusiasm in active participation. It is believed that decentralization will replace the client/server model as dominant model in the new century.

MANETs and P2P systems share similar theoretical foundations. Both break away from the client/server model using multi-hop multicast. Both are dynamic, highly decentralized, self-organized, and self-healed. However, their levels of real world application are polar apart. Cachelogic reported that in January 2006 P2P traffic accounted for approximately 71% of all Internet traffic. On the other hand, only few MANETs applications have been commercially realized. Remarkable research initiatives in the synergy of P2P systems and MANETs have been sparked by this interesting phenomenon. While most focus on routing, the bootstrapping problem remains indispensable for the transplantation of successful approaches in P2P systems into MANETs.

The crucial problem in bootstrapping is topology construction in P2P overlay layer. In this chapter, a novel solution for this problem, i.e. the Ring Ad-hoc Network (RAN) protocol, is introduced. RAN builds effective rings in node ID space of the overlay layer for ring-based P2P systems like Chord, Pastry, and Virtual Ring Routing. With this ring, lengthy stabilization is absolutely unnecessary. It uses only neighbor-based multi-hop primitives. To best of author's knowledge, this is first successful attempt in the area.

Chapter 2 - Providing security and anonymity to users are critical in mobile ad hoc networks (MANETs). In this study, a cluster based security architecture is employed for better management of mobile users and scalability, and a Secure Anonymous Routing scheme for Cluster based MANET termed SARC is proposed. SARC includes intra-cluster routing and inter-cluster routing, where intra-cluster routing uses a common broadcast channel to provide anonymity, and inter-cluster routing uses a sequence of temporary public keys as the trapdoor information. One of the unique features of SARC is that critical network elements, such as the cluster heads, are hidden from adversaries during the routing process. In order to maximize routing efficiency, key indexing is used and symmetric cipher is employed in most part of the proposed scheme to reduce computational complexity. In addition, a technique based on XOR operations for data forwarding is applied to provide anonymity and maximize efficiency during data transmissions. The tradeoff between security/anonymity and efficiency is also addressed. Analytical results are derived using information theoretic measure for anonymity analysis of the proposed scheme. Detailed implementation of SARC is provided and extensive simulations are performed for a large (16 clusters, 800 nodes) network using OPNET. It is observed that SARC has good scalability and it introduces very limited overhead comparing to other cluster based routing protocol which has no security features. Route establish time and packet delivery ratio are also evaluated while taking into account node mobility. Both anonymity analysis (including sender anonymity, receiver anonymity and sender-receiver anonymity) and attack analysis show the effectiveness of SARC against a wide range of strong adversarial attacks.

Chapter 3 - A major trend in next generation wireless networks (NGWN) or fourth generation wireless networks (4G) is the coexistence of diverse but complementary architectures and wireless access technologies. This heterogeneity brings several design and deployment challenges. Among them, integration of existing heterogeneous wireless networks requires the design of efficient mobility management schemes, at IP layer as well as lower layer, to enable seamless roaming of users. IP technology is the best choice for interworking and integration of various radio access technologies and is the main drive of networks evolution towards all-IP core networks. IP mobility management is a crucial issue in heterogeneous mobile environments. Several IPv6-based mobility management schemes have been proposed and the most known of them is Mobile IPv6 (MIPv6). Despite some advantages, MIPv6 and its extensions are hindered by several shortcomings, such as handoff latency, signaling overhead, packet loss. Thus, they fail to fulfil requirements of real-time applications. Moreover, with growing demand for real-time and multimedia applications, quality of service (QoS) support in heterogeneous wireless networks is of primary importance in order to improve system performance and users' satisfaction. However, QoS provision mechanisms like IntServ/RSVP and DiffServ have been developed for wired networks and are not optimized for mobile and wireless environments. This chapter addresses mobility management issues and QoS provision in IP-based NGWN. The authors present recent techniques addressing these problems and discuss their limitations as well as outstanding challenges that still need to be addressed to motivate research activities for the design of efficient protocols and mechanisms in NGWN. Finally, performance analysis of IPv6-based mobility management protocols is provided to show their pros and cons.

Chapter 4 - In the last years, a great research effort has been done to reduce the power consumption of integrated circuits, specially microprocessors. The main advantage of this trend is the fact that battery operated devices may live longer with the same small size cells. However, the use of wireless technologies to transmit information is an energy consuming activity that may exhaust the batteries: the use of high frequency carriers and the need of emitting at least some milliwatts of RF signal consume a relatively large amount of energy that must be invested carefully to ensure that it is not wasted. This article describes which are the main sources of energy waste in medium access control protocols, analyzes the pros and cons of the wireless MAC protocols proposed so far and gives a series of directives to design new and more efficient protocols in the future. The result is an article that any researcher or engineer working with wireless battery operated devices must read in order to ensure that his or her devices are as long-lived as possible.

Chapter 5 - Wireless sensor networks (WSNs) are a recent technology designed for unattended, remote monitoring and control, which have been successfully employed in several applications. WSNs perform environmental data sampling and processing, and guarantee access of the processed data to remote users. In traditional WSN models these tasks consist in transmitting sensed data to a powerful node (the sink) which performs data analysis and storage. However these models resulted unsuitable to keep the pace with technological advances which granted to WSNs significant (although still limited) processing and storage capabilities. For this reason recent paradigms for WSN introduced *data base* approaches to define the tasks of data sampling and processing, and the concept of *data-centric storage* for efficient data access. In this paper, the authors revise the main research contributions on both sides and discuss their advantages with respect to traditional approaches.

Chapter 6 - The advancement of the self-forming, multi-hop Mobile Ad-hoc Wireless Networks (MAWN) created the need for new analysis methods which enable the accurate determination of the reliability and availability of these networked systems. Accordingly, a set of new and innovative methods has been developed and further research is on-going. The need for these methods is because contrary to hardwired networks, the MAWN is a scalable network without infrastructure. Along this line, the MAWN's configuration forms dynamically and probabilistically and as such no singular graphical depiction or mathematical function is able to describe its reliability. It is this feature that precludes the use of traditional methods. These new analytical methods are progressing in parallel with the proliferation of this technology so that reliable performance may be realized. Due to its flexibility, this new network scheme is often deployed in critical applications such as military and first responder applications. In such cases, reliability becomes a paramount system attribute. The primary contribution of this method is to describe and summarize the published methods in a single location. Taken together the tools are now available to a practitioner to analyze and optimize MAWN reliability. Mostly, this work is motivated by the application of MAWN technology in the DoD tactical networks and the import of their reliable operation when employed for this use. Throughout the chapter, DoD network examples will be used to demonstrate the reliability methods developed for the MAWN. The methods include both closed form analysis and Monte Carlo simulation techniques to establish terminal-pair reliability of the MAWN under a random waypoint mobility model; the metrics include two-terminal, k-terminal, and all-terminal reliability.

Chapter 7 - Invasive and implantable biomedical devices used for diagnostic and therapy, ranging from neural prosthesis to video-capsule endoscopy (VCE) systems, are emerging innovative technologies and they are expected to originate significant business activity in the near future. The success of such systems is in part due to the advent of microtechnologies, which made possible the miniaturization of several sensors and actuators, as well their integration with readout and communication electronics.

The new biomedical devices offer the possibility of improved quality of life, as well cost savings associated with health care services. However, one open challenging is to communicate to and from a biomedical device placed inside the human body with devices outside the human body. The lack of antennas, small enough to be integrated with the sensing microsystem, is a difficult task to overcome because such communications must be made at relatively low frequencies, due to live tissue signal attenuation. The straightforward solution is to increase the devices size to dimensions where it becomes possible to integrate an antenna. Up to now solutions, use conventional antennas together with miniaturization techniques to achieve the smallest antennas possible. However, the size of such devices is usually limited by the antenna and, is some cases, also by the batteries size.

Micro-Electro-Mechanical Systems (MEMS) are becoming an available option for RF communication systems since they can offer, simultaneously, devices with improved performance and they use IC-compatible materials, allowing their integration in a silicon chip, side by side with semiconductor circuits. Up to now, MEMS have been used for antenna applications to obtain non-conventional front-ends with improved, or new characteristics. However, some preliminary tests have shown that some MEMS structures could have the ability to operate as an antenna itself and this solution would have the potential to be smaller than the conventional antennas.

In this chapter, it is first discussed the need for small wireless biomedical devices. This requires the use of a microsystem completely integrated, from sensors to communications, thus requiring the use of integrated antennas. The electrical properties of substrates available in integrated circuit technology are very important for antenna design and one method used to characterize wafer materials is presented. Moreover, the antenna integration requires the availability of an electrically small antenna fabricated on materials compatible with the fabrication of integrated circuits. This integration requires the use MEMS techniques, like micromachining and wafer level packaging.

Finally, MEMS structures previously used for non-conventional front-ends will be introduced and investigated, having in mind a new application, the MEMS structure itself will be operating as an antenna. The development of new integrated antennas using MEMS solutions has the potential to make the devices smaller and more reliable, which will make them cheaper and adequate for mass production, resulting in a key advantage for competitors in the RF market. Also, the availability of smaller biomedical wireless devices can lead to new applications not yet fully envisioned. The new solutions envisions power saving, smaller volume, lower cost, and increased system lifetime, which are very important features in biomedical microsystems for diagnosis and therapy.

Chapter 8 - One of the key enablers for business applications in future mobile communication systems is the ability to set up secure channels across the Internet and mobile networks. In this paper, a hybrid transport layer security protocol (HTLS) is described, which sets-up secure channels across different networks, such as the Internet, Bluetooth, and UMTS, using a single protocol. HTLS's sub-protocols and its unique features are explained versos the features of well known security protocols, such as TLS and WTLS, at the OSI transport layer. A comparison of the *i*mplementation results is also presented.

Chapter 9 - In this chapter, the authors develop Markovian models to study the dynamics of real time and elastic calls in a cell served by UMTS/HSDPA and GSM/EDGE systems. The authors first present analytical models for interference and throughputs in GERAN and UTRAN. They then consider different strategies of Joint Radio Resource Management (JRRM) with or without inter-system vertical handovers and show how to calculate the steady-state probabilities and the performance measures (blocking probabilities, mean sojourn times, loads). The authors numerical results compare the different JRRM strategies and show that the best performance is obtained with the strategy where several vertical handovers are allowed all over the communication in order to continuously choose the best system.

Chapter 10 - The popularity of wireless networks makes interference and cross-talk between multiple systems inevitable. This chapter describes techniques for quantifying the effect of the UWB system on the second and third generation mobile communications systems.

Ultra-wideband (UWB) radio signals have characteristics that are different from conventional radios. Of special interest is the ability to spread the transmission power over a sufficiently wide bandwidth to make the signal appear as noise to a narrowband receiver, while still being able to transmit very high data rates over short distances. In this context "narrowband" may actually mean 20 MHz Wide. Ultra Wideband was traditionally accepted as impulse radio, but the FCC and ITU-R now define UWB in terms of a transmission from an antenna for which the emitted signal bandwidth exceeds the lesser of 500 MHz or 20% bandwidth. Thus, pulse-based systems—wherein each transmitted pulse instantaneously occupies a UWB bandwidth, or an aggregation of at least 500 MHz worth of narrow band carriers, for example in orthogonal frequency-division multiplexing (OFDM) fashion—can gain access to the UWB spectrum under the rules. Pulse repetition rates may be either low or very high. Pulse-based radars and imaging systems tend to use low repetition rates, typically in the range of 1 to 10 megapulses per second. On the other hand, communications systems favor high repetition rates, typically in the range of 1 to 2 gigapulses per second, thus enabling short-range gigabit-per-second communications systems. Each pulse in a pulse-based UWB system occupies the entire UWB bandwidth, thus reaping the benefits of relative immunity to multipath fading (but not to intersymbol interference), unlike carrier-based systems that are subject to both deep fades and intersymbol interference.

The aim of this chapter is to present the effect of UWB on UMTS, CDMA-450 , DCS-1800 and GSM-900 on the urban macrocell downlink performance (range and capacity) for a critical distance (distance between the UWB transmitter and the mobile receiver under study) of 1m.

Chapter 11 - This paper presents an efficient fault-tolerant approach for public wireless local access networks (public WLANs). In a public WLAN, multiple access points (APs) are first deployed in the public area to provide wireless communication. For a user in the public WLAN, it must associate with an AP to acquire a wireless communication path before performing data services. If a failure occurs in an AP of the public WLAN, the users under the coverage range of the faulty AP (the affected users) cannot perform data services again. To tolerate the AP failure, previous approaches are based on the hardware redundancy or network planning technique. In this paper, the authors proposed a new fault-tolerant approach which directs each affected user how to move itself to the coverage range of another AP. For quickly reconnecting the wireless communication, the moving distance is considered in the proposed approach. In addition, the proposed approach also considers the overloading problem to avoid causing significant performance degradation on an AP. Finally, extensive simulations are performed to evaluate the performance overhead of the proposed approach.

Chapter 12 - An important key concept of the Virtual Home Environment (VHE) is dynamic service provisioning. In 3G/B3G, the mobile network will have such a capability. The users can dynamically subscribe new services anytime, and the system operator or service provider can dynamically provide services to subscribed users immediately. Based on the UMTS CAMEL (Customized Applications for Mobile Enhanced Logic) architecture, the authors propose an efficient mobile-agent–based platform to provide services dynamically, which can greatly reduce signaling traffic. To demonstrate the efficiency of the authors' platform, the authors used the operations of incoming and outgoing calls to illustrate the operation of mobile agents. In an existing approach, a CORBA agent-based platform was deployed in a distributed processing environment, and it requires a standard, OMG Mobile Agent System Interoperability Facility (MASIF), to be interoperable between agent environments of different vendors or operators. However, there are some problems in this approach, such as problems in the aspects of security and performance. Analysis results have shown that the signaling traffic in the authors CAMEL mobile-agent-based platform can be reduced 40% compared to that in the CORBA agent-based platform. The authors' platform can provide efficient mobility management, and enhance network performance, security and interoperability.

In: Wireless Communications Research Trends
Editor: Tong S. Lee, pp. 1-53
ISBN: 978-1-60021-674-9

Chapter 1

TOPOLOGY CONSTRUCTION FOR BOOTSTRAPPING PEER-TO-PEER SYSTEMS OVER AD-HOC NETWORKS

Wei Ding
University of Maine at Fort Kent

Abstract

As leaders of the decentralization movement, Mobile Ad-hoc Networks (MANETs) and Peer-to-Peer (P2P) systems are hot topics in networking community. Decentralization model puts ordinary network users on the driver's seat, gives them much more control, and stimulates their enthusiasm in active participation. It is believed that decentralization will replace the client/server model as dominant model in the new century.

MANETs and P2P systems share similar theoretical foundations. Both break away from the client/server model using multi-hop multicast. Both are dynamic, highly decentralized, self-organized, and self-healed. However, their levels of real world application are polar apart. Cachelogic reported that in January 2006 P2P traffic accounted for approximately 71% of all Internet traffic. On the other hand, only few MANETs applications have been commercially realized. Remarkable research initiatives in the synergy of P2P systems and MANETs have been sparked by this interesting phenomenon. While most focus on routing, the bootstrapping problem remains indispensable for the transplantation of successful approaches in P2P systems into MANETs.

The crucial problem in bootstrapping is topology construction in P2P overlay layer. In this chapter, a novel solution for this problem, i.e. the Ring Ad-hoc Network (RAN) protocol, is introduced. RAN builds effective rings in node ID space of the overlay layer for ring-based P2P systems like Chord, Pastry, and Virtual Ring Routing. With this ring, lengthy stabilization is absolutely unnecessary. It uses only neighbor-based multi-hop primitives. To best of author's knowledge, this is first successful attempt in the area.

1. Introduction

1.1. Peer-to-Peer Systems and Mobile Ad-hoc Networks

As shown dramatically by YouTube, the power of decentralization is irresistible. In most areas, it is only a matter of time for the decentralization model to replace the prevalent client/server model.

Although the decentralization model is powerful, not every technology of decentralization is successful. For instance, peer-to-peer (P2P) systems and mobile ad-hoc networks (MANETs), two leading technologies in decentralization, are divergent in their commercialization. In case of P2P systems, welcomed applications succeeded in the market, gained popularity, and attracted active research. Research in turn brought in better applications. This is a virtuous cycle. In the case of MANETs, theoretical research dominated the area for more than a decade, but very little has been transfer into commercial application, if we do not include sensor networks as MANETs. Virtually no application has been widely used in real world except Bluetooth-based MANETs.

P2P systems and MANETs share fundamental homogeneity in many aspects. For example, both communicate by multi-hop messaging. Both are characterized by the absence of network infrastructure. Both are against the framework of central controller and instead rely upon self organization. These inherent similarities imply promising probability of successful synergy and transplantation.

The term "P2P system" is used throughout this chapter. However, it is not semantically differentiated with "P2P network."

1.2. Bootstrapping P2P Systems over MANETs

In the synergy of P2P systems and MANETs, a trend has been seen in transplanting achievement in P2P systems into MANETs. In this direction, majority of research has focused on issues of stable status, especially routing. Transplantation hence concentrated on layer substitution which match layer model in wired IP networks to layer model of MANETs. [HGRW2006, LLS2004, HPD2003, PDH2004]

Very limited research has been done in exploring the bootstrapping. Bootstrapping has been largely circumvented using unrealistic assumptions. This has been a repeated characteristic in the research of P2P systems over wired networks. [RD2001, SMKKB2001, CCNOR2006]

Bootstrapping includes two major tasks. The first task is automatic nodes address configuration. If we follow the traditional layer model of MANETs and keep the stiff separation between layers, we need two configurations: the lower in the networking layer and the higher in overlay or application layer. The second task is setting up overlay topology. This chapter focuses on the second task.

In computer networks, topology is frequently used to define qualitative geographic relationships, such as "which node is directly connected to which node," or "which node is neighbor of which node." Certain type of structured P2P systems imposes particular topologies among nodes to form a specific global structure. For a structured P2P system, overlay topology is crucial to its functionality. It lays foundation for other functions like routing, resource sharing, advertising, looking up, retrieval, and data dissemination. It is one of dominant factors that affect primary performance parameters such as efficiency, robustness, scalability, and feasibility. In fact, topology has broader functionality. For example, Jelasity and Babaoglu [JB2005] have shown that problems such as clustering and sorting can be transformed into topology problems and be solved by specific topology construction.

1.3. Current Status

There are two major approaches for bootstrapping a structured P2P system over wired networks. One is to jumpstart a network from one or a few predefined nodes, in which the common way to expand the network is node joining. In wired networks, many structured P2P systems require manual creation of a "seed" network in bootstrapping. Nodes have to be booted one by one in a slow, linear manner, which costs long time. In addition, the jumpstarted network often needs extra long time for stabilization before normal routing could work. In another approach all nodes cooperate concurrently to construct an overlay topology. This approach is distributed and decentralized. The concurrency makes it much faster than joining approach. Furthermore the P2P system could advance to normal working status immediately after bootstrapping.

Remarkable advance has been recently made in topology construction in wired networks. Topology generators can construct topologies such as line, ring, mesh, star, and tree. Generic generator, which could construct any topology if given a mathematical expression, is already available. [JB2005] However, no such construction tool has been reported in MANETs.

Many problems remain unsolved in this area. In wired networks, existing protocols for topology construction and maintenance are usually based upon unrealistic assumptions. Most of them assume the existence of a specific initial topology. Some protocols demand that the network remains in an ideal topology all the time as a necessary condition for normal operations. The second problem is the ignorance of network merger and partition. Some systems even require each node keep and monitor global state of the entire network. Another problem is: many schemes for topology construction still employ centralized strategy. Some require central coordinators; some follow a network-wide top-down view in protocol design. Some have deficiencies in fault-tolerance and recovery. For structured P2P systems over MANETs, some approaches can not keep up with the rate of change.

1.4. RAN – A New Solution

In this chapter, the Ring Ad-hoc Network (RAN) protocol is introduced. It builds a ring topology in P2P ID space. To best of author's knowledge, it is the first successful attempt in topology construction for MANETs. The ring is used to bootstrap ring-based P2P systems, such as Chord [SMKKB2001, DBKKMSB2001], Pastry [RD2001], and Virtual Ring Routing [CCNOR2006], over MANETs. On this ring, ring-based P2P systems could be put into normal operation immediately without lengthy stabilization. As in many literatures, Chord is used in demonstration. Three patterns are tested in RAN: distributed exhaustive pattern, virtual centralized exhaustive pattern, and random pattern. Simulation shows that the distributed exhaustive pattern has best overall performance. So this pattern is essentially representative of RAN.

The rest of this chapter is organized as follows. Section 2 introduces P2P systems, especially Chord. Section 3 depicts three previous research projects — T-Man, T-Chord, and Ring Network. They are most successful approaches for Chord ring construction in wired networks. Section 4 describes P2P systems in MANETs. Section 5 outlines the RAN protocol suite. Section 6 gives algorithms of RAN in AP notation. Simulation results are given in Section 7. Section 8 concludes the chapter.

2. Peer-to-Peer Systems

2.1. Peer-to-Peer Paradigm

P2P overlay systems provide fast, accurate, and scalable resource discovery, resource sharing, and storage services without a central controller. The concept of P2P systems first appeared in mid 1990s. As file sharing platforms, especially to distribute music over Internet, P2P systems became a hot topic in the late 1990s. A traditional P2P system is built upon IP. It uses IP as the communication platform. An IP capable host can reach anything attached to the Internet or other IP networks like IEEE 802 family by an IP address. However, IP layer could not tell a host how and where to find given content or another host. This is done by P2P overlay systems. The basic task of P2P overlay systems is to connect to other peers and find out interesting content.

P2P systems are distributed and self-organized. A host is called peer, because all hosts usually have same status, share same responsibility, and the relationship among them is characterized by equality. Unused bandwidth, storage, CPU cycles are shared among peers. Peers enjoy great freedom and privacy. Usually consumers are also producers, so aggregate resources grow exponentially with utilization. P2P systems have excellent fault tolerance, because there is no single point of failure in a P2P system.

The emergence of P2P systems was a revolution against long time dominance of client/server model in computing and communication. In the client/server model, powerful, reliable servers provide data and services. Clients request data and services from servers. The client/server model has proved extremely successful by its famous offspring, such as World Wide Web, database systems, and FTP. However, it has following inherent defects:

- need central controller
- dictation in which clients look like slaves
- presents a single point of failure
- unused resources through out the network
- poor scalability

2.2. Peer-to-Peer Systems

P2P systems address above defects of client/server model. At large P2P computing aims at sharing and exchanging resources and services between terminals. These resources and services include information (file or data structure), CPU cycles, storage (memory, cache, and disk), I/O devices, etc. P2P paradigm takes advantage of superfluous computing capacity, storage, and network bandwidth, so end users can unite and leverage their collective power to carry out huge task or achieve mutual benefits.

In a P2P system, all nodes are clients, servers, and routers at same time. All provide and consume data and services. No centralized data source endangers the system as the single point of failure. Nodes collaborate directly with each other. Any node can initiate a connection. All nodes are totally free: they may enter and leave the network arbitrarily and frequently. It will be "the ultimate form of democracy on the Internet" as well as "the ultimate threat to copy-right protection on the Internet." [Kaashoek2003]

P2P systems have following advantages: [Muthusamy2003]

- Efficient use of resources
- Unused bandwidth, storage, CPU cycles at the edge of the network become available to any user
- Scalability
  - Consumers of resources also donate resources. If remarkable consumers turn into producer, aggregate resources will grow with utilization.
  - Self-scaling
- Reliability
  - No single point of failure
  - Geographic distribution
  - Replicas
  - Built-in fault tolerance
  - Fault tolerance
- Easy administration
  - Nodes self organizing
  - No need to deploy servers
  - Load balancing

Besides file sharing, P2P paradigm could be applied in collaborative Internet (e.g. ICQ, shared whiteboard), distributed computing and grid computing (e.g. UC Berkley Seti@Home Project), multiplayer network games (e.g. Doom) and many other fields. However, P2P systems, especially those for file sharing, remain to be the oldest and most sophisticated P2P application. In a typical file sharing network, a user makes files (music, video, etc.) on her computer available to others. Then another user connects to the network, searches for the files, finds the first user's computer, and downloads files directly from first user's computer.

P2P systems fall into two categories: unstructured P2P systems and structured P2P systems. [Muthusamy2003] An unstructured P2P system does not have a fixed topology for routing. By the existence of central index servers, unstructured P2P systems are divided into three subgroups: centralized with a central index server, like Napster; semi-centralized with local index servers, like KaZaA; decentralized without any index server, like Gnutella. [Clip2, Ivkovic2001]

Structured P2P systems use fixed topologies like ring or grid for routing. They impose specific local relationships between peers, which finally generate global structures. These topology structures can be used for efficient data placement, search, and retrieval. They have guaranteed scalability — hops in routing is not linear with number of nodes. Most of them could reach the logarithm. They are self-organized, fault-tolerant, and they support load balancing. Structured P2P systems are usually implemented via Distributed Hash Table (DHT). Typical systems include Chord [DBKKMSB2001, SMKKB2001], Pastry [RD2001], CAN [RFHKS2001], BitTorrent [Cohen2003I, Cohen2003B], and Virtual Ring [CCNOR2006].

2.3. Unstructured P2P Systems

Napster was devoted to sharing music files on Internet. Providers upload their list of files and IP addresses to Napster server. Downloaders send queries to Napster server for files of their interest in the format of keyword search. Keywords could be artist, song, album, even bit rate. Napster server replies with IP address of users with matching files. Downloaders connect directly to the provider's computer to download file. Using a central directory/index server and a central query database, Napster guarantees correct results. At same time, the central server forms a single point of failure and bottleneck for scalability. Napster is Susceptible to denial of service attack and mischief from malicious users.

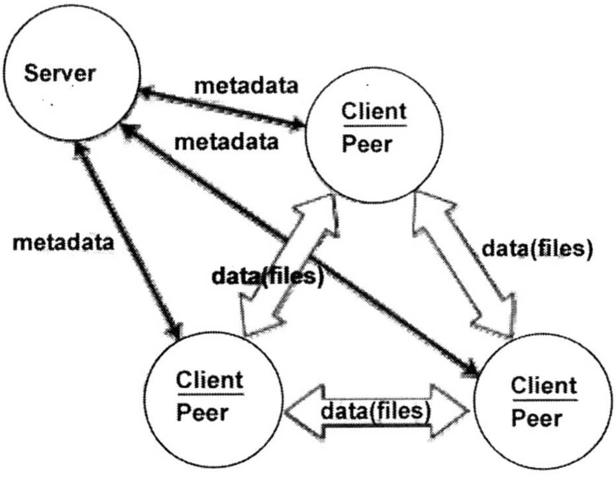

Figure 1. Centralized architecture of Napster

Gnutella enables sharing any type of files, not just MP3. It employs decentralized search. In Gnutella a user A asks her neighbors for files of interest, those neighbors ask their neighbors, and so on. Finally either users with matching files reply to A's query, or the packet is destroyed after a preset Time To Live (TTL). Each message has a parameter which sets the max number of hops the packet can "live". Search is distributed by the means of queries flooding. Comparing to Napster, Gnutella is decentralized and robust to denial of service attacks, for it has no single point of failure. Nevertheless, it can not guarantee correct results for every query. Gnutella is still not scalable. [Clip2, Ivkovic2001]

KaZaA is a hybrid of centralized and decentralized structures, where super-peers act as local central nodes and local search hubs. Each super-peer is similar to a Napster server in a smaller scale. Super-peers are automatically chosen by the system based on their capacities (storage, bandwidth, etc.) and availability (connection time). Users upload their list of files to a super-peer, which periodically exchange file lists with neighbor super-peers. When the query reaches a super-peer for files of interest, the file is transferred back to requesting node following the reverse path. [Muthusamy2003]

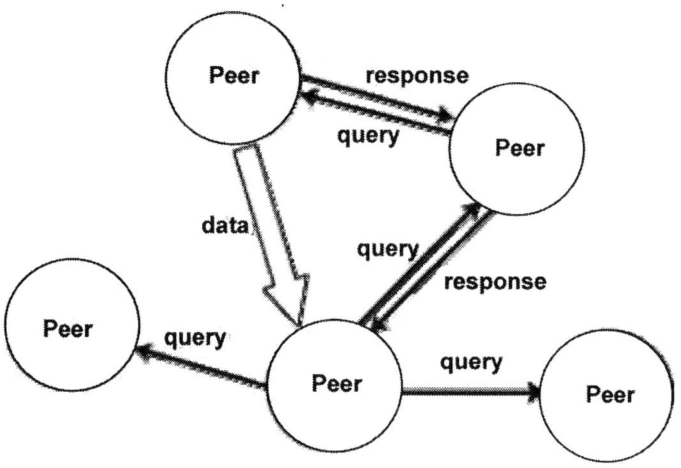

Figure 2. Flooding style search in Gnutella

2.4. Structured P2P Systems

Structured P2P systems are also called second generation P2P overlay networks. [Muthusamy2003] They are self-organized and fault-tolerant with balanced load. Scalability is guaranteed on numbers of hops to answer a query. One frequently cited difference with unstructured P2P systems their DHT interface.

A DHT stores (key, value) pairs. Each peer stores a subset of (key, value) pairs. Core functions of DHT API include insert, lookup, and delete. Insert function stores a (key, value) pair at the node responsible for the key. Lookup function returns value associated with a key from the host peer of the pair. Basic operation is to find node responsible for a key. A key need to be mapped to a node before insert, lookup, or delete functions could be used for this node. DHT maps Keys evenly to all nodes in the network. Each node maintains information about only a few other nodes. Messages can be routed to a node efficiently. Arrival or departure of one node only affects a few nodes.

Many services can be built on top of a DHT interface, like file sharing, archival storage, databases, naming, service discovery, chat, rendezvous-based communication, publish and subscribe. There are several implementations of DHT generic interface, for instance, Chord from MIT, Pastry from Microsoft Research in UK and Rice University, Tapestry from UC Berkeley, Content Addressable Network (CAN) also from UC Berkeley, SkipNet from Microsoft Research and University of Washington, Kademlia from New York University, Viceroy from Israel government and UC Berkeley, P-Grid from EPFL in Switzerland, Freenet developed by Ian Clarke. These systems are also called P2P routing substrates.

Routing in Chord is based upon a ring, on which nodes are organized according to their node IDs. Keys are assigned to their successor node in the ring. The consistent hash function ensures even distribution of nodes and keys on the ring. In a system with N nodes and K keys, lookups are resolved with $O(\log N)$ hops as well.

Pastry has a similar interface to Chord, however, it has good network locality to minimize hop traveling distance. To achieve locality new node needs to know a nearby node. Each routing hop matches the target identifier by one more digit. There are many choices in each hop, called possible locality.

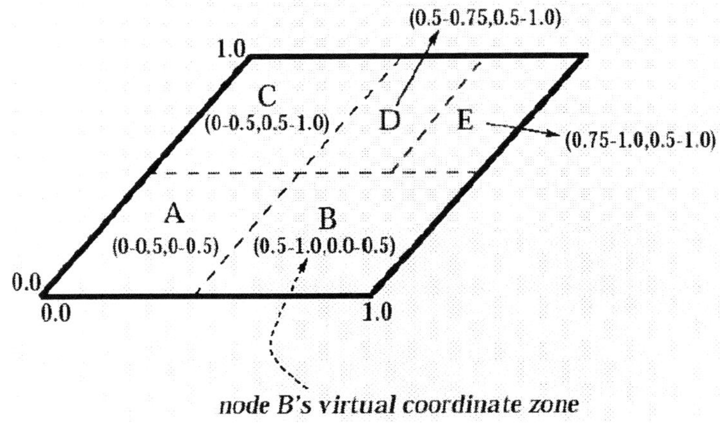

Figure 3. CAN network with 5 nodes in 2-d space

CAN uses a d-dimensional Cartesian coordinate space on a d-torus. Each node occupies a distinct zone in the space. Each key is hashed to a point in the space.

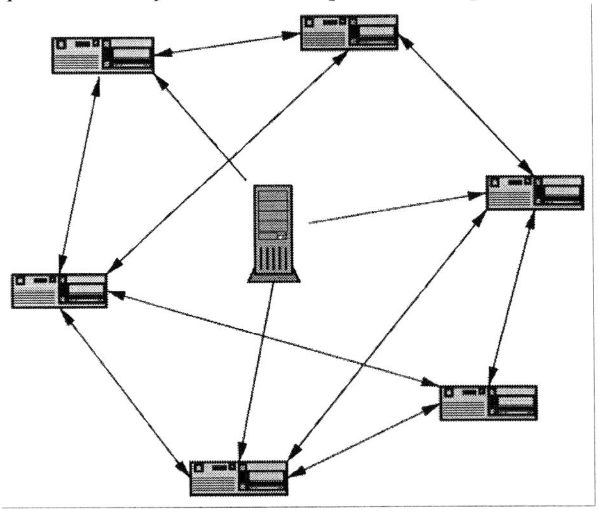

Figure 4. BitTorrent nodes upload pieces of common file to each other

BitTorrent has a highly connected ring topology with a center, like a bike wheel. BitTorrent uses economic methods in file sharing. It is faster and more reliable than most P2P approaches. BitTorrent forces concurrent downloaders of a same file to share the cost of upload. By using BitTorrent, they have to upload pieces of the file to each other. [Cohen2003I, Cohen2003B]

2.5. Chord

2.5.1. Consistent Hashing

Chord employs consistent hashing to assign ID to nodes and keys. The consistent hashing uses SHA-1 cryptographical hash as its base hash function. The compositive effect of two functions provides fast distributed hash computation. The consistent hashing has three attractive idiosyncrasies.

First, like other DHT, consistent hashing helps routing in Chord remain scalable to network size, that is, node number in the network. Unlike many proactive routing algorithm, Chord does not need its nodes keep tracking of every other node. A Chord node just need track $O(\log N)$ other nodes in its finger table. Each node resolves the hash function by communicating with other nodes. A lookup search for a key in Chord DHT only requires $O(\log N)$ messages to be exchanged.

Second, it has superb load balancing and map keys evenly to nodes with uniform random distribution. This character is very important to Chord's success. It provides solid foundation for Chord's scalability, that is, the scalability to base. Many calculations in Chord involve modular operation. The scalability to base makes Chord calculations independent of base. No matter how big a base you chose, this feature will keep Chord at similar performance level.

Third, consistent hashing is very stable. With help of consistent hashing, Chord could smoothly absorb disturbance from joining and ungraceful leaving (leaving without handling problems arising from the leave). In Chord ID space, a joining or leaving node only affects $O(1/N)$ existing keys in network which need move to other nodes to maintain the network-wide load balance. This is almost theoretical optimum.

2.5.2. Routing in Chord

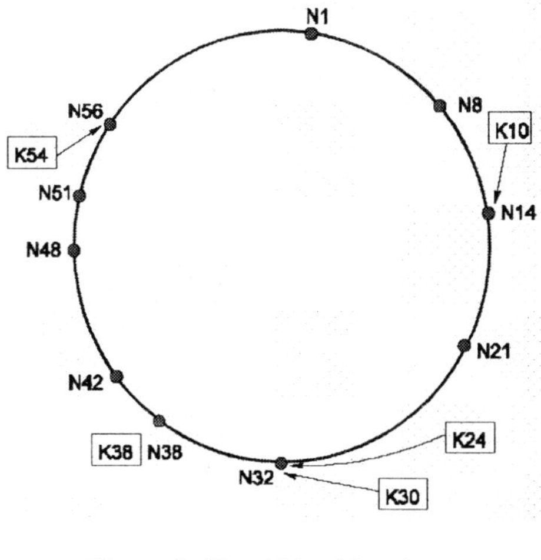

Figure 5. Chord identifier ring.

Routing in Chord is implemented by Chord identifier ring, as shown in Figure 5 and 6, on which nodes are organized according to node IDs. Keys are assigned to their successor node in the ring. The Hash function ensures even distribution of nodes and keys. In an *O* (log *N*) size Chord finger table associated with an *N* size node set, *i*th finger points to the first node that succeeds *n* by at least 2^{i-1}.

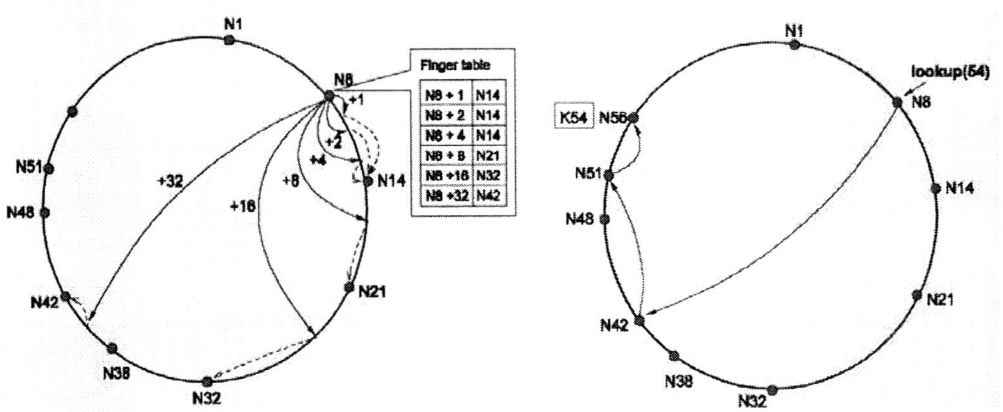

Figure 6. Looking up a key in Chord

To look up a key n, we first locate the furthest node that precedes the key in the finger table. Chord queries could find the target's home address in *O* (log *N*) hops. In a system with *N* nodes and *K* keys, with high probability, each node receives at most *K*/*N* keys. Each node maintains information about *O* (log *N*) other nodes. And lookups are resolved with *O* (log *N*) hops. However, the efficiency comes with a loss in accuracy. In Chord, there is no guaranteed delivery and no guaranteed consistency among replicas. Hops have poor network locality, nodes close on ring can be amazingly far in the physical network.

2.5.3. Chord Algorithm

In Chord, a node ID is a unique m-bit identifier, hashed from IP address or other unique ID. A key is an *m* bit identifier, hashed from a sequence of bytes. A value is sequence of bytes. Chord API includes following functions:

```
// node n finds the successor of id
n.find_successor(id)
    if (id ∈ (n, successor])
        return successor;
    else
        p= closest_preceding_node(id);
        return p.find_successor(id);

// search the local table for the highest predecessor of id
n.closest_preceding_node(id)
    for i = m downto 1
        if (finger[i] ∈ (n, id))
```

```
        return finger[i];
    return n;

// create a new Chord ring
n.create()
    predecessor = nil;
    successor = n;

// join a Chord ring containing node p.
n.join(p)
    predecessor = nil;
    successor = p.find_successor(n);

// called periodically. verifies n's immediate successor, and tells the successor about n.
n.stabilize()
    x = successor.predecessor;
    if (x ∈ (n, successor))
        successor = x;
    successor.notify(n);

// n thinks p might be predecessor.
n.notify(p)
    if (predecessor is nil or p ∈ (predecessor, n))
        predecessor = n;

// called periodically. It refreshes finger table entries. next stores the index of the next finger
to fix.
n.fix_fingers()
    next = next + 1;
    if (next > m)
        next = 1;
    finger[next] = find_successor(n + 2^(next-1) );

// called periodically. It checks whether predecessor has failed.
n.check_predecessor()
    if (predecessor has failed)
        predecessor = nil;
```

3. Previous Works on Bootstrapping in Wired Networks

3.1. T-Man — A Gossip-Based Approach

Based upon popular gossip communication model [LMM2000] in distributed computing, T-Man [JB2005] is designed as a general purpose protocol for building and maintaining network topology. T-Man targets large scale and highly dynamic networks. It is simple, scalable, robust, and flexible. It may be used as a standalone program, a bootstrapping component, or a recovery component in other protocols. It is mainly used in P2P community, but has an application range far beyond. With the aid of its original concept — the ranking function, T-Man controls self-organization of topologies in a straightforward, intuitive, and adaptive manner. T-Man follows a stepwise refining procedure with a short asymptotic time. T-Man is completely distributed. Each node relies solely upon local communication to increase the quality of the current set of neighbors. Its fast convergence and high robustness in dynamic environments have attracted considerable follow-up research.

T-Man is so adaptive and flexible that it allows for topology change on-the-fly at run time without any change in protocols. All previous approaches have to revise protocol for each possible topology to achieve the same objective. As a general abstraction, topology can be used to solve problems or to enhance and support other solutions. Therefore changing topology on-the-fly will have significant benefit in both theory and practice. It may drastically increase the efficiency of distributed applications as well as the efficiency in deploying such applications. With the support for quick topology change, we can derive best topology for a certain scenario by progressive evolution of topologies.

3.1.1. Gossip Protocol

The Gossip protocol [BEGH2004, JHB2001, LMM2000, MMA2000] provides a scheme for performing probabilistically reliable network broadcasts. In the Gossip protocol nodes send a message to some instead of all neighbors (usually only one). The recipients are often selected randomly, but deterministic algorithms are used as well. Due to the redundancy in links, most nodes received the packet in limited hops. Gossip minimizes amount of transportation, and hence reduce communication overhead. Gossip has much better performance than flooding. Gossip can be used to deliver multicast messages with less overhead and enhanced efficiency than normal flooding style broadcasting.

3.1.2. Ranking Function

Key concept of T-Man is *ranking function*, which specifies the preference for a node to choose its neighbors in the target topology. A node uses the ranking function to order any set of nodes according to the preference. This simple abstraction results in an effective algorithm which generates various topologies with preciseness and efficiency. The ranking function is the source of effectiveness, versatility, and flexibility of T-Man.

Suppose a network contains nodes, all connected to each other. Each node has an address sufficient for sending messages to it. Each node maintains addresses of other nodes through a partial view, which is a set of node descriptors. Besides a node address, a node descriptor contains a profile, which contains topology related properties, such as ID, geographical location, etc. Links of topology are determined by addresses in partial views descriptors.

Following the selected ranking function, T-Man use local gossip messages and gradually evolves the current topology towards the target. According to its simulation report, the convergence is fast and scalable. Convergence time grows as the logarithm of the network size. The high speed guarantees that T-Man can build divergent topologies on-the-fly. This feature makes T-Man a perfect fit for dynamic systems in which the nodes and their properties change rapidly.

Here gives a formal description. Suppose N is the node set of a network. Each node x maintains addresses of other nodes through a partial view, denoted as $view_x$. c is the maximal size of partial views in the network. Ranking function R has following parameters as its input.

- x, base node
- $S = \{y_1, y_2, \ ,y_m\}$, a set of nodes

The output of R is an m-tuple, which is a re-ordered S. The task is to construct views of all nodes such that the view of node x, $view_x$, contains exactly the first c elements of $R(x$, {all nodes except x}), which is output of R over the entire node set. That is,

$$R(x, \text{view}_x) = R(x, N - \{x\})$$

One convenient way to get a ranking function is through a distance function, which is derived from a metric space over the node set. The ranking function measures the Euclidean or other distance from the base node. Here are few examples of defining distance function. For lines, the profile of a node is a real number. The distance function is

$$d(a, b) = |a - b|$$

Its variant can be extended to a ring. For example for a Chord ring with range [0, N], node profile is an integer in [0, N]. Here distance is directional, that is, $d(a, b)$ is not necessarily equal to $d(b, a)$. The distance function is defined as

$$d(a, b) = (a - b) \bmod (N+1)$$

Extending one dimensional distance function for line to two dimensions, we can derive distance function for a mesh. The profile for node is two-dimensional real vector. The distance for the mesh is the Manhattan distance, which is the sum of two one dimensional distances on two coordinates. Use the same transformation from line to ring, we can get profile and distance function for tube from those for mesh.

3.1.3. T-Man Protocol

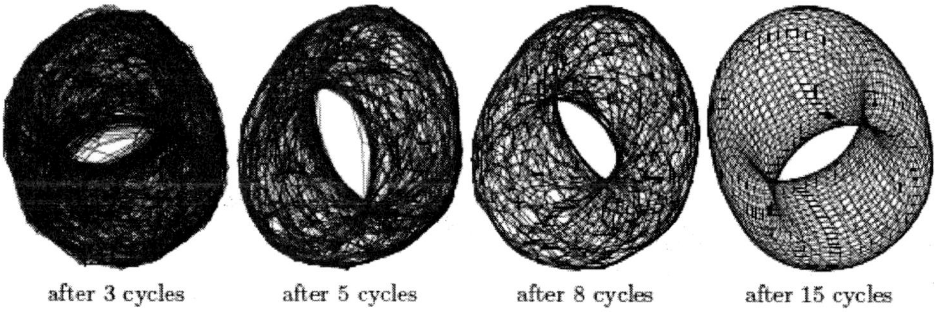

Figure 7. Constructing a torus over 50×50 nodes starting from a uniform random distribution of nodes with c = 20.

Given an arbitrary overlay network, constructing a target topology is realized via connecting all nodes to the right neighbors. T-Man's basic idea is there is a general relationship of nodes behind a given topology, which is expressed by a ranking function. The relationship between nodes could be geographical location, semantic description of stored data, storage capacity, etc.

T-Man is based on the gossip communication scheme. After initialization, each node executes the same protocol concurrently. No synchronization or coordination is needed. Nodes' running is not synchronous. The protocol consists of two threads: an active thread initiating communication with other nodes; a passive thread waiting for and processing incoming messages.

Initialization

view ← rnd.view ∪ {(myAddress, myProfile)}

Active Thread

do at a random time once in each consecutive interval of T time units
p ← selectPeer()
myDescriptor ← (*myAddress*, *myProfile*)
buffer ← merge(*view*, {*myDescriptor*})
buffer ← merge(buffer, rnd.view)
send *buffer* to *p*
receive $buffer_p$ from *p*
buffer ← merge($buffer_p$, view)
view ← selectView(*buffer*)

Passive Thread

do forever
receive $buffer_q$ from *q*
myDescriptor ← (*myAddress*, *myProfile*)
buffer ← merge(*view*, {*myDescriptor*})
buffer ← merge(*buffer*, *rnd.view*)
send *buffer* to *q*
buffer← merge($buffer_q$, *view*)
view ← selectView(*buffer*)

As described above, each node maintains a view. The view is a set of node descriptors. Function merge($view_1$,$view_2$) returns the union of $view_1$ and $view_2$. In above protocol, two key functions are selectPeer() and selectView(*buffer*). Function selectPeer() uses the current view to return an address. First, it applies the ranking function to order the elements in the view. Then it returns the first descriptor that belongs to a live node. Function selectView(*buffer*) applies the ranking function to order elements in the buffer. Then it returns first *c* elements of the buffer. By using views of their current neighbors, all nodes improve their views, so that their new neighbors will be closer to the target topology. Neighbors will become closer and closer.

3.2. T-Chord — An Application of T-Man

3.2.1. Advantages of T-Chord

T-Chord efficiently bootstraps Chord from a random unstructured overlay using T-Man. It is one of most successful Chord ring building approaches in terms of thoroughness, speed, and efficiency. Simulation proved that T-Chord is able to create a perfect Chord ring in O(log(*N*)) steps where *N* is network size. It also shows optimized message latency. The generated network is immediate operable and could be handed over to the Chord protocol right away.

T-Chord completely breaks away from the old pattern of bootstrapping structured P2P system — that is, using a jumpstart node and node joining procedure. The joining based method is very inefficient. It is unable to make nodes' bootstrapping concurrent. Nodes have to be booted one by one in a linear manner, which is very unrealistic for large network. [DBKKMSB2001] Some require booting nodes in a fixed order, which will not only need linear run time but also need complicated synchronization and coordination. Without the constraint of single jumpstart node, in T-Chord every node starts its own topology building and optimization concurrently. Furthermore, unlike many other attempts to bootstrapping Chord, T-Chord does not need any a prior configured initial network or jumpstart node.

3.2.2. T-Chord Protocol

T-Chord starts from a connected unstructured overlay network with a random topology. In T-Chord simulation, the unstructured random network is generated by a lightweight membership protocol called NEWCAST. [JGKS2004] Bootstrapping of T-Chord does not include node ID automatic generation. Nodes are a priori configured and unique IDs are assigned to nodes from a circular ID space. T-Man ranking function just needs minor revision to be adapted for T-Chord. In T-Man's running procedure, not only direct successor and predecessor are located as outcome of ring topology, many encountered nodes are also remembered. These buffered nodes are very useful in building Chord finger table.

3.2.3. Deficiencies of T-Chord

The most notable problem with T-Chord is its requirement for a priori configuration of Chord IDs. It ruins its good reputation and great prospective due to its ability to unconditionally bootstrap from arbitrary initial topology. Another short coming is its distance function, which inherited from T-Man. Its definition

$$d(u,v) = \min\{(v-u) \bmod 2^m, (u-v) \bmod 2^m\}$$

is not compatible with the distance defined in Chord, which is

$$d(u,v) = (v-u) \bmod 2^m$$

3.3. Ring Network

3.3.1. Features of Ring Network

The Ring Network (RN) protocol is an asynchronous message-passing distributed protocol, which fits well the autonomous behavior of peers in a P2P system. [SR2005] Peers do not need to be informed of any global network state. They are not required a grace leave, i.e. to assist in repairing the network topology caused by their leave.

RN protocol is not gossip based. RN uses message passing, a traditional distributed computing technique. Another notable difference is initial condition. RN requires the presence of a weakly connected initial network called minimum bootstrapping system to be able to return a Chord ring, while T-Chord can start at any condition and find any connected component. Two nodes are weakly connected means that there is a directed path between them no matter which direction the path is. For author's RAN protocol and T-Chord, differentiating weakly connected components from strongly connected components does not make much sense, since we do not have any preliminary requirement about connectivity. In addition, since most devices in MANETs support duplex mode, there is no much pragmatic significance to find this difference. RN does not specify the scale of the bootstrapping system and how the system is configured. From the Proposition 2.1 in [SR2005], we guess the bootstrapping system is a subset of all nodes to which every node is connected with at most one hop distance.

3.3.2. RN Protocol

The RN protocol is fully distributed. It can quickly adapt to churns in the network. All peers independently and asynchronously run a same set of procedures while they exchange asynchronous messages. Periodically each peer calls the Closer Peer Search procedure to search a closer predecessor in ID space, by which a closer successor candidate is also returned. As shown later in Section 3.3.5, authors of [SR2005] confuse successor and predecessor in the RN algorithm. But the pseudocode is still consistent and correct if we ignore the textual description.

Peers that participate in this search record information in any message they received. After collecting information returned by the predecessor search, returned by bootstrapping process, or gleaned from message propagation, each peer selects a currently closest successor. This process repeats till a complete consistent ring is formed. Local information stored by each peer includes:

- Γ: the set of current neighbors of the peer.
- W: the set of peers returned by Closer Peer Search.
- B: the set of peers that the peer has learned by the Search Monitor while propagating search request messages on behalf of other peers.
- s: a peer selected randomly from the current successor, and peers returned by the bootstrapping system.
-

Three steps of the protocol are described below in more detail.

Closer Peer Search

Each peer x periodically initiate a search for the successor candidate to which it is closer than to its current successor in the ID space. Current node first finds the closer predecessor. Current node x randomly chooses a peer s, which is either its current successor $x.\Gamma_0$ or a peer returned by the bootstrapping system, and sends s a *CloserPeerSearch* message. s forwards the message to one of its neighbors to which x is closest. The receiver of this request propagates this request in a similar manner. This way x gets closer and closer to the target. When a receiver u finds that the initiator x is closer to itself than any of its neighbors, the search is terminated. u then sends to x the address and ID of its successor $u.\Gamma_0$, which x adds to its set $x.W$.

The result of the Closer Peer Search depends on the current network topology. If the network is already in a ring topology, the search will not be really launched. Note that the search does not necessarily returns the closest node of x in ID space, because the ending node of the search may have a unvisited descendent node, which is more than one hop away, and x is closer to it. Furthermore, since the search is actually for a closer predecessor, it does not ensure of finding the successor to which x is closest. No matter x is closest to $u.\Gamma_0$ or not, since u is closest to x, x will be always between u and $u.\Gamma_0$. So $u.\Gamma_0$ is a promising candidate for x's successor. The frequency of this search only affects the speed of the protocol, not its correctness.

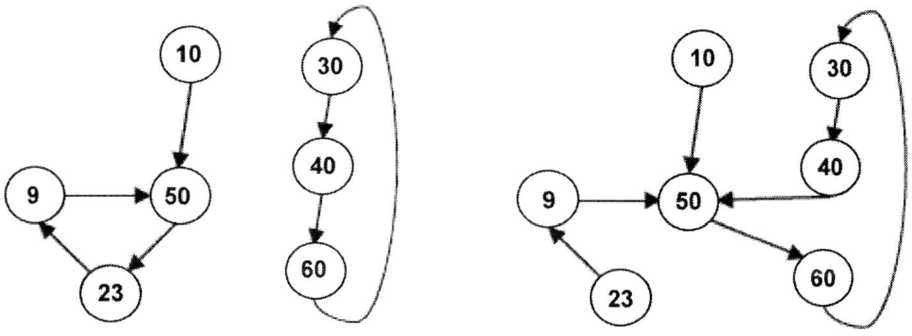

Figure 8. Closer Peer Search

Figure 8 illustrates the Closer Peer Search. Left-hand side is the starting situation; right-hand side is the ending situation. Node 50 starts this search at node 30. The search terminates at node 40, which notifies node 50 its successor 60. Node 50 then sets 60 as its new successor. Actually the exact next step for node 50 is adding node 60 to its successor candidate set W. To make it clearer, node 60 is assumed to be selected as new successor of node 50.

Every peer u records each received Closer Peer Search message. If a search is initiated by $x \neq u$ and is terminated at u, then x is closer to u than $u.\Gamma_0$. u then adds the address and ID of x to its set B. In Figure 8, peer 40 adds 50 to its set B.

Neighbor Update

Periodically every peer u checks if it has found a closer successor than its current successor $u.\Gamma_0$. It examines its current list of neighbors, a bootstrapping peer returned by the bootstrapping system, its set W, and its set B. The peer closest to u from among the union of these is chosen as the new $u.\Gamma_0$. In figure 9, after *W* and *B* have been updated, nodes 40 and 50 update their successors as well.

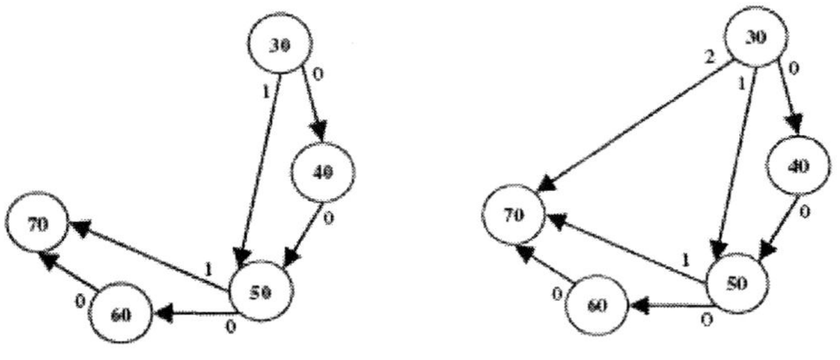

Figure 9. Neighbor update in RN

At the same time *u* updates all its other neighbors $u.\Gamma_1$, $u.\Gamma_2$, and so on. *u* sends a message to neighbor $u.\Gamma_i$ asking it to return the ID of $u.\Gamma_i$'s *i*th neighbor *v*. If the ID of *v* is between $u.\Gamma_i$ and *u*, *u* sets it as new $u.\Gamma_{i+1}$. Similar to finger table used in Chord, the purpose of such an update process is to minimize the number of hops and improve the search speed. [HGS1987] In Figure 9 peer 30 updates its third neighbor. Since the order number starts at 0, the third is actually its No.2 neighbor. It first asks its No. 1 neighbor, peer 50, for the No.2 neighbor. Peer 50 sends back 70. Peer 30 then sets peer 70 as its No.2 neighbor. This is because peer 70 is between peer 50 and peer 30 if we look at them in a ring. Eventually peer 30 has discovered a closer peer that is 4 hops away from it, using two messages.

3.3.3. AP Notation

AP notation is tailored pseudocode format for expressing network protocols. [Gouda1998] AP notation is very instrumental for correctness analysis. This analysis model has been proved useful and widely adopted by the distributed computing community. It ignores the execution order of interleaving of actions of nodes in a protocol by assuming arbitrarily random order. It is especially suitable for asynchronous protocols, for it expresses asynchronous protocols clearer by eliminating the need for interrupts.

In AP notation a distributed protocol consists of a series of procedures associated with nodes in a network. A node is the carrier of protocols. Data structures of a node *p* are classified into three categories: constants, inputs, and variables, denoted by keywords **const**, **input**, and **var** respectively. The operation procedures of a protocol is put in actions section denoted by (a< *i* >), where *i* is the order number of procedures. Actions are delimited by two square brackets. An action is expressed in syntax

<guard> → *<statement>*

The statement of an action can be executed only if the corresponding guard condition is evaluated to true. At the beginning of every round of running of a protocol, all guards of all actions of all peers are evaluated. Then only one statement of an action whose guard evaluated to true is executed. When there are more than one statements whose guards are evaluated to true, a true guarded statement is selected for execution at random. Every enabled action will eventually be executed, but the order and frequency of execution are arbitrary. RN and RAN protocols are written in AP notation.

3.3.4. RN Algorithm

Below is the algorithm for the RN protocol in AP notation. [SR2005]

Peer u

const

T : set of bootstrapping peers

input

w : a peer (successor candidate)

x : peer being searched for

c : index of received neighbor

z : new neighbor

s : a bootstrapping peer

var

S : Set of peers

B : Set of successor candidates

W : Set of successor candidates

Γ_i : ith neighbor

(a1) *true* $\rightarrow$

$S := \{s\} \cup W \cup B \cup \Gamma$

$\Gamma_0 :=$ argmin $_{k \in S}\, d(u, k)$

$B := W := \varnothing$

[]

(a2) *true* $\rightarrow$

s:= Get random peer from $\{T \cup \Gamma_0\}$

send *closerPeerSearch*(u) **to** s

[]

(a3) **receive** *closerPeerSearch*(x) **from** $q \rightarrow$

if x is closer to u than any neighbor $\in \Gamma$

then

$B := B \cup \{x\}$

send *successorCandidate*(Γ_0) **to** x

else

send *closerPeerSearch*(x) **to** argmin $_{k \in \Gamma}\, d(k, x)$

[]

(a4) **receive** *successorCandidate*(w) **from** $q \rightarrow$

$W := W \cup \{w\}$

[]

(a5) *true* $\longrightarrow$

for each $h \in \Gamma$ **do**

send *getNeighbor*(*index*(h)) **to** h

[]

(a6) **receive** *getNeighbor*(j) **from** $q \longrightarrow$

if Γ_j exists

then send *neighbor*(Γ_j, j) **to** q

[]

(a7) **receive** *neighbor*(z, c) **from** $q \longrightarrow$

if $\Gamma_c \leq z < u$

then $\Gamma_{c+1} := z$

else $\Gamma_{c+1} :=$ **NIL**

[]

Note that function argmin $_{k \in S}\, d(u, k)$ returns a k, instead of $d(u, k)$ or (u, k), which gives minimum $d(u, k)$.

3.3.5. Problems with RN

The most serious problem with RN is the minimum bootstrapping system required as a necessary condition to apply RN protocol. RN does not specify: (1) scale of the minimum bootstrapping system; (2) whether and how the minimum bootstrapping system is generated? manually or automatically by a program? from an arbitrary network topology or an a priori configured topology? (3) how many hops away from the minimum bootstrapping system is any node outside the minimum bootstrapping system? (4) how the RN is interfaced with the minimum bootstrapping system?

Second, RN is not guaranteed to converge to the ideal Chord ring within finite time. When a connected network has more nodes it is getting more difficult for RN to converge to the ideal ring. Situation in wired network is similar.

Third, as the direct reason for above problem, the basic strategy of RN in searching closer node to the target node — continuously choosing closer neighbor at each step — has no logical support at all. The common sense reasoning is against this strategy. The distribution of node IDs is totally random. The proximity of one node has nothing to do with the proximity of its children nodes. No proof of correctness of RN is presented in [SR2005].

Fourth, in [SR2005], the authors confused some basic concepts and logic. For instance, they mixed up distance from node u to node v, i.e. $d(u, v)$ with distance from v to u. A subsequent mix-up is the concept u is closer to v when $d(u, v)$ is smaller. Because the distance is directional and modulus based, suppose here the modulus is m, the following equation always holds

$$d(u,v) + d(v,u) = m$$

Obviously, by definition, when $d(u, v)$ gets smaller, distance from u to v becomes smaller, so u is closer to v. At the same time, v is getting farther to u. However, in [SR2005], the "smaller the value of $d(u, v)$ the closer v is said to be to u." It is not just a trivial issue as chopping logic. This mistakes leads to a more serious misuse of concept in following part of the paper. For example, the loser peer search is actually a search for closer predecessor of the current node by interpreting the pseudocode of their algorithm; however, they describe it as a search for closer successor in Section 4.1, which cause a lot more confusion and logical mess-up in RN protocol and algorithm.

Next, the procedure and result of simulation of RN is not very convincing. (refer to [SR2005] Section 5) The simulator used for RN simulation is NetLogo. [Wilensky1999] Not many models and functions for network simulation are included in libraries of NetLogo. For networking simulation the choices and possibilities are limited. In its latest version, i.e. Version 3.1, no model in the integrated library is ready for use for simulation in scenarios like RN. More important, authors of [SR2005] did not mention anything about how the simulation is implemented. No information for following questions is provided in [SR2005]: (1) whether and how the program is designed? (2) how the RN is terminated in the simulation? what is the ending condition of entire RN protocol? RN has already given the ending condition of the closer successor search, but nothing has been said about terminating the whole protocol.

Last, in simulation of RN described in [SR2005], no convergence time data or any other data about performance of RN is provided. The simulation is about the quality of Chord ring generated. Authors of [SR2005] used a concept "perfect Chord ring", however, the perfect ring does not perform best in their simulation. By definition given in Chord position paper [SMKKB2001], it is clear that there could be only one perfect Chord ring, in which all nodes in the networks are linear sorted. No other ring should be target of Chord topology construction, unless Chord is revise to a better version.

4. Previous Works on Structured P2P Systems over MANETs

4.1. Special Issues on P2P Systems over MANETs

In wired networks like Internet, neighbor is defined on overlay layer and low layers such as network layer. We can say being neighbor is equivalent to knowing address. Two nodes u and v, we say v is u's neighbor only if u knows v's network address and be able to send a message to v. By this definition, neighbor relation is unidirectional and not commutable. When u knows v's address, we have no clue if v knows u's address.

On the contrary, in MANETs, neighbor is only defined on lowest layer, e.g. physical layer or MAC layer. Defining layer could be expanded to network layer. In most cases, it is define by radio range. From this point of view, it has nothing to do with Chord ID space or overlay layer. In both wired networks and MANETs, the distance function is defined in the same way. From above property, a natural extension is: in MANETs, a node's neighbor set is fixed at a given time, while for a node in a wire network, it could have countless variation. Therefore, in RN protocol in Section 3.3, the neighbor update procedure can only be applied to wired networks. It is not applicable to MANETs.

For Chord or any other structured P2P systems built on wired networks like Internet, all nodes are actually connected. Even though two nodes can not connect to each other or do not know the existence of the other if they do not know the network address of the other, they are still connected. This is not the case in MANETs. Nodes in MANETs are strictly constrained by the radio range in physical layer. If there is no path from no node to another formed by neighborhood relations in a MANET, these two nodes are not reachable to each other unless their movement establish a path later. A MANET is consisted of a set of connected components, which are disjoint to each other. A component could contain only one node if the node is isolated. If a MANET has more than one component, there is no way to have one comprehensive Chord ring which includes every node like what always happen in Chord over Internet. The best scenario is we can find a Chord ring for each connected component.

Both P2P over wired networks and P2P over MANETs have proximity concerns, but in MANETs this issue is has more serious impact. The reason is still from the physical layer characteristics. A hop in MANETs is more costly than in Internet. Hence Proximity optimization has more urgent, more realistic significance in MANETs.

Substituent of IP address is necessary in MANETs for the purpose of building a P2P overlay, for example, source route in DPSR [HPD2003]. The reason is intuitive: overlay layer only makes sense or semantically correct if an underlay layer exists.

A P2P system over wired network, especially one over Internet, usually does not cover intermediate nodes of its path on the Network layer. Otherwise the P2P system may cover too many unrelated nodes. In a P2P system over a MANET, the situation is poles apart: all intermediate nodes should be included to secure connectivity on the overlay layer.

4.2. Cramer and Fuhrmann's Pessimistic Verdict

In Cramer and Fuhrmann's [CF2006], several serious problems could be found.

First, the whole paper is built upon some unrealistic, far-fetched assumptions. For example, they assume that all nodes can reach a common bootstrap node (which is called joint point) immediately after they power up. To make it possible, either all nodes in the MANET have to be only one hop away — which requires very small network or very powerful transmitter/receiver; or every ordinary node already has a route to that super node before power up, which is almost same as assuming that all nodes already have a pre-configured Chord successor and finger table — so the network is already initialized, why does it need bootstrapping? Another example is the assumption of single bootstrap node, which is against the definition of MANET and cause the single point of failure.

The most unrealistic assumption is at the time of power up, that is, in their own words, in the first stabilization cycle, a Chord ring has been set up and all nodes have already joined the this ring in ID space. A minor assumption, which is not serious as other assumptions, is every node knows the size of the network n. Another untenable assumption is all nodes on the ring are in a complete sequential order, from 0 to n.

4.3. Out of IP Box: Strength of RAN

In wired networks, core operations in T-Man and Ring Network are based upon IP. Traditional approach to transfer anything in wired IP networks to MANETs is replacing lower IP layers with existing routing protocols and MAC protocols of MANETs. Successful examples include DPSR [HPD2003] and Ekta [PDH2004]. If we follow this train of thoughts, the MANET version T-Man would be very complicated and inefficient, therefore not feasible in MANETs. However, the traditional approach has not been rigorously tested, even though it has been so prevalent in MANETs community. The author believes that the dominance of traditional approach is largely from the historical monopoly of IP. It is probably neither valid nor necessary.

RAN abandoned the dominant IP model and integrates the overlay layer into network and MAC layers. This avoids complicated mapping of overlay layer onto lower layers and seamlessly integrates dynamic source routing into ring-based DHT routing and tremendously reduces cost in setting up T-Man over MANETs. The resulted RAN has advantages of T-Man, T-Chord, and RN in a simplified MANET model.

The general strategy of RAN is distributed stepwise refinement. Three patterns are designed, namely distributed exhaustive pattern, virtual centralized exhaustive pattern, and random pattern. A spanning tree called component tree is used to simplify the model of connected component. A node sets itself as the root of its component tree. All nodes in the component are included in this tree. In its distributed construction procedure, the component tree goes through nodes on the fly. In two exhaustive patterns, all nodes of the tree are passed, while only nodes on one root-to-leaf path are passed in the random pattern. No node keeps its component tree in storage. Only some parts of the tree exist in memory when searching for next closer successor. This statelessness considerably increases flexibility and robustness.

Distance from node A to node B is defined as

$$(ID_B - ID_A) \bmod max$$

where *max* is a relatively huge modulo. The ring topology is determined by the successor relation among nodes of a connected component. At each step, the current successor is compared with a selected node. If the selected node has smaller distance from root than current successor, it is assigned as new successor. The process repeats till the tree is parsed.

5. RAN — an Optimal and Realistic Approach

5.1. Introduction

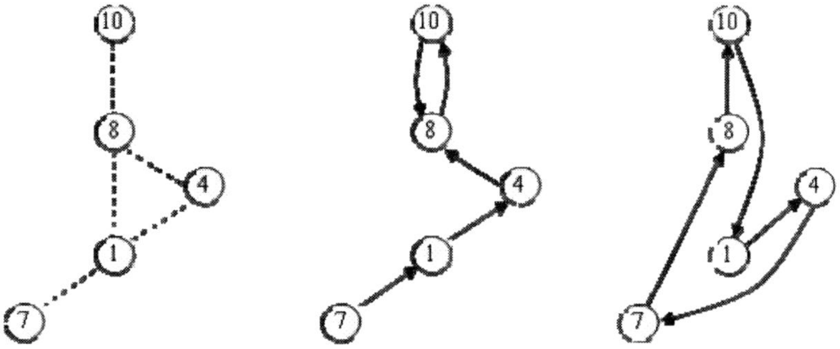

Figure 10. RAN Examples. Left is a network with only neighbor relation. Middle shows original successor relation. Right is successor relation after running RAN

Ring Ad-hoc Network (RAN) is a protocol to build a ring topology over MANETs. RAN has integrated merits of T-Man, T-Chord, and Ring Network and adapts well to MANETs. RAN is completely distributed. It uses only neighbors and local information. RAN builds an ideal ring topology for each connected component in the node ID space of a MANET. Upon this ring, ring-based P2P systems could run immediately without any stabilization. For instance, entire Chord protocol can run immediately. No stabilization is necessary unless large scale disturbance occurs. RAN integrates automatic non-IP address configuration into bootstrapping. To best of author's knowledge, it is the first successful try in the filed of bootstrapping ring-based P2P systems over MANETs.

The basic algorithm in RAN is distributed stepwise refinement. Each node treats its connected component as a tree, called component tree. All nodes in the component are included in this tree. It sets itself as the root. If the depth of a node in the tree is i, the node is said at level i. At each step, we compare the current successor with a random chosen node, all nodes in current sub tree, or all nodes in current level, depending on the pattern of the algorithm. If a chosen node in current level has shorter distance to root, we use this node as new successor. The process repeats till the tree is traversed. Here the distance function is exactly same as define by Chord, also same as that of RN.

Chord ring is determined by the successor relation among nodes in a connected component. Unlike RN, in RAN the successor of a node is not always its neighbor. If the depth of node n's component tree is $p > 1$, the successor of n is n's neighbor only at the first round of RAN execution. As RAN runs into deeper levels, the successor may change. The distance between n and its successor may be the depth of current level at most.

5.2. Design Goals and Assumptions

RAN is designed to achieve following goals:

- Generate an ideal Chord ring for each connected components, which will guarantee the quality of ring-based P2P systems running on the ring.
- Compatible to any MANET routing protocols, that is, routing independent.
- No any kind of a priori bootstrapping node or bootstrapping
- Pure distributed and decentralized
- Have all capability of T-Chord and RN except those incompatible with nature of MANETs
- Asynchronous, only use message passing
- Scalable to MANET size
- Good proximity and optimized for MANETs

RAN integrates automatic non-IP address configuration into bootstrapping, which is often deliberately ignored in previous approaches by assuming that an ideal IP address configuration has been a priori established from the very beginning. A non-IP node ID configuration is assumed. It generates unique random ID in structured P2P layer. No network layer address is needed. Routing in low layers uses this node ID as well.

5.3. Component Tree

A component tree is one spanning tree of the connected graph which is derived from a connected component. The rule of construction is:

(1) Select the searching node, which is looking up the closest successor, as the root.
(2) Add all neighbors of the root to the first level of the component tree.
(3) For all following levels, construct the next level according to the direct neighborhood relation.
(4) Delete all edges which connect a lower level node to an upper level node, no matter if the former is a descendent of latter.

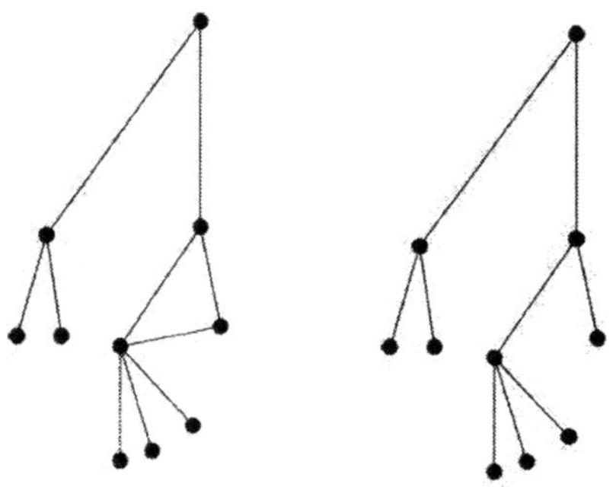

Figure 11. Convert a connected component to a component tree

For a complete component tree with N nodes, uniform downward degree k, and the depth d, following equations hold.

$$N = 1 + k + k^2 + k^3 + \cdots + k^d$$

$$N = \frac{k^{d+1} - 1}{k - 1} \tag{1}$$

Equivalently,

$$d = \log_k (kN - N + 1) - 1 \tag{2}$$

$$k^{d+1} = kN - N + 1 \tag{3}$$

Most performance parameters share following scenarios. With regard to number of nodes involved, we have one node parameters versus network parameters, which involve all nodes in the network. With regard to number of rounds in searching, we have one round parameters versus life-long parameters, which cover all rounds. We can mix up these two categories using simple combinations. We may have one node one round parameters, one node life-long parameters, network one round parameters, and network life-long parameters.

An ideal network is defined as a connected MANET, which has only one connected component. In an ideal network, the component tree has to be a complete tree, in which all leaf nodes are at depth d or d-1, and all leaves at depth d are toward the left. In an ideal network, one node one round message complexity M is the number of messages sent in one round of searching in this node's component tree. In unicast mode, M is also the number of messages received.

Theorem 1

In an ideal network, by symmetry, all nodes have same one node message complexity. The one round network message complexity M^{Net} is the sum of one node message complexities of all nodes.

$$M^{Net} = M \times N \tag{4}$$

Theorem 2

In an ideal network, the network time complexity T^{Net} is equal to the one node time complexity T.

It applies to both one round time complexity and life-long time complexity. Obviously, all nodes run concurrently and spend same amount of time in one round of searching. Let T_d be downward time, and T_u upward time

$$T = T_d + T_u \tag{5}$$

From (2) we have

$$d = O(\log N) \tag{6}$$

5.4. Three Patterns

To compare performance and find out intrinsic mechanism which determines the performance, we designed three patterns for RAN protocol. The primary concern is the balance between effectiveness and efficiency, to be specific, the trade-off between the completeness of generated ring and the time, message, and storage complexities of construction.

Three patterns are studied in length. Two of them are exhaustive patterns, namely, distributed exhaustive pattern and virtual centralized exhaustive pattern. In virtual centralized exhaustive pattern the searching node acts as a central controller and coordinates the searching procedure. Two exhaustive patterns use unicast in message exchange, exhaustive search at each level of component tree. Output ring is guaranteed to be ideal Chord ring for every connected component. Because it compares all node identifiers in the component, the finding the closest successor is ensured. However, this exhaustion may suffer from high cost in time, message, and storage. We need measures to mitigate the overhead. These two exhaustive patterns are equally excellent in effectiveness. Both keep 100 percent nodes of connected component in ring constructed. The distributed exhaustive pattern has better performance than the virtual centralized exhaustive pattern due to the fact that nodes in distributed exhaustive pattern only exchange messages with parents and children.

The third pattern is random pattern, which has its root in Ring Network [SR2005]. To adapt to MANETs environment, the minimum bootstrapping system is eliminated. A breadth-first search scheme is used in lieu of it. The search scheme traverses the component tree of searching node in a cascading manner to make up the poor effectiveness of RN.

5.4.1. Distributed Exhaustive Pattern

In the distributed exhaustive pattern, the searching node (root) sends a *getClosestCandidate* message to each of its k child in its component tree. Each child concurrently forwards the message to its k children at next level. At following levels, nodes keep forwarding the request message to their own children until leaf nodes are reached. Then, from leaf nodes up, the closest successor candidate of the root in the subtree is calculated at the root of the subtree and is returned to the parent node in a *closestCandidate* message. The calculation is done by comparing distances from the root to returned candidates of the children of the root of subtree. Obviously, each node sends out $k + 1$ messages except root and leaves.

$$M = (k+1)N - k^{d+1} - 1$$

Here k^{d+1} is the number of skipped *getClosestCandidate* messages from k^d leaf nodes, for they have no children. Similarly, 1 is the number of skipped *closestCandidate* messages from root. Plug in (3), we get

$$M = 2N - 2$$

In an ideal network with uniform downward degree k, the one node one round message complexity M of the distributed exhaustive pattern is independent of k. It is only decided by the size of network N.

$$M^{Net} = 2N^2 - 2N$$

$$T_d = dk$$

$$T_u = d$$

$$T = d(k+1)$$

$$T = (k+1)(\log_k (kN - N + 1) - 1) = O(k \log N)$$

In an ideal network, the distributed exhaustive pattern has time complexity $O(k \log N)$.

In multicast option, each node sends out 2 messages except root and leaves.

$$M = 2N - k^d - 1$$

Plug in (3),

$$M = (1/k + 1)(N - 1)$$

$$M^{Net} = MN = (1/k + 1)(N - 1)N$$

When k is a big number,

$$M \approx N - 1$$

$$M^{Net} \approx N^2 - N$$

That is,

$$M \approx O(N)$$

$$M^{Net} \approx O(N^2)$$

The messages complexity is about half of that in plain option.
The *getClosestCandidate* message needs 1 time unit to move from one level to next,

$$T_d = T_u = d$$

$$T = 2d$$

$$T = 2\log_k(kN - N + 1) - 2 = O(\log N)$$

5.4.2. Virtual Centralized Exhaustive Pattern

In the virtual centralized exhaustive pattern, all direct children nodes of the root form the first level. Direct children of nodes in first level form the second level, and so on. The root sends every node in current level a *getAllNeighbors* message. Then these nodes send their children set to root in an *allNeighbors* message, so root gets the next level. Then root sets the next level as new current level and repeats the same procedure till leaves are reached. This algorithm is most expensive in terms of overhead. However, it gives the root node tremendous power to control whole process upon a distributed network. Individualized services could be implemented this way.

If we count a multi-hop message as one message, both M^{Net} and M are independent of k and d; they only depend on N. The simple fact is: the root sends each node except itself a message; then each of these nodes sends a message back to the root.

$$M = 2(N - 1) = 2N - 2 = O(N)$$

$$M^{Net} = 2N^2 - 2N = O(N^2)$$

It looks like same as in the distributed exhaustive pattern. However, unlike in the distributed exhaustive pattern, almost all messages have to go through multi-hops. To get precise comparison with the distributed exhaustive pattern, the per hop message complexity M_{hop} should be used. At first level, there are k nodes. Each needs two one hop messages. So there are $2k$ *getAllNeighbors* and *allNeighbors* messages sent to and from this level. It takes $k + 1$ time units to transfer all of them. At second level, there are k^2 nodes. Each needs two two-hop messages. There are $2 \times 2 \times k^2$ messages sent to and from this level. It takes $k^2 + 2$ time units to transfer all of them. Similarly, level i needs $2 \times i \times k^i$ messages, which need $k^i + i$ time units to transfer.

$$M_{hop} = 2k + 4k^2 + 6k^3 + \cdots + 2dk^d = 2\sum_{i=1}^{d} ik^i$$

Suppose $S = \sum_{i=1}^{d} ik^i$

$$S = \frac{dk^{d+2} - (d+1)k^{d+1} + k}{(k-1)^2} = \frac{dk^{d+1}(k-1)}{(k-1)^2} - \frac{k^{d+1} - k}{(k-1)^2}$$

From (1),

$$N - 1 = \frac{k^{d+1} - k}{k - 1}$$

Plug it in, we have

$$S = \frac{dk^{d+1}}{k-1} - \frac{N-1}{k-1} = \frac{dk^{d+1} - N + 1}{k-1}$$

Plug in (3),

$$S = dN + \frac{d - N + 1}{k - 1} = O(N \log N)$$

When $N \to \infty$,

$$S \approx dN$$

$$M_{hop} = O(N \log N)$$

$$M_{hop}^{Net} = NM_{hop} = 2NS = O(N^2 \log N)$$

$$T = (k+1) + (k^2+2) + (k^3+3) + \cdots + (k^d+d)$$

$$T = N + d(d+1)/2 - 1 = O(N)$$

5.4.3. Random Pattern

Random pattern is the variant of RN in MANETs. The minimal bootstrapping system is replaced with the breadth-first traversal of all nodes in the component. Each round of search does begin with the starting node. The searching node gets the starting node and its route from the queue for the breadth-first traversal. It sends the node a *closerNodeSearch* message. Note that this message may need multi-hop to reach its destination when the traversal proceeds beyond neighbors of the searching node. In worst case, this message needs d hops. On average, this first message needs

$$\bar{h} = \frac{k + 2k^2 + 3k^3 \cdots + dk^d}{N-1}$$

$$\bar{h} = \frac{S}{N-1} = d + \frac{d}{N-1}\frac{k}{k-1} - \frac{1}{k-1}$$

When $N \to \infty$,

$$\bar{h} \approx d$$

$$\bar{h} = O(d) = O(\log N)$$

This means when M is very big, the average hop number is close to the worst case hop number.

Along the same path of the received message, the starting node first sends back its neighbor set to the searching node to feed the traversal. Then it passes the *closerNodeSearch* message to one of its unvisited children, whose ID is closest to ID of the searching node. The receiver passes the message to one of its unvisited children, and so on, until one receiver finds that no neighbor is closer than itself. M is the number of messages exchanged between the searching node and the starting node, plus the number of messages in the search, which is d in worst case and $d/2$ on average.

$$\overline{M} = \frac{d}{2} + 2 = O(\log N)$$

$$\overline{M}_{hop} = \frac{d}{2} + 2\bar{h} \approx 2.5d = O(\log N)$$

The computation time is negligible comparing to communication time, so

$$T = T_{start} + T_{search} + T_{return}$$

T_{start} is the time to pass message to the starting node; T_{search} is the time to find the closest node; T_{return} is the time to transfer the found node ID back to the searching node.

$$T_{return} = T_{search} + T_{start}$$

On average,

$$T = 2(d + d/2) = 3d = O(\log N)$$

Due to the limit of length, detailed algorithms for three patterns are not given. Please refer to full paper if interested.

5.5. Three Options

Besides three patterns just described, three auxiliary options are defined to improve the efficiency, especially time complexity and message complexity. Plain option means no additional operation and the search should ends with a complete ring. Other two options are explained below.

5.5.1. Approximation Option

The approximation option could be applied to all three patterns. Approximation pattern does not change underlying algorithm. It works by changing the end condition of all patterns. End condition in the approximation option is much looser than normal scenario. Normally, all patterns set the ideal ring as their objective. With approximation option, a small fraction of nodes are allowed to be left out of the final rings if they are in very short line segments attached to rings.

After first running of check_rings() function in the simulator, a connected component breaks down to a ring and a set of lines which are attached to the ring at only one node. With running of the simulation, the lines gradually shrink and are absorbed by the ring. Finally with sufficient running of our simulator only ring exists.

In the plain pattern, we require that all lines are absorbed by the corresponding ring of the connected component. However, this approach becomes so resource demanding when network size increases over 100 nodes. To reduce overhead in time, message, and storage, we revise the ending condition to allow a small fraction of nodes of a component to remain in short lines. Usually the fraction is set to 10 percent, or 15 percent. This approximation tremendously reduced the complexity in time, storage, and message. The growth function of time versus network size dropped from sub exponential to linear. Similar improvement happened to the growth function of the number of sent or received messages versus network size

5.5.2. Multicast Option

Another option is multicast option, in which a node sends message to all direct downward neighbor nodes (its children) at next level by multicasting one message instead of unicasting multiple messages. It considerably improves time, message, and storage complexity.

However, as we mentioned above, multicast option can not be applied to any random pattern.

Please refer to Section 7 for detailed algorithm and simulation.

5.6. Mobility

Mobility and its complication have been one of major difficulties in MANETs. Mobility will cause leave of neighbors. It also causes repartition of connected components, which is the primary concern in topology construction. Excessive mobility may cause general failure of a MANET because it makes multi-hop communication impossible. In real world, only moderate mobility needs to de addressed.

When mobility is moderate, especially when its stable period is significantly longer than the search cycle of RAN protocols, RAN suite could solve the mobility problem with simple refreshing and dynamic variable component tree. For two exhaustive patterns, a search may miss the closest node if it just joins a node's neighbor set after this node sends out its closest successor; but second search will cover it. Refreshing is very effective and reliable. No error or exception needs to be handled. The only cost is one more round of search. For random pattern, however, the optimal successor may be missed in simple refreshing, because the traversal is not stateless. More sophisticated approach needs to be found.

Other scenarios will be addressed in the Mobile RAN protocol MRAN, which is not covered by this chapter. Please check following publications of author or contact the author by email at *weiding@ieee.org*.

6. Algorithms

6.1. Distributed Exhaustive Pattern

6.1.1. Message Format

getClosestCandidate(originator, sender, receiver, sequence_num)

originator: the ID of searching node (root of component tree)
sender: sender of this message, not necessarily the searching node
receiver: receiver of this message
sequence_num: a random number user to identify this message

closestCandidate(originator, candidate, sender, receiver)

originator is the ID of searching node (root of component tree)

candidate: closest candidate returned
sender: sender of this message, not necessarily the searching node
receiver: receiver of this message

alreadyReceived(*originator*, *sender*, *receiver*)

originator: the ID of searching node (root of component tree)
sender: sender of this message
receiver: receiver of this message

6.1.2 Algorithm

peer *u*

constant
maximum: upper bound of ID

input
init: initialization flag, set to **true** at beginning
size: number of nodes in the connected MANET
in-que-len: length of incoming message queue
out-que-len: length of outgoing message queue

var
in-queue: incoming message queue
out-queue: outgoing message queue
Γ: set of one-hop neighbors
Γ_0: successor
reply_received: number of responses returned to *getClosestCandidate* messages sent by this node
originator: ID of searching node (root of component tree)
closest_candidate(*originator*): current closest candidate for *originator*
closest_candidate_distance(*originator*): ID space distance from *originator* to *closest_candidate*(*originator*)
already_received_neighbor: number of neighbors that already received the *getClosestCandidate* message from me (this node)
roots_info: map from node index of another root node to the node index of candidate of closer successor at this node for the other root node. NOT for itself.
search_finished: indicate the search for this node's closest successor is finished

Library Function
lookfor(*x*, *y*): Return *x* if $x \in y$. Return<*x*, *> if <*x*, *> $\in$y. Otherwise return **NIL**.

Action
(a1) *init* $\rightarrow$

```
    construct Γ
    if Γ = ∅
    then
        search_finished := true
        return

    search_finished := false
    init := false
    reply_received := 0
    for each h ∈ Γ do
        send getClosestCandidate(u, u, h) to h
    []

(a2) receive getClosestCandidate(originator, q, u) from q →
    if originator ∈ roots_info
    then
        send alreadyReceived(originator, u, q) to q
        return
    else
        roots_info := roots_info + {<originator, q>}
        closest_candidate(originator) := u
        closest_candidate_distance(u) := (u – originator) MOD maximum
        already_received_neighbor(originator) := 0
        for each h ∈ (Γ - {q}) do
            send getClosestCandidate(originator, u, h) to h
    []

(a3) receive closestCandidate(originator, cd, q, u) from q →
    if lookfor(originator, roots_info) = NIL
    then
        error(closestCandidate message does not have a root entry)
        exit

    <originator, x> := lookfor(originator, roots_info)
    reply_received ++
    d := (cd – originator) MOD maximum
    if d < closest_candidat_distance e(originator)
    then
        closest_candidate(originator) := cd
        closest_candidate_distance(originator) := d
    if u ≠ originator
    then
        if reply_received = |Γ| - 1
        then send closestCandidate(originator, closest_candidate(originator), u, x) to x
    else
```

if *reply_received* = |*Γ*|
then *search_finished* := **true**
[]

(a4) **receive** *alreadyReceived*(*originator*, *q*, *u*) **from** *q* →
if *lookfor*(*originator*, *roots_info*) = **NIL**
then
error(*closestCandidate* message does not have a root entry)
exit
<*originator*, *x*> := *lookfor*(*originator*, *roots_info*)
reply_received ++
already_received_neighbor ++
if *u* ≠ *originator*
then
if *reply_received* = |*Γ*| - 1
then send *closestCandidate*(*originator*, *closest_candidate*(*originator*), *u*, *x*) **to** *x*
else
if *reply_received* = |*Γ*|
then *search_finished* := **true**
[]

Note: For **receive** primitives, actual triggering event: message is taken out from *in-queue*.

6.2. Virtual Centralized Exhaustive Pattern

6.2.1. Message Format

The format of *getAllNeighbors* message is

getAllNeighbors(*originator*, *sender*, *receiver*, *route*)

originator: the root node.
sender: sender of this message, not necessarily the searching node
receiver: receiver of this message
route: the route from the *originator*

The format of a*llNeighbors* message is

allNeighbors(sender, originator, neighbors, broute)

originator: *originator* of received corresponding *mGetAllNeighbors* message
neighbors: all neighbors except the sender of corresponding *mGetAllNeighbors* message
broute: route from current node to *originator*

6.2.2. Algorithm

peer *u*

input
init: initialization flag, set to **true** at beginning
size: number of nodes in the connected MANET
in-que-len: length of incoming message queue
out-que-len: length of outgoing message queue
msg-rate: message processing rate

Note: Assuming rates for incoming messages and outgoing messages are same.

var
in-queue: incoming message queue
out-queue: outgoing message queue
T: external timer, simulator by discrete counter
L: set of peers at the current level
N: set of neighbors, including all hops
level: current level of hops from u
Γ: set of one-hop neighbors
Γ_0: successor
R: *u*'s routing records for this algorithm
r: a route in *R*
s: a node in a set with smallest node ID
AN_received: number of received allNeighbor messages
AN_in_queue: number of allNeighbor messages in in_queue
nodes_last_level: node number in previous level
current_completed: if all searching is completed at current level

Library Function
route(*a*, *b*): return a route from node *a* to node *b*. Actually a route is a string or vector. *route*(*a*, *a*) returns *a*.
distance(*u*, *h*): RAN distance function
reverse(*r*): return the reverse path of route *r*

Action
(a1) *init* →
    construct *Γ*
    if $\Gamma \neq \varnothing$
    then
        find Γ_0
        $T := 0$
        init := **false**
        $R := \varnothing$

```
        for each h in Γ do
            R := R + {uh}
    N := L := Γ
    []
```

Note: Here *route*(u, h) = *uh*. *uh* is a series, like string, vector in C++, ArrayList in Java.

```
(a2) current_completed →
    level ++
    current_completed := false
    T := 0
    for each h ∈ L do
        if route(u, h) ∈ R
        then send getAllNeighbors(u, u, h, route(u, h)) to h
    L := ∅
    []

(a3) receive getAllNeighbors(o, q, u, r) from q →
    br := reverse(r)
    send allNeighbors(u, o, Γ- {all nodes in r}, br) to o
    []

(a4) receive allNeighbors(q, o, S, br) from q →
    for each h in (S – N) do
        route(u, h) := route(u, q) + h
        R := R + { route(u, h)}
    L := L + (S – N)
    N := N + (S – N)
    []
(a5) true →
    Local
    timeout_small: lower bound of ending time
    timeout_big: upper bound of ending time
    timeout_small := 2 × level × nodes_last_level / msg-rate
    timeout_big := max{4 × level × nodes_last_level / msg-rate, 6 × log(size)}
    if (AN_received = nodes_last_level) or ((T >= timeout_small) and (AN_in_queue = 0))
                or (T >= timeout_big)
    then
        current_completed := true
        for each allNeighbors(q, u, S, br) message still in in-queue
            take out allNeighbors(q, u, S, br)
    s := argmin_{k∈L} d(u, k)
    if d(u, s) < d(u, Γ0)
    then Γ0 = s
    []

(a6) receive a message destined for another node →
```

send the message to next node on the route
[]

Note: For **receive** primitives, actual triggering event is: message is taken out from *in-queue*.

6.3. Virtual Centralized Exhaustive Pattern with Multicast Option

6.3.1. Message Format

Suppose a node always sends multicast messages at low frequency, so there is no need of an out queue for sending multicast messages. Only in-queue is needed for receiving multicast messages from other nodes. We also suppose multicast messages have priority over normal messages, they could use all msg-rate to process multicast in queue if needed.

There is only one kind of multicast messages, that is, *mGetAllNeighbors*. The format of *mGetAllNeighbors* message is

mGetAllNeighbors(originator, serial_number, sender, depth, back_route)

originator: the root node.
serial_number: a random number used to find out later repeated coming of a same multicast message from a same originator.
sender: the forwarding node of the message.
depth: *sender*'s depth in the broadcasting tree, which is a spanning tree converted from the current connected component with root at the querying node.
back_route: the route back to the *originator*

The format of a*llNeighbors* message is

a*llNeighbors(sender, originator, neighbors, serial_number, depth, route)*

originator: *originator* of received corresponding *mGetAllNeighbors* message
neighbors: all neighbors except the sender of corresponding *mGetAllNeighbors* message
serial_number: *serial_number* of corresponding *mGetAllNeighbors* message
depth: depth of current node
route: route from current node to *originator*

6.3.2. Algorithm

peer *u*

Const
init: initialization flag, set to **true** at beginning
in-que-len: length of incoming message queue
out-que-len: length of outgoing message queue

input

size: number of nodes in the connected MANET
msg_rate: message processing rate
x: the querying node at the root of the broadcasting tree
max_depth: maximum depth, usually log(*size*), at most *size*
timeout: upper bound of running time of whole procedure

Note: Assume rates for incoming messages and outgoing messages are same

var
in_queue: incoming message queue
brd_in_queue: incoming multicast message queue
out_queue: outgoing message queue
T: external timer, simulator by discrete counter
N: set of neighbors, including all hops
Γ: set of one-hop neighbors
Γ_0: successor
R: *u*'s routing records for this algorithm
r: a route in *R*
s: a node in a set with smallest node ID
current_completed: if all searching is completed at current level
rnd: a random number
received_brdcst: set of all received multicast messages

Library Function
route(*a*, *b*): return a route from node *a* to node *b*. Actually a route is a string or vector. *route*(*a*, *a*) returns *a*.
distance(*u*, *h*): RAN distance function
reverse(*r*): return the reverse path of route *r*

Action
(a1) init →
    construct Γ
    if $\Gamma \neq \varnothing$
    then
        find Γ_0
        T := 0
        init := **false**
        rnd := *getRandomNum*();
        back_route := *route*(*u*, *u*)
        multicast *mGetAllNeighbors*(*u*, *rnd*, *u*, 0, *back_route*)
        R := $\varnothing$
        for each *h* in Γ **do**
            R := *R* + {*uh*}
        N := Γ
    []

Note: Here route(u, h) = uh. uh is a series, like string, vector in C++, ArrayList in Java.

(a2) **receive** *mGetAllNeighbors* →
    if (*<originator, serial_number>* $\notin$ *received_brdcst*) **and** (*depth* < *max_depth*)
        back_route := *back_route* + route(*sender*, *u*)
        received_brdcst := *received_brdcst* + {*<originator, serial_number>*}
        multicast *mGetAllNeighbors*(*originator*, *sn*, *u*, *depth* + 1, *back_route*)
        route := reverse(*back_route*)
        send a*llNeighbors*(*u*, *originator*, Γ - {*sender*}, *serial_number*, *depth* + 1, *route*) **to** *originato*
    []

(a3) **receive** a*llNeighbors*(*q*, *u*, *S*, *sn*, *d*, *rt*) **from** *q* →
    for each *h* in ($S - N$) **do**
        route(u, h) := route(u, q) + h
        $R := R + \{$ route(u, h)$\}$
        if *distance*(u, h) < *distance*(u, Γ_0) **then** $\Gamma_0 := h$
    $N := N + (S - N)$
    []

(a4) **receive** a non-multicast message destined for another node →
    send the message to next node on the route
    []

(a5) T > *timeout* →
    disable (a1), (a3), and (a5)
    stop the whole procedure for *u*
    []

Note: For all **receive** primitives, actual triggering event is: message is taken out from *in-que*.

6.4. Random Pattern

The basic objective of random pattern is to seek high efficiency, faster convergence time, i.e. build topology faster, instead of completeness. In another word, random pattern prefers speed to the quality of ring. In this pattern, at each level of the component tree, not every node is searched like in exhaustive patterns. We pick up only one.

One way to do it is: always pick up the closest neighbor of current node. However, since node ID is assigned randomly from a huge ID space, which has nothing to do with other properties of a node, such as neighborhood. The implied strategy behind this approach — a node *x* closer to *u* may has neighbor even more closer to *u* — is not tenable. However, if we use pure random selection, we will lose the ending condition. Maybe we can use the depth of searching path as an end condition.

From this point of view, RN may have chosen a better approach. Closer Peer Search in RN actually searches a node similar to the predecessor of *u*. *u* is the current node searching closer successor. The underlying logic is: the successor of *u*'s predecessor definitely has better chance to be close to *u*.

6.4.1. Random Pattern Message Format

closerNodeSearch(*starting_node*, *master*, *serial_number*, *sender*, *depth*, *route*, *back_route*)

starting_node: boolean variable, indicating if this is the first *closerNodeSearch* message for the starting node.
master: the root node.
sender: the real creator and sender of the *closerNodeSearch* message.
receiver: the destination node.
serial_number: a random number used to find out later repeated coming of a same broadcast message from a same master.
sender: the forwarding node of the message.
depth: sender's depth in the broadcasting tree, which is a spanning tree converted from the current connected component with root at the querying node.
route: Route from sender to receiver.
back_route: the route back to the master.

successorCandidate(*sender*, *successor*, *receiver*, *serial_number*, *route*)

sender: the sender, i.e. current node.
successor: the successor of current node.
receiver: receiver of this successorCandidate message. It should be the master of its received closerPredecessorSearch message.
serial_number: serial_number.
route: the route from sender node to the master.

newCNNeighbors(*sender*, *receiver*, *serial_number*, *route*, *neighbor_set*)

sender: the sender of message, i.e. current node.
receiver: receiver of this message. It should be the master of its received *closerNodeSearch* message.
serial_number: serial_number.
route: the route back to the master.
neighbor_set: starting node's neighbors

6.4.2. Random Pattern Algorithm

Version 5

type (class)
component_node: element of variable *component_queue*
component_node.nodeIdx: node index, internal expression, not node ID.

component_node.traversed: Indicates if a node has been traversed in current searching node's Random execution.
component_node.in_route: record route from current searching node to this node.

input .
max_depth: maximum depth, usually log(*size*), at most *size*
timeout: upper bound of running time of whole procedure

var
init := **true**: mark the very beginning of algorithm
in_queue: incoming message queue
out_queue: outgoing message queue
S: Set of nodes
B: Set of successor candidates obtained while forwarding other nodes' *closerNodeSearch* messages
W: Set of successor candidates obtained from own *closerNodeSearch* messages
w: a node (successor candidate)
x: peer being searched for
T: external timer, simulator by discrete counter
Γ: set of one-hop neighbors
Γ_0: successor
R: *u*'s routing records for this algorithm
s: a node
rnd: a random number
new_round: indicates this round should stop and a new round should be started
component_queue: The BFS connected component queue, used as the source set to feed the *closerNodeSearch*
dumped_component: dumped_component contains those nodes poped out from component_queue

Library Function
route(*a*, *b*): return a route from node *a* to node *b*. Actually a route is a string or vector. *route*(*a*, *a*) returns *a*.
distance(*u*, *h*): RAN distance function
reverse(*r*): return the reverse path of route *r*
empty(*x*): return true if set or series x is empty, otherwise return false

Action

(a1) *init* →
Local
cn: component node
init := **false**
construct Γ
in_queue := ∅

```
out_queue := ∅
if Γ = ∅
then
    new_round := false
    return
else
    find Γ0
    new_round := true
    component_queue := ∅
    dumped_component := ∅
    for each h ∈ Γ
        cn := new(component_node)
        cn. nodeIdx := h;
        cn.traversed := true
        cn.in_route := {u} + {h}
        push_back (component_queue, cn)
[]

(a2) new_round and (not empty(component_queue)) →
    new_round := false
    B := W := ∅
    cn1 := pop_front(component_queue)
    push_back (dumped_component, cn1)
    rnd := getRandomNum()
    route := cn1.in_route
    back_route := reverse(route)
    send closerNodeSearch(true, u, rnd, u, 0, route, back_route) to cn1
    []

(a3) receive closerNodeSearch(starting_node, x, sn, q, depth, route, back_route))
        from q →
    Local
    send_candidate := false
    next

    if (depth >= max_depth)
    then
        send_candidate := true
        new_round := true
    else if u is closer to x than any u's neighbor h ∈ u.Γ
    then
        send_candidate := true
        B := B ∪ {x}
    if send_candidate = true
    then
```

send *successorCandidate*(u, Γ_0, x, sn, *back_route*) **to** x
else
$next$:= $\text{argmin}_{k\in\Gamma}\, d(k, x)$
rt := *route*(u, *next*)
bk_route := *reverse*(rt) + *back_route*
send *closerNodeSearch*(**false**, x, sn, u, *depth* + 1, rt, *bk_route*) **to** *next*
if *starting_node* = **true**
send *newCNNeighbors*(u, x, sn, *back_route*, $u.\Gamma$)
[]

(a4) **receive** *successorCandidate*(q, w, *org*, sn, *route*) **from** $q \rightarrow$
if u = *org*
then
$W := W \cup \{w\}$
$S := W \cup B \cup \Gamma$
$\Gamma_0 := \text{argmin}_{k \in S}\, d(u, k)$
new_round := **true**
else send the message **to** next node on the route
[]

(a5)**receive** *newCNNeighbors*(q, *org*, sn, *route*, *nb_set*) **from** $q \rightarrow$
if u = *org*
then
for each $h \in$ *nb_set*
if $h \not\in$ (*component_queue* $\cup$ *dumped_component*)
then
cn2 := *new*(*component_node*)
cn2. nodeIdx := h
cn2.traversed := **false**
cn2.in_route := $\{u\} + \{h\}$
push_back (*component_queue*, *cn2*)
[]

7. Simulation

To simplify programming, the simulation is based upon static network. As mentioned before, the mobile situation will be addressed in WRAN protocol, which is not covered by this chapter. It is recommended that bootstrapping is launched at a relatively less mobile setting; since mobility will change the composition of connected components. All discussion involving mobility is based upon the premise that the change in components caused by mobility disturbance should be limited to a reasonable range. The premise validates that early research could be based upon static assumption.

This simulator is not for one single connected component; instead it is for a MANET randomly generated on a 100 × 100 two–dimensional square. The length unit is meter. Each node is independently generated with node x coordinate and y coordinate uniformly distributed in range [0, 100], and node ID uniformly distributed in range [0, 65535]. The direct connectivity between two nodes is purely decided by their Euclidean distance and the uniform radio range for all nodes in the MANET. It is much more realistic than generating only one connected component.

Basic parameters tested include completeness, time, number of sent messages, and number of received messages. The completeness examines the effective of algorithm by checking completeness of rings generated. Time is the time used to construct rings. Two messages measure the message complexity.

7.1. Completeness

Completeness is the ratio of number of nodes in generated rings to number of all nodes. The simulation shows that all nodes are either in rings, or in lines. Each line is connected to one and only one ring. If more than one rings exist in one component, only nodes in the biggest ring is counted in the completeness calculation as nodes in rings. Isolated nodes are regarded as rings, so they are always counted as constructed. This is rational for some MANETs which are unfortunately initialized with considerable isolated nodes. Each connected component has one or more rings which are connected by lines. The bottom line is: at any time of the construction, even before anything is done for the construction, all nodes in same component should always remain in the same component. The construction only changes the number of rings and the nodes in rings in the component. It does not change the component, as the component is defined by the neighborhood relation among nodes, which remains identical in a static network.

Table 1. Completeness of Algorithms

Network Size	20	40	60	80	100
Random Pattern	0.93	0.705	0.547	0.395	0.343
Distributed Exhaustive Pattern	1	1	1	0.98	0.97
Centralized Exhaustive Pattern (Plain)	1	1	1	1	0.99
Centralized Exhaustive Pattern (Approximation 0.85)	0.94	0.89	0.87	0.855	0.86

7.2. Time

Time is defined as the algorithmic time used in one run of simulation, from beginning to end. The critical question is how the end condition is defined in simulation. As we mentioned in Section 5.5.1, normally the end condition is defined by the formation of ideal Chord ring, which is unique and fixed for a given MANET. For a pattern that needs too much time to finish, the end condition could be adapted by using the approximation option. For all combinations of patterns and options, the algorithmic time unit is set as virtually synchronized discrete time unit. In each unit, all nodes are supposed to complete the processing of m incoming messages and n outgoing messages. Except in multicast option, m

is assumed to equal to n. m and n are determined by the simulation parameter message processing rate.

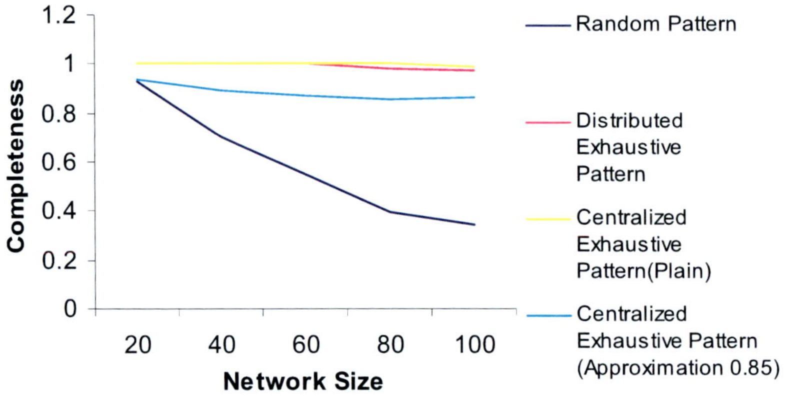

Figure 12. Completeness of algorithms.

Table 2. Time used in ring construction

Network Size	20	40	60	80	100
Random Pattern	11.8	15	32	63.6	91
Distributed Exhaustive Pattern	8.3	21.8	43	87.4	98.5
Centralized Exhaustive Pattern (Plain)	5.8	56.4	1091.4	2552	21409
Centralized Exhaustive Pattern	6	49.6	313.8	632	910

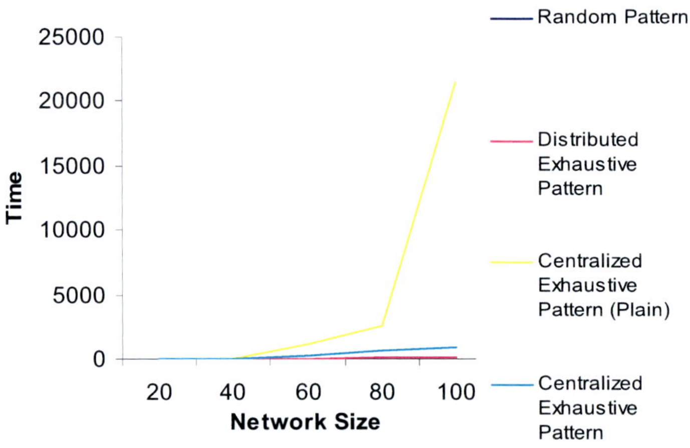

Figure 13. Time used in ring construction

7.3. Message Complexity

Table 3. Messages Sent

Network Size	20	40	60	80	100
Random Pattern	72.8	1117.8	4721	13839	25117.3
Distributed Exhaustive Pattern	103.6	1355.1	5089.8	16077	26853.5
Centralized Exhaustive Pattern (Plain)	412	15863.4	162705.6	465754.6	2581659.6
Centralized Exhaustive Pattern (Approximation 0.85)	226	5556.8	62625.4	186703	364112.2

Messages complexity is measured by two parameters: messages sent and messages received. The message sent is defined as total number of messages sent by all nodes in the network during the simulation. The message received is defined as total number of messages received by all nodes in the network during the simulation. These messages are not user data related. They are pure control messages used to create the ring.

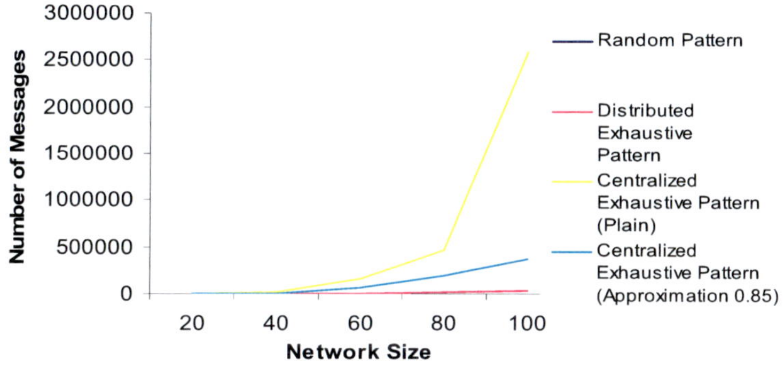

Figure 14. Messages sent

Table 4. Messages Received

Network Size	20	40	60	80	100
Random Pattern	72.8	1117.8	4721	13839	25117.3
Distributed Exhaustive Pattern	98	1312.8	5045	15938.3	26783
Centralized Exhaustive pattern	394.6	15639	161335.2	460179.4	2539312.2
Centralized Exhaustive Pattern (Approximation 0.85)	209.8	5452.2	61693.8	183749.6	358354

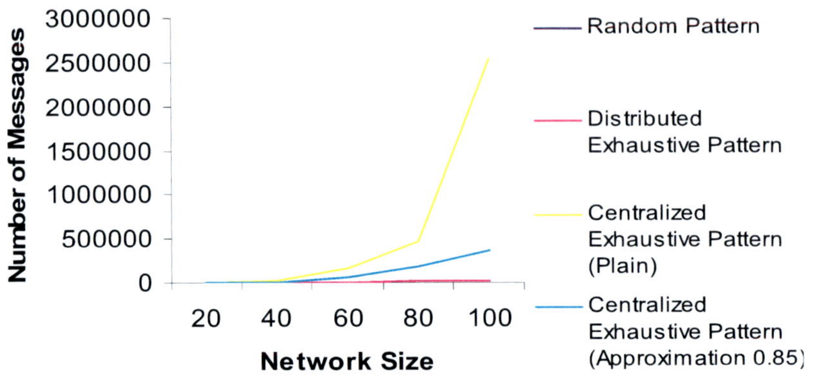

Figure 15. Messages received.

7.4. Analysis of Simulation Results

Obviously, the winner is the distributed exhaustive pattern. It shows perfect effectiveness, at the same time and unlike other two patterns, it has no serious side effect. Other two, however, suffer from different fatal problems. For random pattern, it is the effectiveness. For virtual centralized exhaustive pattern, it is efficiency.

As shown in preceding sections, it is clear that in RAN family, random pattern has the best overhead cost in both time complexity and message complexity. However, it is also the worst approach in terms of quality of ring constructed. The reason is behind its searching strategy. The closest first criterion does not make sense in a pure stochastic uniform distribution of ID space. The closest successor could be hidden anywhere in the component tree. It could be child of any node. It may be child of current closest successor, or child of current farthest successor, or child of any other node. The lesson is: in face of such complete randomness, exhaustion in search is necessary. It has been illustrated clearly. Exhaustive approaches almost always returns the best rings, unless we intentionally prohibit it from doing so with an approximation.

The simulation demonstrates poor efficiency of centralized approach. The virtual centralized exhaustive pattern is worst in terms of cost in time and message. On the other hand, its twin approach, the distributed exhaustive pattern shows tremendous divergent performance. It proved that exhaustion is not necessarily a synonym of expenditure. The distributed exhaustive pattern has almost same efficiency as the random pattern.

The third enlightenment is at a high level of abstraction, it is kind of philosophy. As the author always advocate, the benefit of decentralization has shown by the distributed exhaustive pattern. This result also raises a question: could centralization be implemented above decentralized infrastructure?

8. Conclusion

This chapter introduces a novel approach for bootstrapping P2P overlay ring over MANETs. Originally it could be used to all ring-based P2P systems, like Pastry and recent Virtual Ring

Routing. This approach benefits from successful P2P topology construction methods in wired networks. To the best of the author's knowledge, this approach is the first successful attempt in this field.

RAN protocol is proposed for ring topology construction. RAN builds perfect ring in P2P ID space. It integrates upper layer into lower layer. No underlying routing protocol is needed. Ring-based P2P systems could be immediately put into normal operation upon the ring. RAN includes a variety of algorithms. Pros and cons of these algorithms are shown, both in theory and in simulation. Simulation shows that the distributed exhaustive pattern is the best in terms of effectiveness and efficiency.

This is a new area of study; many questions are unanswered; many research topics could be developed. Here we only give one example. As explanation for poor effectiveness of both Ring Network and RAN-Random, the very idea to keep tracing the closest node at every round of closer successor search, does not yield best successor, nor does using its other varieties like searching for closest predecessor. Mathematical analysis and simulation results both show the weakness of this approach: its guideline of finding closest node limits its range of choice. To improve this but not going to another extreme of exhaustive search, another approach could be tried. That is: keep the random itinerary but discard the closest standard. However, a follow-up question would be immediately raised, that is: without the closet criterion, what can be our end condition? Simplest answer is search depth or search time, or quality of returned node. In this direction, we guess the biggest gold mine may be under the way. It could be the most prospective follow-up research for RAN.

References

[BEGH2004] M. Brahami, Patrick Th. Eugster, Rachid Guerraoui, Sidath B. Handurukande: BGP-Based Clustering for Scalable and Reliable Gossip Broadcast. Global Computing 2004, pp. 273-290, Rovereto, Italy, 2004

[BRAN2006] TC (Technical Commitee) BRAN, "ETSI HIPERLAN/2 Standard," http://portal.etsi.org/radio/HiperLAN/HiperLAN.asp, June 2006 (Last updated: 2006-06-05 15:13:24).

[Boleng2002] J. Boleng, "Efficient Network Layer Addressing for Mobile Ad Hoc Networks," Proceedings of the International Conference on Wireless Networks (ICWN'02),
pP 271–277, Las Vegas, NV, June 2002.

[Borg2003] Joseph Borg, "*A Comparative Study of Ad Hoc & Peer to Peer Networks*", M.S. Thesis, University College London, August 2003.

[CAG2005] S. Cheshire, B. Aboba, and E. Guttman, *"Dynamic configuration of IPv4 link-local addresses,"* Proposed Standard, Internet Engineering Task Force, draft-ietf-zeroconf-ipv4-linklocal, May 2005.

[CCL2003] Imrich Chlamtac, Marco Conti, Jennifer J.-N. Liu, "*Mobile ad hoc networking: imperatives and challenges*", Ad Hoc Networks 1, pp 13–64, Elsevier Press, 2003.

[CCNOR2006] Matthew Caesar, Miguel Castro, Edmund B. Nightingale, Greg O'Shea, and Antony Rowstron, "*Virtual Ring Routing: Network Routing Inspired by DHTs,*" Proc. ACM SIGCOMM 2006.

[CF2006] Curt Cramer and Thomas Fuhrmann, *"Bootstrapping Chord in Ad Hoc Networks: Not Going Anywhere for a While,"* Fourth Annual IEEE International Conference on Pervasive Computing and Communications Workshops (PERCOMW'06), pp. 168-172, Pisa, Italy, 2006.

[CWLG1997] C.-C. Chiang, H.K. Wu, W. Liu, M. Gerla, *"Routing in clustered multihop, mobile wireless networks with fading channel,"* Proceedings of IEEE SICON97, pp. 197–211, April 1997.

[Clip2] Clip2, "*The Gnutella Protocol Specification v0.4*," Document Revision 1.2, http://www9.limewire.com/developer/ gnutella_protocol_0.4.pdf.

[Cohen2003B] 33. Bram Cohen, "*BitTorrent Economics Paper*," May 2003 http://bitconjurer.org/BitTorrent/bittorrentecon.pdf

[Cohen2003I] Bram Cohen, "*Incentives Build Robustness in BitTorrent*," May 2003. http://www.bittorrent.com/bittorrentecon.pdf

[DBKKMSB2001] Frank Dabek, Emma Brunskill, M. Frans Kaashoek, David Karger, Robert Morris, Ion Stoica, Hari Balakrishnan, *"Building Peer-to-Peer Systems with Chord, a Distributed Lookup Service,"* In the Proceedings of the 8th Workshop on Hot Topics in Operating Systems (HotOS-VIII), Schloss Elmau, Germany, May 2001.

[DVH2003] Nitin Desai, Varun Verma and Sumi Helal, *"Infrastructure for Peer-to-Peer Applications in Ad-Hoc Networks"*, 2nd International Workshop on Peer-to-Peer Systems(IPTPS), Berkeley, CA, February 2003.

[EE2000] J. Elson and D. Estrin. An Address-free Architecture for Dynamic Sensor Networks. Technical Report 00-724, Computer Science Department, USC, January 2000.

[FL2001] James A. Freebersyser, Barry Leiner, *"A DoD perspective on mobile ad hoc networks"*, in: Charles E. Perkins (Ed.), Ad Hoc Networking, pp. 29–51, Addison Wesley, Reading, MA, 2001.

[Gast2002] M. S. Gast, "*802.11 Wireless Networks – The Definitive Guide*," O'Reilly & Associates, California 2002.

[Gouda1998] M. G. Gouda, "*Elements of Network Protocol Design*," John Wiley and Sons, 1998.

[HDN2003] R. Hinden, S. Deering, E. Nordmark, "*RFC3587: IPv6 Global Unicast Address Format*," Proposed Standard, Internet Engineering Task Force, August 2003.

[HDVL2003] Sumi Helal, Nitin Desai, Varum Verma and Choonhwa Lee, "*Konark - A Service Discovery and Delivery Protocol for Ad-Hoc Networks*", Proceedings of the Third IEEE Conference on Wireless Communication Networks(WCNC), New Orleans, Louisiana, March 2003.

[HGRW2006] T. Heer, S. Gotz, S. Rieche, and K. Wehrle, "*Adapting Distributed Hash Tables for Mobile Ad Hoc Networks,*" *Proceeding of Fourth Annual IEEE International Conference on Pervasive Computing and Communications Workshops*, pp.173 – 178, 2006.

[HGS1987] W. D. Hillis, J. Guy, and L. Steele, "*Data Parallel Algorithms*," *Communication ACM*, **30**(1), pp.78–78, 1987.

[HPD2003] Y. C. Hu, H. Pucha, and S. M. Das, "*Exploiting the Synergy between Peer-to-Peer and Mobile Ad Hoc Networks,*" Proceedings of HotOS-IX: Ninth Workshop on Hot Topics in Operating Systems, Lihue, Kauai, Hawaii, May 2003.

[Henson2003] Val Henson, "*An Analysis of Compare-by-Hash*," Proceedings of the 9th Workshop on Hot Topics in Operating Systems, Lihue, Hawaii, May 2003

[Heritage2000] American Heritage. "*The American Heritage Dictionary of the English Language*", Fourth Edition, Houghton Mifflin Company, Boston, MA, January 2000.

[Ivkovic2001] Igor Ivkovic, "*Improving Gnutella Protocol: Protocol Analysis and Research Proposals*," Prize-Winning Paper for LimeWire Gnutella Research Contest, September 2001

[JB2005] M. Jelasity and O. Babaoglu, "*T-Man: Gossip-based overlay topology management*," In Engineering Self-Organising Applications (ESOA'05), 2005.

[JGKS2004] M. Jelasity, R. Guerraoui, A.-M. Kermarrec, and M. van Steen, "*The Peer Sampling Service: Experimental Evaluation of Unstructured Gossip-Based Implementations*," In Middleware 2004, volume 3231 of Lecture Notes in Computer Science, pp. 79–98, Springer-Verlag, 2004.

[JHB2001] K. Jenkins, K. Hopkinson, and K. Birman, "*A Gossip Protocol for Subgroup Multicast*," In International Workshop on Applied Reliable Group Communication (WARGC), April 2001.

[JM1996] D.B. Johnson, D.A. Maltz, "*Dynamic source routing in adhoc wireless networks*," in: T. Imielinski, H. Korth (Eds.), Mobile Computing, pp. 153–181, Kluwer Academic Publishers, Dordrecht, 1996.

[KLW2003] A. Klernm, C. Lindemann and O. Waldhorst, "*A special Purpose Peer-to-Peer File Sharing System for Mobile Ad Hoc Networks*", Proc. Workshop on Mobile Ad Hoc Networking and Computing (MADNET 2003), Sophia-Antipolis, France, pp 41-49, March 2003.

[Kaashoek2003] Frans Kaashoek, "*Peer-to-peer computing research: a fad?*" 2003 http://project-iris.net/talks/dht-toronto-03.ppt

[Kortuem2001] Gerd Kortuem. "*Proem: A Peer-to-Peer Computing Platform for Mobile Ad-hoc Networks*", Online proceedings of Advanced Topic Workshop in Middleware for Mobile Computing, November 2001.

[LJLQC2004] Sei-yon Lee, Ju-wook Jang, Kyung-Geun Lee, Lan Quan, Tae-kyoung Cho, "*A Peer-to-Peer Search Scheme over Mobile Ad hoc Networks*," (ISPC) International Scientific-Practical Conference 2004, Institute of Mathematics of National Academy of Sciences (IM NAS, Bishkek, Kyrgyz Republic), 2004.

[LLS2004] Mei Li, Wang-Chien Lee, Anand Sivasubramaniam, "*Efficient peer to peer information sharing over mobile ad hoc networks*," the Second WWW Workshop on Emerging Applications for Wireless and Mobile Access (MobEA'04), New York City, NY, May 2004.

[LMM2000] Meng-Jang Lin, Keith Marzullo, and Stefano Masini, "*Gossip versus Deterministically Constrained Flooding on Small Networks*," Proceedings of the International Symposium on Distributed Computing (DISC), Toledo, Spain, October 2000.

[LW2002] C. Lindemann and O. Waldhorst, "*A Distributed Search Service for Peer-to-Peer File Sharing in Mobile Applications*," Proceeding of. 2nd IEEE Conference on Peer-to-Peer Computing, 2002.

[MDMD2001] Archan Misra, Subir Das, Anthony McAuley, and Sajal K. Das, "*Autoconfiguration, Registration, and Mobility Management for Pervasive Computing*", IEEE Personal Communication, pp 24-31, August 2001

[MG1996] S. Murthy and J.J. Garcia-Luna-Aceves, "*An Efficient Routing Protocol for Wireless Networks*", ACM Mobile Networks and Application, Special Issue on Routing in Mobile Communication Networks, pp. 183-97, Oct. 1996.

[MMA2000] K-C Mei, R Mathur, S. K. Agarwal, "*Gossip Style Data Stability in Networks*," Project Report, Boston University, December 2000.

[MP2002] Mansoor Mohsin and Ravi Prakash, "*IP Address Assignment in Mobile Ad Hoc Networks,*" Proceedings of IEEE MILCOM, September 2002.

[Muthusamy2003] Vinod Muthusamy, "*An Introduction to Peer-to-Peer Networks*," October 2003. http://www.eecg.toronto.edu/~jacobsen/mie456/slides/p2p-mie.pdf

[Naugle1998] Matthew Naugle, "*Illustrated TCP/IP – A Graphic Guide to the Protocol Suite*," John Wiley & Sons, Inc., November 1998.

[OSL2003] B. Oliveira, I.G. Siqueira, A.A. Loureiro, "*Evaluation of ad-hoc routing protocols under a peer-to-peer application*," IEEE Wireless Communication and Networking Conference, 2003

[OSMLWN2005] Leonardo B. Oliveira, Isabela G. Siqueira, Daniel F. Macedo, Antonio A. F. Loureiro, Hao Chi Wong, Jose M. Nogueira, "*Evaluation of Peer-to-Peer Network Content Discovery Techniques over Mobile Ad Hoc Networks*," IEEE International Symposium on a World of Wireless, Mobile and Multimedia Networks (WOWMOM'05), pp. 51-56, Taormina, Italy, June 2005.

[PB1994] C.E. Perkins, P. Bhagwat, "*Highly dynamic destination-sequenced distance-vector routing (DSDV) for mobile computers*," Computer Communications Review (October 1994), pp. 234–244, 1994.

[PC1997] V.D. Park, M.S. Corson, "*A highly adaptive distributed routing algorithm for mobile wireless networks*," in: Proceedings of INFOCOM 97, April 1997.

[PDH2004] Himabindu Pucha, Saumitra M. Das, Y. Charlie Hu, "*Ekta: An Efficient DHT Substrate for Distributed Applications in Mobile Ad Hoc Networks*," Sixth IEEE Workshop on Mobile Computing Systems and Applications, pp. 163-173, 2004.

[PMWBS2001] C. Perkins, J. Malinen, R. Wakikawa, E. Belding-Royer, and Y. Sun, "*IP Address Autoconfiguration for Ad Hoc Networks*", IETF Internet-Draft, November 2001 (work in progress).

[PR1999] C.E. Perkins, E.M. Royer, "*Ad-hoc on-demand distance vector routing*," in: Proceedings of 2nd IEEE Workshop on Mobile Computing Systems and Applications, February 1999.

[PS2001] M. Papadopouli and H. Schulzrinne, "*Effects of power conservation, wireless coverage and cooperation on data dissemination among mobile devices*," Proceedings of ACM Interational Symposiumon Mobile Ad Hoc Networking and Computing (MobiHoc), pp 117–127, October 2001.

[RD2001] A. Rowstron and P. Druschel, "*Pastry: Scalable, distributed object location and routing for large-scale peer-to-peer systems*," Proceeding of IFIP/ACM International Conference on Distributed Systems Platforms (Middleware), pp 329-350, Heidelberg, Germany, November 2001.

[RFHKS2001] S. Ratnasamy, P. Francis, M. Handley, R. Karp, S. Shenker, "*A Scalable Content-Addressable Network*," Proceedings of the SIGCOMM, pp 161-172, 2001

[RR2002] R. Ramanathan and J. Redi, "*A Brief Overview of Ad Hoc Networks: Challenges and Directions*," IEEE Communications, 50th Anniversary Commemorative Issue, pp. 20-22, May 2002.

[RS1998] Ram Ramanathan, Martha Steenstrup, "*Hierarchically organized, multihop mobile wireless networks for quality of service support*", Mobile Networks and Applications, 1998 No.3, pp101–119, 1998.

[RT1999] Elizabeth M. Royer, Chai-Keong Toh, "*A Review of Current Routing Protocols for Ad Hoc Mobile Wireless Networks*", IEEE Personal Communications, Vol. 6, No. 2, pp. 46-55, April 1999.

[SB2003] Y. Sun and E. M. Belding-Royer, "*Dynamic Address Configuration in Mobile Ad hoc Networks*," Technical Report 2003-11, Computer Science Department, UCSB, March 2003.

[SMKKB2001] Ion Stoica, Robert Morris, David Karger, M. Frans Kaashoek, and Hari Balakrishnan, "*Chord: A Scalable Peer-to-peer Lookup Service for Internet Applications*," In Proceeding of ACM SIGCOMM 2001, pp. 149-160, San Diego, CA, August 2001.

[SR2005] A. Shaker and D. S. Reeves, "*Self-stabilizing structured ring topology P2P systems*," Technical Report 2005-25, Department of Computer Science, N.C. State University, 2005.

[VM2003] John Viega and Matt Messier, "*Secure Programming Cookbook for C and C++: Recipes for Cryptography, Authentication, Input Validation & More*," 1 edition, O'Reilly Media, Inc, July 2003

[Vaidya2002] N. H. Vaidya, "*Weak Duplicate Address Detection in Mobile Ad Hoc Networks*", Proceedings of the ACM International Symposium on Mobile Ad Hoc Networking and Computing (MobiHoc' 02), pp 206–216, Lausanne, Switzerland, June 2002.

[Wilensky1999] U. Wilensky, "*NetLogo*," http://ccl.northwestern.edu/netlogo/, Center for Connected Learning and Computer–Based Modeling, Northwestern University, Evanston, IL., 1999.

[ZNM2003] H. Zhou, L. Ni, and M. Mutka, "*Prophet Address Allocation for Large Scale MANETs*," Proceedings of the IEEE Conference on Computer Communications (INFOCOM 2003), San Francisco, CA, March 2003.

In: Wireless Communications Research Trends
Editor: Tong S. Lee, pp. 55-81
ISBN 978-1-60021-674-9

Chapter 2

SARC: SECURE ANONYMOUS ROUTING FOR CLUSTER BASED MANET

***Lijun Qian*[1]*, *Ning Song*[1]† and *Xiangfang Li*[2]‡**
[1]CeBCom Research Center, Department of Electrical and Computer Engineering, Prairie View A&M University, Prairie View, TX 77446, USA
[2]WINLAB, Rutgers University, Piscataway, NJ 08854, USA

Abstract

Providing security and anonymity to users are critical in mobile ad hoc networks (MANETs). In this study, a cluster based security architecture is employed for better management of mobile users and scalability, and a Secure Anonymous Routing scheme for Cluster based MANET termed SARC is proposed. SARC includes intra-cluster routing and inter-cluster routing, where intra-cluster routing uses a common broadcast channel to provide anonymity, and inter-cluster routing uses a sequence of temporary public keys as the trapdoor information. One of the unique features of SARC is that critical network elements, such as the cluster heads, are hidden from adversaries during the routing process. In order to maximize routing efficiency, key indexing is used and symmetric cipher is employed in most part of the proposed scheme to reduce computational complexity. In addition, a technique based on XOR operations for data forwarding is applied to provide anonymity and maximize efficiency during data transmissions. The tradeoff between security/anonymity and efficiency is also addressed. Analytical results are derived using information theoretic measure for anonymity analysis of the proposed scheme. Detailed implementation of SARC is provided and extensive simulations are performed for a large (16 clusters, 800 nodes) network using OPNET. It is observed that SARC has good scalability and it introduces very limited overhead comparing to other cluster based routing protocol which has no security features. Route establish time and packet delivery ratio are also evaluated while taking into account node mobility. Both anonymity analysis (including sender anonymity, receiver anonymity and sender-receiver anonymity) and attack analysis show the effectiveness of SARC against a wide range of strong adversarial attacks.

*E-mail address: liqian@pvamu.edu
†E-mail address: nsong@pvamu.edu
‡E-mail address: xfli@winlab.rutgers.edu

1. Introduction

There has been great interest in mobile ad hoc networks (MANETs) recently since they have tremendous military and commercial potential. In order to deploy MANET to cover a large area and to support a large amount of users, scalability of such networks is one of the main concerns. It has been shown that proper clustering in MANET reduces the complexity of link-layer and routing protocol design significantly and improves the scalability of the protocols [1], [2]. In addition, clustering increases the network capability of supporting Quality-of-Service (QoS) [3]. Clustering is also desirable because of practical reasons. For instance, in a battlefield deployment, a cluster may be naturally formed by a set of soldiers equipped with wireless communication devices and a tank serving as cluster head (CH). Hence, a cluster based architecture is employed in this work.

Security is of paramount importance in cluster based MANET. Routing security is one of the main concerns because routing is needed for both intra-cluster and inter-cluster communications, and the adversary may perform various attacks on the routing traffic. Usually active attacks are easy to detect and there are many research proposals on security architecture and secure routing in cluster based MANET, such as [10], [11], [19], [20], that addressed active attacks. On the contrary, passive attacks are very difficult to detect and they will provide opportunities for effective active attacks when critical network elements (such as the CHs) are located [9]. In addition, MANET is extremely vulnerable to passive attacks based on eavesdropping and traffic analysis. Thus, providing anonymity to users (including users' identities, locations, data, etc.) is critical, especially in a hostile environment, such as in a battlefield. For example, it is important to protect the CHs by making them indistinguishable from other nodes in the network, thus keeping them anonymous to the enemy. Hence, the focus of this work is on assurance of mobile users' anonymity during the routing process rather than just routing security itself.

Secure anonymous routing is one of the primary counter-measures to various attacks (especially passive attacks) on the routing traffic. It is very challenging to provide both security and anonymity in MANET, because of the infrastructure-less nature of the network, limited network resources and numerous possible attacks. Although there are many secure anonymous routing proposals for MANET recently (a detailed review is given in Section 7. of this chapter), none of them address the secure anonymous routing design in cluster based MANET (except our previous work [36] and HANOR [35]). However, critical network elements such as the CHs are not hidden in HANOR, thus they may be located and attacked by adversaries. Furthermore, since the inter-cluster routing packets will be broadcast and processed by every node rather than just the gateways (GWs), the overhead is still high for HANOR. We are motivated to provide an efficient secure anonymous routing scheme for cluster based MANET that expected to cover a large geographic area and manage a large amount of users. In this study, a novel Secure Anonymous Routing scheme for Cluster based MANET (SARC) is proposed to provide both security and anonymity, and to prevent various attacks on the routing traffic. One of the unique features of SARC is that CHs are hidden from adversaries during the routing process. Specifically, the proposed scheme will provide privacy for mobile users, including both identity privacy and location privacy as defined in [6]. In other words, no one should know the real identities of the source and the destination of a route except themselves, and the source and the destination have no information about the real identities of intermediate nodes en route. In addition, no one should know the exact location of the source or the destination except themselves, and other nodes, typically intermediate nodes en route, should have no information about their distance from either the source or the destination [6]. The proposed scheme will also protect routing and data traffic from traffic analysis and packet analysis attacks. Active attacks such as fabrication and modification of packets, and Denial-of-Service (DoS) attacks are not treated in this study since they can be easily detected and thwarted by an Intrusion Detection System (IDS) [9].

In order to maximize routing efficiency, the technique of key indexing is used and the tradeoff between security/anonymity and efficiency is also discussed. Analytical results are derived using information theoretic measure [24], [25] for anonymity analysis of the proposed scheme. The attack model defined by [22] is adopted for attack analysis. Both anonymity analysis (including sender anonymity, receiver anonymity and sender-receiver anonymity) and attack analysis show the effectiveness of the proposed scheme against a wide range of strong adversarial attacks (except for a global adversary, dummy traffic may be injected to thwart traffic analysis attacks). Detailed implementation of SARC is provided and extensive simulations are performed for a large (16 clusters, 800 nodes) network using OPNET. It is observed that the proposed scheme introduces very limited overhead comparing to other cluster based routing protocol which has no security features. Route establish time and packet delivery ratio are also evaluated while taking into account node mobility. In this study, only network-layer security and privacy are considered. Security and privacy issues at other layers such as the physical-layer are out of the scope of this work.

The chapter is organized as follows. Section 2. provides details of the cluster based architecture and assumptions. Secure anonymous routing schemes for intra-cluster and inter-cluster traffic are proposed in Section 3.. Secure anonymous data transmissions are discussed in Section 4.. Section 5. provides anonymity analysis and attack analysis. Simulation results of protocol overhead, route establish time and packet delivery ratio are presented in Section 6.. Related works are discussed in Section 7.. Section 8. contains the concluding remarks.

2. Architecture and Assumptions

In this study, it is assumed that all the nodes are grouped into a number of overlapping or disjoint clusters in a distributed manner. A cluster head (CH) is elected for each cluster to maintain cluster membership information and perform other administrative functions. There are also multiple gateways (GWs) within each cluster. We further assume that key distribution has done and each node has one or more public-private key pairs, which might be pre-installed or generated by itself, or using a scheme such as the one proposed in [10]. It is also assumed that the CHs have similar physical characteristics as the GWs and the cluster members (CMs). In other words, the CHs do not have much higher processing capabilities than other nodes, and they may be compromised as well.

2.1. Cluster Affiliation

It is assumed that each cluster has an asymmetrical key pair KUc/KPc, where the public key KUc is signed by a root Certificate Authority (rCA), and private key KPc is held and maintained by the CH. A CH is designated initially, and it holds the private key of the cluster in order to authenticate all the members. A new CH (if needed) might be re-designated when the current CH relinquishes its role, or when it is broken down.

Each node typically affiliates with one cluster when the network is deployed. Each cluster member (CM) has the public key of the cluster, but not the private key. The CMs that belong to the same cluster should share a secret with that cluster, and one possible implementation is a signature of a random number using the cluster's private key. For example, if node A belongs to cluster x, it may manually install the < NA, KPcx(NA) > pair during initialization, where NA is a random number, and KPcx is the cluster's private key. KPcx(NA) is a signature of the cluster. A more efficient implementation is using the hash value of KPcx instead of KPcx(NA), i.e., node A initially has the pair <NA, H(KPcx, NA) >. A node may share multiple secrecies with different clusters at any time so that it may join different clusters.

GWs are automatically determined by each node rather than designated. For example, nodes that locate at the border of a cluster may act as GWs and perform the corresponding functions. It is expected there will be sufficient number of nodes that qualify as GWs when the network is dense and nodes are uniformly distributed in a cluster.

2.2. Nodes Join or Leave a Cluster

When a node wants to join a new cluster, it needs to be authenticated by the CH. Suppose that node A initially has the pair <NA, H(KPcx, NA) >, it generates a temporary session key Kses, and broadcasts an Authentication Request (AuRQ),

[ARQ, KUc(Kses), Kses(NA, H(KPcx, NA))]

where ARQ is the request ID. When the CH receives AuRQ, it will obtain Kses with KPc, then verify H(KPcx, NA) after decrypting it with Kses. If succeeded, the CH will send an Authentication Response (AuSP) attaching its Cluster Name (CN) encrypted by Kses

[ASP, Kses(CN, IV)]

where ASP is the response ID. Note that CN might change periodically by the CH to keep the cluster anonymous. IV is a 32-bit increasing number maintained by CH. Each time CH updates CN, it will increase IV by one, which is used to defend against replay attacks. If authentication failed, CH will send an error message to specify the reason of failure, such as error decryption, wrong secrecy, etc. Node might try to select other public key and secrecy for authentication when obtaining an error message.

CH will keep a list of all CMs. After each successful authentication, CH will add an entry to its member list (Table I)

CH will periodically check whether its CMs in the list are present. This procedure may also thwart Denial-of-Service (DoS) attacks since repeated AuRQs can be easily detected by comparing the obtain random number with the list.

If a CM leaves a cluster, it may not need to send any notification. CH will delete a CM from its list when that node is found not present for certain time during periodic checks. However, if a CH plans to leave a cluster, it needs to claim an election for a new

Table 1. Cluster members' table

Random Number	Valid Time
NA	valid time A
NB	valid time B
..	..

CH, which might be based on a specified security policy such as the one discussed in [19]. After a new CH is designated, the private key KPc will be securely transferred to it from the original CH. In some extreme occasions, CH might break down before the private key can be transferred. The (n, k) threshold scheme [4] can be adopted as a backup scheme to protect KPc. A suitable value of k may be chosen to guarantee security of KPc.

2.3. Key Management

In the proposed cluster based architecture, CN acts as the group key for a cluster. It is used to identify the current cluster and should be only known by the CMs. CN should be periodically updated by the CH, since CN might be divulged because of node movements. To update CN, the CH simply broadcasts an update [CNUP, IV, CNc(CNn), KPc(H(IV, CNn))] where CNUP is the ID of the update. The new cluster name CNn is encrypted by the current cluster name CNc. Meanwhile a signature by the CH is used to guarantee both integrity and authority. We assume that in most cases a divulged CN is out-of-date since CN is updated periodically. In case that a valid CN is known by an adversary or a CM is compromised, point-to-point updates are needed (discussed in Section 5.2.).

3. Secure Anonymous Routing

In this study, all the nodes are assumed stationary or have low mobility during the routing process such that routing will not become meaningless. However, node mobility may not be neglected during data transmissions, i.e., a route may be broken due to node mobility during a traffic session. Links between nodes are assumed bi-directional because most routing protocols and wireless Medium Access Control (MAC) protocols (such as the MAC protocol in 802.11) require symmetric links.

3.1. Intra-cluster Secure Anonymous Routing

Single-hop communication is assumed within each cluster, i.e., every node can directly communicate with any other node in the same cluster. This can be achieved by only allowing strongly received nodes (during authentication) to join a cluster.

Three steps are included in the proposed intra-cluster secure anonymous routing: Key broadcasting, Intra cluster routing request (Intra-RREQ), and Intra cluster routing response (Intra-RRSP) (see Fig. 1).

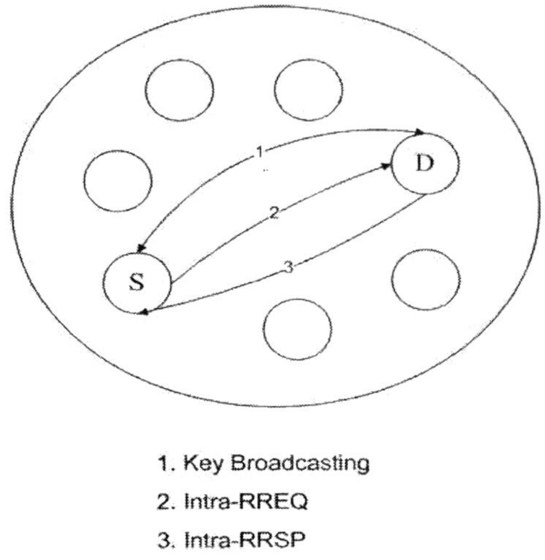

Figure 1. intra-cluster routing.

In the step of key broadcasting, each node will randomly generate a pseudo name, and broadcast the pseudo name and the corresponding public key (KU) with the format

[pseudonym $\bigoplus$ CN, KU $\bigoplus$ CN, H(CN, pseudonym, KU)] ,

where $\bigoplus$ represents XOR operation. The use of pseudonym $\bigoplus$ CN and KU $\bigoplus$ CN guarantees that only the current CM can get the pseudonym and KU pair of other CMs in the same cluster (by performing XOR operation using the current CN) because only CMs in the same cluster have the knowledge of CN. Here we use the hash value of the CN, pseudonym and KU rather than the CN itself. The strong collision resistance of the hash function guarantee the uniqueness of the hash value, thus prevent replay attacks. The integrity of the message is also assured by checking the hash value. All the nodes in a cluster need to build a table to map public keys and node names (pseudo names) of all the CMs, see for example, Table II. Because one-hop communication is assumed within each cluster, all other CMs can receive the broadcast and keep the message in its local mapping table. In order to improve anonymity, all the CMs will periodically (but randomly) update their public keys and pseudo names by key broadcasting. For example, each CM choose to broadcast a new public key and pseudo name every m minutes. It may choose a random number uniformly distributed in $[lm + m/2,\ (l+1)m]$ as the time for its lth key broadcasting. It will prevent the link-ability of two (public key, pseudo name) pairs from the same CM. The local timestamp helps to keep track of the validity of the public keys. Entries will be deleted when their corresponding timestamps expire.

Table 2. Name-PubKey mapping table

Name	Key	Local Timestamp
A	Key1	time1
B	Key2	time2
..	..	..

Because of the high computational complexity of the public key schemes, they are only applied to identify the designated receiver and help to deliver a symmetric session key. For example, if node S wants to communicate with node D, they need to negotiate a symmetric session key first. Node S simply broadcasts a routing request (RREQ) packet that is encrypted by node D's public key. Although all nodes of that cluster will receive the RREQ, only node D has the corresponding private key and thus can decrypt it. Therefore, it guarantees receiver anonymity. Node D will send a routing response (RRSP) and encrypt it with node S's public key, which will guarantee sender anonymity. Furthermore, the pseudonyms of the source and destination nodes will guarantee sender-receiver anonymity. After node S decrypts the RRSP, node S and node D will have a shared session key for secure data transmissions. In order to thwart packet analysis attacks, each packet needs to have the same packet size (by added padding).

Note that a CM may have multiple public/private key pairs. It is computationally very expensive for the CM to try all its private keys when receiving a packet. A technique called key indexing is proposed in [21]. A similar key indexing technique may be applied here and the tradeoff between efficiency and anonymity is discussed in detail in Section 3.3..

The format of the Intra-RREQ (without key index) is

[KU_D(Ks), $Ks(RREQ \| Req\_ID \| PN_S)$, H(CN, KU_D(Ks)), padding], and the format of the Intra-RRSP (without key index) is

[KU_S(Ks'), $Ks'(RRSP \| Req\_ID \| Kses)$, H(CN, KU_D(Ks')), padding], where PN_S is the pseudonym of S; KU_D and KU_S are the public keys of node D and node S, respectively; K_s and K'_s are temporary symmetic keys; K_{ses} is the symmetric session key for data transmissions. Req_ID is an identifier of the request and it is also used to defend against replay attacks. The hash values in the Intra-RREQ and Intra-RRSP are used to maintain the integrity of those messages.

The Intra-RREQ has the same format as Intra-RRSP so that attackers are unable to distinguish them by packet analysis. Hence attackers can not correlate the source and the destination by packet format. Furthermore, since Intra-RREQ and Intra-RRSP are encrypted by the public keys of destination and source separately, attackers can not obtain the pseudonym of the source or the destination, and can not feign others' pseudonym to communicate.

Note that each Intra-RREQ and Intra-RRSP are only broadcast once in intra-cluster secure anonymous routing and they do not propagate to other clusters. Hence, high bandwidth efficiency can be achieved. Furthermore, since each node (including the CH and GWs) behaves exactly the same, no special function need to be performed by the CH and GWs in the intra-cluster routing process. Thus, critical network elements can be hidden from the attackers.

3.2. Inter-cluster Secure Anonymous Routing

In the proposed inter-cluster anonymous routing, we extend the method in [6] to cluster based wireless ad hoc networks. The tradeoff between bandwidth efficiency, computational complexity, and the level of anonymity achieved is the main concern.

It is assumed that there exists a security association between any source and destination node pairs. The shared keys may be distributed by a Key Distribution Center (KDC) or manually.

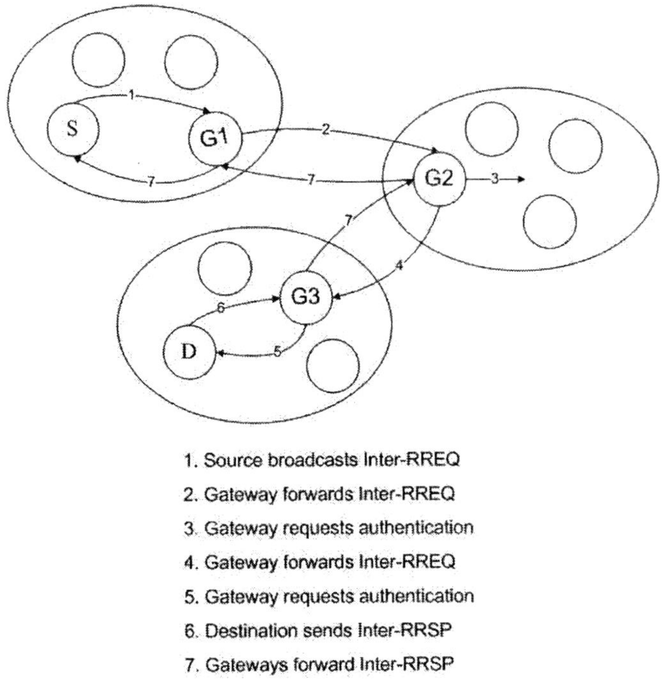

Figure 2. inter-cluster routing

The procedures of inter-cluster anonymous routing are outlined in Fig. 2.

3.2.1. Source Broadcasts Inter-Cluster Routing Request

Source node S generates an inter-cluster routing request (Inter-RREQ), and broadcasts Inter-RREQ in its cluster. Here we require that only GW nodes take part in inter-cluster routing. Other CMs simply ignore this request to avoid packet propagations (thus avoid wasting bandwidth). The format of this request is

[RREQ, Req_ID, H($K_{sd}||Req\_ID$), $K_{sd}(K_{ses})$, K_{ses}(Req_ID), PK0] .

- Req_ID: identifier of the request;
- K_{sd}: the shared key between node S and node D;
- K_{ses}: a session key (will be used to verify response later);
- PK0: a temporary public key of node S.

The hash value of $K_{sd}||Req\_ID$ acts as a key index and is used for locating a key quickly. If none of the symmetric key (stored locally) match the hash value, the node is not the destination. K_{sd} is used for authentication between the source node S and the destination node D. To prevent possible mistakes when multiple keys have the same hash value, K_{ses} is used by intermediate node to verify whether it is the destination node, because only destination node D has K_{sd} to obtain K_{ses} and is able to verify that it is indeed the destination by decrypting the fourth field in Inter-RREQ and comparing it with Req _ID.

Note that the above procedure is only needed when the hash values match. PK0 is kept by its next hop node (GW) to encrypt routing response. Since only node S has the corresponding private key and the public key is temporary, it can guarantee both security and anonymity in this step.

3.2.2. Gateway Forwards Inter-RREQ

The Inter-RREQ will be forwarded by GWs to neighboring clusters. Before forwarding Inter-RREQ, the GW firstly keeps the public key of the sender and replaces it with the public key of the current GW. For example, in step 2 of Fig. 2, G1 will keep PK0, and replace it with PK1 (a temporary public key of G1). The Inter-RREQ changes to

[RREQ, Req_ID, H($K_{sd}\|Req\_ID$), $K_{sd}(K_{ses})$, K_{ses}(Req_ID), PK1] ,

Similarly, in step 4, the Inter-RREQ changes to

[RREQ, Req_ID, H($K_{sd}\|Req\_ID$), $K_{sd}(K_{ses})$, K_{ses}(Req_ID), PK2] ,

where PK2 is a temporary public key of G2.

When a GW receives a fresh Inter-RREQ, it will save Req_ID and the corresponding K_{ses}(Req_ID) for identifying duplicate Inter-RREQs and later verification, and forward the Inter-RREQ to GWs in neighboring clusters. When a foreign GW receives a fresh Inter-RREQ, it will also broadcast an authentication request in its local cluster to check whether the destination is there. For example, the packet format in step 3 is [AREQ, Req_ID, H($K_{sd}\|Req\_ID$), Ksd(Kses), Kses(Req_ID), PK2, H(CN, Ksd(Kses))] , where AREQ is the authentication request ID, and the hash value is used to identify the cluster and maintain the integrity of the message. Because it is an intra-cluster request, nodes in other clusters will ignore it.

The GW may wait until a node replies and stop forwarding Inter-RREQ, or a timer expires and then forward Inter-RREQ to GWs in neighboring clusters. However, there are two concerns with the above design. Firstly, this may incur excessive delay in inter-cluster routing. Secondly, anonymity may be sacrificed if the GW stop forwarding the Inter-RREQ. For example, an attacker can figure out the cluster of the destination node although not the exact location of the destination. In order to avoid these problems, in our design the GW will not wait for responses and will forward the Inter-RREQ immediately after step 3 in Fig. 2. Of course, additional bandwidth is needed since each GW will re-broadcast the Inter-RREQ exactly once.

3.2.3. Destination Sends Inter-cluster Routing Response

When a CM receives an authentication request, it checks whether it is the destination. If it is, it will generate an inter-cluster routing response (Inter-RRSP), such as step 6 in Fig. 2. In this example, the destination uses a pseudonym T_4, and encrypts T_4 by sender's public key (PK3) such that the intermediate GWs and the source can authenticate the destination. It also includes the encrypted (by T_4) session key K_{ses} and Req_ID. The packet format of Inter-RRSP is [RRSP, PK3(T_4) , $T_4(K_{ses} \parallel Req\_ID)$].

3.2.4. Gateway Forwards Inter-RRSP

When an intermediate GW receives a routing response, it decrypts the pseudonym Tx by using its corresponding private key. Then it uses the obtained Tx to decrypt the session key K_{ses} and verify the destination, because the original Req_ID and the corresponding K_{ses} (Req_ID) in the routing request has been saved by intermediate GWs. If the verification is successful, the intermediate GW will perform the same operation as that of the destination, i.e., it will generate a new pseudonym and encrypt it by last sender's public key. Then it will encrypt K_{ses} and Req_ID with the new pseudonym. For example, the packet format in step 7 is

[RRSP, PK2(T_3) , $T_3(K_{ses} \parallel Req\_ID)$].

Therefore, after the Inter-RRSP reaches the source, an inter-cluster route is formed as S:T_1:T_2:T_3:T_4(D).

The proposed inter-cluster secure anonymous routing implements two different packet formats at a GW for forwarding Inter-RREQ and authentication within its local cluster. Thus an adversary may distinguish GW nodes from other nodes. However, since each GW re-broadcasts exactly twice for each Inter-RREQ (one for forwarding Inter-RREQ and the other for local authentication), it is not possible for the adversary to locate the cluster of the destination node unless key indexing is applied. The tradeoff between provided anonymity and routing efficiency is discussed later in Section 3.3..

Note that GWs may use the same packet format for forwarding Inter-RREQ and authentication within its local cluster. However, this approach violates the semantics of clusters. For example, every node will have to examine every routing packets (local or not) which results in much higher overhead.

Furthermore, the proposed scheme ensures location privacy because nodes do not reveal their real identity to other nodes, and their pseudonyms are changed dynamically. Therefore, an attacker can trace a node to a certain cluster at the most. Moreover, since source and destination identifiers are never disclosed during route discovery, the relationship anonymity between the source and the destination is guaranteed.

3.3. Efficiency Analysis

In the secure anonymous routing process, each packet is encrypted by either a symmetric or an asymmetric key, and the intended receiver is identified by the key. However, one problem (as pointed out in [21]) is that the receiving node may have many keys and does not know which key to use. Therefore, each node has to try to decrypt any packet received with *all* its keys in order to identify whether it is the intended receiver. This process causes very low efficiency and high cost on computation and runtime.

One way to solve this problem is to add a key index for each encrypted packet. Each node only needs to compare the key index to identify whether it is the intended receiver and which key to use instead of performing many decryptions. Consequently, the cost on computation and runtime will be greatly reduced. If a hash algorithm is used to generate the key index, then only hash operation will be performed rather than decryption. Hash algorithm such as MD5 is almost a thousand times faster than the RSA asymmetric algorithm and is ten times faster than DES. For example, see test results in [26].

For intra-cluster routing request and response, the key index could be H(KU, CN). CN is used to prevent non-CMs from analyzing the packet. Thus the Intra-RREQ will change to

[H(KU_D, CN), KU_D (Ks), Ks (RREQ || Req_ID || PN_S), H(CN, KU_D (Ks)), padding].

Similarly, key indexing may be applied to inter-cluster routing as well. For example, the Inter-RRSP may be modified as

[RRSP, H(PKi), PKi(T_{i+1}), $T_{i+1}(Kses \| Req\_ID)$], where H(PKi) acts as the key index.

However, use of key indexing might weaken the anonymity of the system. For example, during inter-clustering routing, an attacker may correlate Inter-RREQ and Inter-RRSP by recording the public keys (PKi) in the Inter-RREQ and comparing with the key index H(PKi) in the Inter-RRSP. However, it may not affect the anonymity of the mobile users seriously. Although the attacker may divulge a few links on the path, it is almost impossible for the attacker to discover the entire path unless many attackers at different segment of the path collude. In addition, data transmission is impossible to track even if the attacker has discovered the entire path, because the data packet format will be different per hop. Therefore, it is possible to use key indexing without jeopardizing the mobile users anonymity too much.

4. Data Transmission

Intra-cluster data transmissions can be achieved by the source node broadcasting data encrypted with the negotiated session key K_{ses} from the intra-cluster route discovery. The packet format is

[DATA, H(K_{ses}), K_{ses}(data)]

where DATA is the packet type. Each node within the same cluster will first check whether it is the destination by verifying the hash value of its session keys. Because only the destination need to decrypt the data, computational complexity is low for all other nodes.

Inter-cluster data transmissions rely on the sequence of symmetric keys generated during Inter-RRSP. After inter-cluster routing is done, each node i on the path will keep a mapping [T_i, T_{i+1}]. T_i is the symmetric key generated by itself and transmitted upstream (to node $i-1$) as a part of the routing response. T_{i+1} is the symmetric key received in the routing response from downstream node $i+1$. During inter-cluster data transmission, the hash value of T_{i+1} is used to identify the downstream node. The packet format (from node i to node $i+1$) is [DATA, H(T_{i+1}), $T_{i+1} \bigoplus K_{sd}$(data)], where $\bigoplus$ represents XOR operation, data is encrypted by the shared key K_{sd} between the source and the destination if data security is required. Each intermediate node will first verify whether it is the downstream node by checking H(T_{i+1}). If it is (and hence has the [T_{i+1}, T_{i+2}] pair), it will change H(T_{i+1}) to H(T_{i+2}), and perform the following operation: $T_{i+2} \bigoplus T_{i+1} \bigoplus T_{i+1} \bigoplus K_{sd}$(data) = $T_{i+2} \bigoplus K_{sd}$(data). Then the updated packet [DATA, H(T_{i+2}), $T_{i+2} \bigoplus K_{sd}$(data)] will be re-broadcast. Since hash value is used to identify the next hop, and the data field changes from hop to hop, the attacker can not track the data flow. Thus sender-receiver anonymity can be maintained.

5. Anonymity Analysis and Attack Analysis

5.1. Anonymity Analysis

An anonymity metric based on entropy (proposed in [24], [25] and adopted by [18]), is used to analyze the anonymity level of the source and the destination. The entropy of a wireless network may be defined as $H(X) = \sum p_i log(1/p_i)$, where X is a discrete random variable with probability function $p_i = P(X = i)$. Suppose the size of the network is N, an attacker can discover node i's identity with probability p_i. $H(X)$ (uncertainty) is maximized when the node is equally likely to be any node, i.e., $H_{max} = logN$ when $p_i = 1/N$. Then the degree of anonymity can be defined as $\eta = H(X)/H_{max}$.

In this study, we adopt the attack model defined by [22]. Specifically, attack-C-M means that there are C compromised nodes and M (outside) malicious nodes in the network. They may perform traffic analysis or packet analysis on the routing traffic and data traffic, and they may collude. We will focus on the source/destination "pseudonym anonymity" for intra-cluster routing, and source/destination "cluster anonymity" for inter-cluster routing, where "pseudonym anonymity" is defined as the uncertainty of mapping a pseudonym to a specific node, and "cluster anonymity" is defined as the uncertainty of mapping a source or destination to a specific cluster, respectively. Moreover, we assume it is very hard for the attacker to distinguish which pseudonyms belong to the same node. In other words, we assume that the attackers do not possess the capabilities of observing the signal-to-noise ratio of a transmitting device or observing the transmitting signal's watermarks.

5.1.1. Pseudonym Anonymity of Intra-cluster Routing

In intra-SARC, key broadcasting is protected by CN so that (outside) malicious nodes can not obtain the public key and the pseudonym of any CM. Therefore, the anonymity pseudonym anonymity is infinite under Attack-0-M, which means Attack-C-M has the same effect as Attack-C-0 in terms of pseudonym anonymity. Consequently, only Attack-C-0 is considered.

Suppose that a cluster has N nodes and C of them are compromised nodes ($C < N$), and each node has equal probability to send and receive routing request. The Intra-RREQ and Intra-RRSP are encrypted by the public keys of the source and the destination, which means other CMs (including the compromised nodes) are not able to obtain the pseudonym of the source and the destination, except for themselves. Firstly we consider the source anonymity. If the destination is one of the compromised node, the source will be revealed; otherwise the probability is $1/(N - C)$. Let Y be a discrete random variable, and $p_0 = P(Y = 0) = C/N$ represents the probability that the destination node is compromised, $p_1 = P(Y = 1) = 1 - C/N$ represents the probability that the destination is a legitimate node. Therefore the entropy under Attack-C-0 is

$$\begin{aligned} H(X|Y) &= p_0 H(X|Y=0) + p_1 H(X|Y=1) \\ &= p_1 \sum [P(X=i|Y=1) log 1/P(X=i|Y=1)] \\ &= (1 - C/N) log(N - C) \end{aligned} \quad (1)$$

The anonymity degree of the source/the destination is $\eta = H(X|Y)/H_{max} = (1 -$

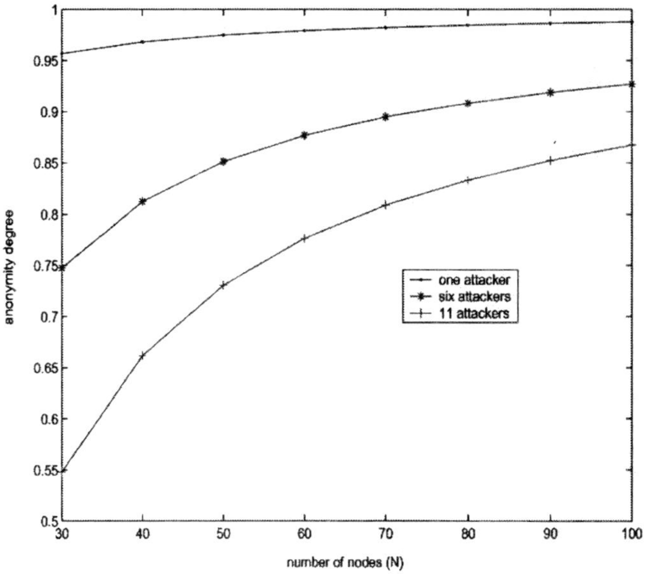

Figure 3. The anonymity degree of intra-cluster routing.

$C/N)log(N - C)/logN$. Figure 3 shows how this quantity varies with N and C. The anonymity degree increase with the number of nodes within the cluster and decrease with number of compromised nodes within the cluster. One compromised node can hardly do any harm, however, when 6 out of 30 nodes are compromised, the anonymity degree drops to 75%.

5.1.2. Cluster Anonymity of Inter-cluster Routing

In inter-cluster routing, no real identity, pseudonym or the corresponding public key is used, thus even compromised nodes can not identify which node is the source or the destination. What they can do is try to locate the source or the destination down to their clusters. Hence, it is only meaningful to consider cluster anonymity. In other words, how accurate the attackers may locate the cluster where the source or the destination node resides. Furthermore, since all the GW nodes perform the same operation no matter where they are, and the packet of inter-cluster routing is transparent for both compromised and malicious nodes, the compromised node can be treated the same as the malicious node with respect to cluster anonymity.

Cluster anonymity analysis may become very complicated because there are many factors (such as cluster distributions, number of attackers, attacker distribution, etc.) that will affect the cluster anonymity. Here we only consider the case of a single attacker. Suppose that there are P clusters in the network, and X is a random variable representing which cluster the source or the destination resides. The maximal cluster anonymity of the network is $H_{max} = logP$.

Assume that each cluster has N nodes, $N1$ of them in the area that do not overlap with other clusters. We also assume the average overlapping degree is D, which means a node in an overlapping area can sense the signals from D clusters on average. For example, in

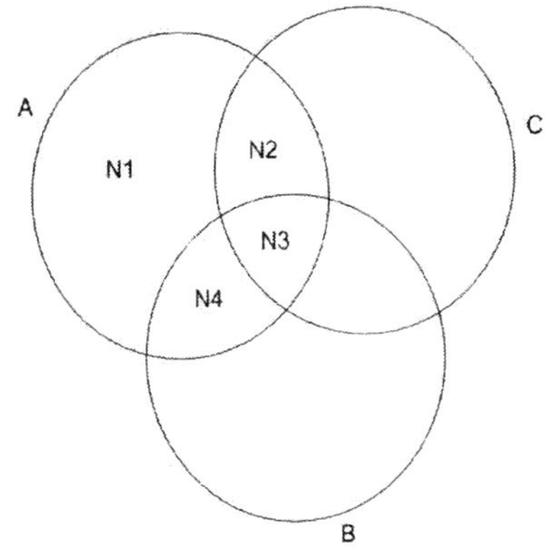

Figure 4. An example of inter-cluster node distribution.

Figure 4, there are three clusters A, B, and C. In cluster A, $N1$ nodes are in the area that do not overlap with B and C; $N3$ nodes are in the area that overlap with both B and C. $N2$ and $N4$ nodes are in the area that overlap with either B or C. $N = N1 + N2 + N3 + N4$. The average overlapping degree is $D = (N2 \times 2 + N3 \times 3 + N4 \times 2)/(N2 + N3 + N4)$

Let us consider the destination cluster anonymity. If a node observes a Inter-RRSP, there are 4 possibilities (represented by a discrete random variable Y):

1. the node is not in any overlapping area, and it resides in the destination cluster with probability $p_0 = P(Y = 0) = \frac{N1}{N}\frac{1}{P}$. In this case, the destination cluster will be revealed.

2. the node is not in any overlapping area, and it resides outside the destination cluster with probability $p_1 = P(Y = 1) = \frac{N1}{N}(1 - \frac{1}{P})$. The destination cluster may be any of the other clusters with probability $1/(P - 1)$.

3. the node is in an overlapping area, and it resides in the destination cluster with probability $p_2 = P(Y = 2) = (1 - \frac{N1}{N})\frac{D}{P}$. The destination cluster may be any of the overlapped clusters with probability $1/D$.

4. the node is in an overlapping area, and it resides outside the destination cluster with probability $p_3 = P(Y = 3) = (1 - \frac{N1}{N})(1 - \frac{D}{P})$. The destination cluster may be any of the other clusters with probability $1/(P - D)$.

Hence, the destination cluster anonymity is

$$\begin{aligned} H(X|Y) &= \sum\sum p_j P(X = i|Y = j) log \frac{1}{P(X = i|Y = j)} \\ &= \frac{N1}{N}(1 - \frac{1}{P}) log(P - 1) + (1 - \frac{N1}{N})\frac{D}{P} log D \\ &\quad +(1 - \frac{N1}{N})(1 - \frac{D}{P}) log(P - D) \end{aligned} \tag{2}$$

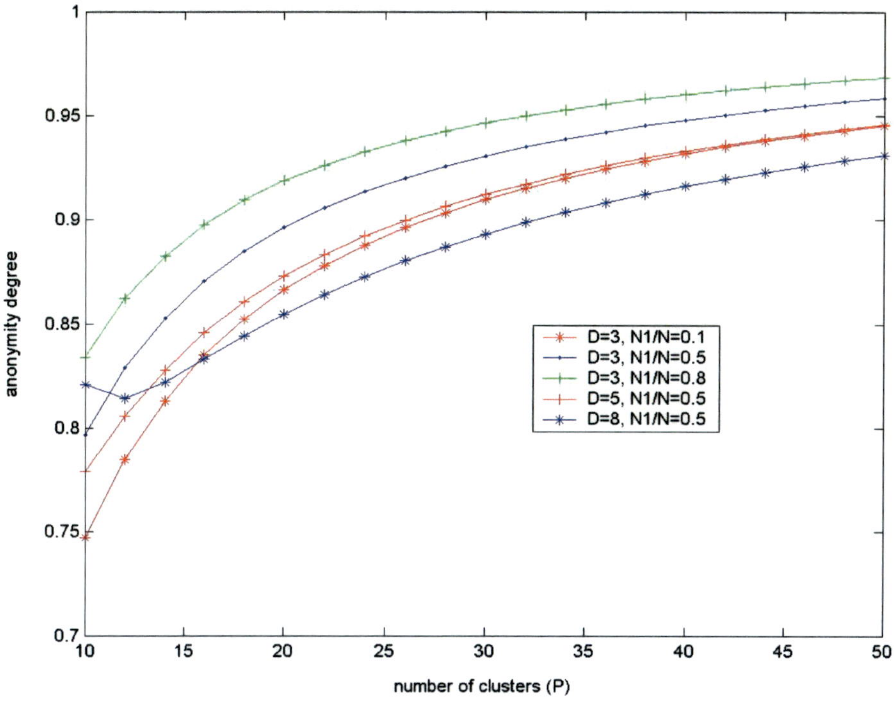

Figure 5. The anonymity degree of inter-cluster routing.

The cluster anonymity degree is

$$\eta = [\frac{N1}{N}(1-\frac{1}{P})log(P-1) + (1-\frac{N1}{N})\frac{D}{P}logD + (1-\frac{N1}{N})(1-\frac{D}{P})log(P-D)]/logP \quad (3)$$

The source cluster anonymity can be obtained similarly.

The cluster anonymity degree with respect to the number of clusters P, the average overlapping degree D, and $N1/N$, is shown in Figure 5. It is observed that the cluster anonymity degree increases with the number of clusters P, as expected. It is also observed that the cluster anonymity degree increases with $N1/N$. In the right-hand side of equation (2), the first term is dominant. Thus, the cluster anonymity degree η is higher when $N1/N$ becomes bigger. This represents the typical case when there are a lot of nodes in the non-overlapping area and they are not in the destination's cluster, and the number of clusters is not too small (larger than 10 in this example). Another observation is that the cluster anonymity degree increases when D decreases and P is large. When there are a lot of clusters, less overlapping (small D) reduces the chance that the observer is within the destination cluster. Therefore, the uncertainty of the destination cluster increases. Thus the cluster anonymity degree improves. However, when P is not very large, there could be a case where two cluster anonymity degree curves with parameters D_1 and D_2 cross at $D_1 = P - D_2$. In this example, it happens at $D_1 = 5$, $D_2 = 8$, and $P = 13$.

5.2. Attack Analysis

The active attacks such as the denial-of-service (DoS) attacks are usually easy to detect because they cause abnormal traffic patterns under many circumstances [9]. Intrusion detection systems can act as one of the counter-measures against such active attacks. Hence, active attacks are not addressed in this work. However, it worths pointing out that the integrity of the routing packets are guaranteed in the proposed scheme, although routing packets are not encrypted (in order to keep the overhead low). The attacker will not be able to alter any field in the routing packets without being detected. In addition, secure routing in cluster based ad hoc networks is much more resistant to active attacks than routing in pure ad hoc networks. The main reason is the existence of an on-line authority (CHs) capable of controlling traffic and monitoring node behavior [18].

On the contrary, passive attacks such as eavesdropping and traffic analysis are difficult to detect. Once locating certain critical nodes through overheard routing information, passive attackers can perform active attacks on the critical network elements. Therefore, passive attackers are more dangerous than active attackers because they are difficult to detect [9]. Such passive attacks are the main concern of this work.

In anonymous communications, two main passive attacks are packet analysis attack and traffic analysis attack. In packet analysis attack, the attackers try to deduce routing information by analyzing the packet length, type, content, etc. In traffic analysis attack, the attackers try to deduce routing information by analyzing the amount of traffic flow among nodes and correlating eavesdropped traffic information to actual network traffic patterns.

In cluster based wireless ad hoc networks, CH plays an important role as the central controller and the trusted authority in a cluster. Thus, one of the main tasks of secure anonymous routing is to hide CH from attackers. In the proposed secure anonymous routing scheme, CH acts exactly the same as the other nodes throughout the routing procedures in both intra-cluster and inter-cluster anonymous routing, which makes it indistinguishable from the other nodes in the network. Consequently, the CHs are safe from both packet analysis attacks and traffic analysis attacks.

Note that the attackers may be able to identify GWs from other nodes. However, since each cluster typically has more than one GW node, it is not as critical as the CH. Furthermore, it is feasible to allow some nodes perform GW functions from time to time. This will shuffle the routing traffic and make traffic analysis attack more difficult to succeed.

In case a global adversary exists, who can monitor *all* traffic flows in the entire network, dummy traffic and/or local mixing of messages may be applied to prevent traffic analysis. For example, a limited flooding technique is proposed in [28] to address this issue.

If a node other than the CH is compromised, its CH should update the group shared secret (CN). Since the compromised node has the old CN, the CH can not broadcast the update request. Instead, it should send the request to each CM using point-to-point mode. The packet format is

[CNUPP, IV, Nx(CN), KPc(H(IV, CN))]

where CNUPP is the packet identifier, Nx is the corresponding random number of node x. Note that IV should be the same for all CMs. Since the compromised node may have pre-installed signatures of multiple clusters, the CH should notify other CHs. It is assumed that all CHs can identify each other by sharing a secret key.

If a CH is compromised, all the signatures for that cluster should be revoked and every node include other CHs should be notified. In order to guarantee authority, the CH revoke message should be signed by the root Certificate Authority (rCA). The message could be a Certificate Revoke List (which is updated periodically). Furthermore, this signed message should be dispatched to at least one trusted CH manually or through a special signaling channel. The trusted CH obtaining the revoked information will send a notification to other CHs; moreover, each CH also needs to broadcast the revoked message to all CMs.

6. Performance Evaluations

6.1. Implementation Overhead Analysis

One routing design for cluster based wireless ad hoc networks is the Cluster Based Routing Protocol (CBRP) [17]. CBRP does not contain any security features. In this study, CBRP is used as a baseline for overhead comparison analysis.

Suppose that 3DES and RSA-512 are employed as the symmetric and public key algorithms, and MD5 is adopted as the hash algorithm. The detailed packet fields of intra-cluster routing and inter-cluster routing are shown in Fig. 6 and Fig. 7, respectively. In intra-cluster routing, public key is only used to deliver a symmetric key, thus the computational complexity is low. The overhead is also low due to the use of hash function.

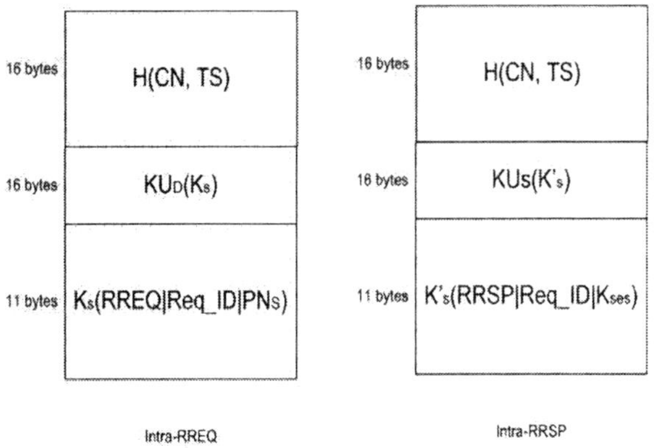

Figure 6. Intra-cluster routing: packet fields.

In inter-cluster routing, the size of Inter-RREQ packet is 85 bytes. The packet size for authentication request from GW to cluster member is 101 bytes. The packet size for Inter-RRSP is 27 bytes. Since the routing packets' sizes are fixed in the proposed SARC, while in CBRP the routing packets' sizes grow with the hop count of the route, the overhead between them becomes close as the obtained route becomes longer (more hop counts). A simulation is performed to demonstrate this effect and the result is shown in Fig. 8. It is assumed that there are 20 clusters in the network and each node in each cluster want to communicate with any other node in a different cluster. The result shown is the average overhead over all obtained routes. It is observed that the overhead of SARC is 26.3% higher than that of CBRP when the average number of hops in the obtained routes is 4 (source and destination are in neighboring clusters). This drops to only 16.7% when the average number of hops

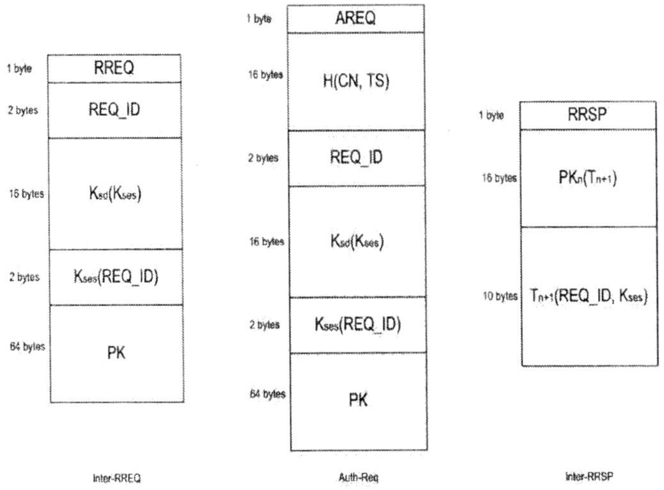

Figure 7. Inter-cluster routing: packet fields.

in the obtained routes increases to 10. When the average number of hops in the obtained routes is more than 16, SARC has lower overhead than CBRP. These observations are in agreement with our expectation.

6.2. Route Establish Time

The routing protocol is implemented within OPNET. The network is $400m \times 400m$ square field with 800 nodes uniformly distributed. Sixteen non-overlapping clusters are formed in the system with equal size of a fixed $100m \times 100m$ area. The GWs are chosen as the nodes who locate at the border of the cluster, to be exact, whose distance to the border is less than 10 meters. An example is given in Fig. 9, where the solid line is the edge of cluster and the nodes outside the dashed line are GWs. Note that Fig. 9 is for illustration purpose only, and has far less nodes than the one (800 nodes) used for simulations.

The inter-cluster route establish time with and without key index is studied in this part of the simulation. During the routing process, it is assumed that the nodes in the network are either stationary or have negligible mobility. In other words, the time scale of routing is much less than the time scale of mobility, such that routing will not become meaningless. The route establish time are collected when the network has one, ten, thirty and fifty source/destination pairs. And the results are averaged over 100 runs. The delay of cryptographic operation is evaluated based on the test results given by [26].

The inter-cluster route establish time with and without key index is shown in Fig. 10. It is observed that the route establish time is much less with key index than that without key index because using key index reduces the large delay caused by decryptions. It is also observed that the route establish time increases linearly with hop counts when using key index, but it is almost unchanged with increased source/destination pairs. The reason is that with key index, each hop causes almost the same amount of delay, thus the route establish time increases linearly with hop counts. On the other hand, each node searches the key based on a hash algorithm when using key index, and hash algorithm is highly efficient and will not be affected much by the number of source/destination pairs. In other words, the

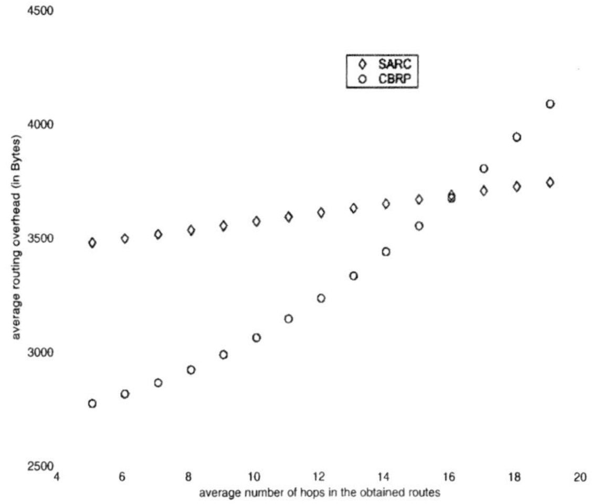

Figure 8. Routing overhead of SARC and CBRP for Inter-cluster routing.

queueing delay at each node will not be affected much by the number of source/destination pairs when using key index. On the contrary, the route establish time increases dramatically with the number of source/destination pairs when key index is not implemented. The main reason is that the delay of cryptographic operation (decryption rather than hash algorithm) is significant at each node and the queueing delay at each node will increase as well. These delays will grow dramatically with the number of source/destination pairs.

6.3. Packet Delivery Ratio

In this part of the simulation, the effects of offered load (in number of flows across the network) and node mobility on packet delivery ratio is investigated. The Random Waypoint mobility model [27] is adopted, with the pause time fixed to 10 seconds and the maximum speed varies from 0 (node is stationary) to 30 m/s. The link capacity is 1Mbps. The data generating rate is 4 packets per second with the packet size exponentially distributed with mean 1000 bits. The simulation time is 10 minutes.

It is observed that the packet delivery ratio decrease as the node speed and the offered load increase, as expected. When all the nodes are stationary, the network is capable of supporting 30 simultaneous traffic flows. Higher mobility is the main reason for the drop in packet delivery ratio because it causes more paths broken and more packet loss. The offered load is less a factor than the node mobility in this simulation since the link capacity is high comparing to the data generating rate.

6.4. Scalability Study

It is expected that the proposed cluster based anonymous routing protocol will have better scalability than flat routing protocols. In this part of the simulation, ASR [6] is tested using the same simulation setup as in Section 6.2. but with no clustering (flat topology with the same number of nodes and node distribution). In all the trials for ASR with 50 source/destination pairs, the computer runs out of memory and the simulation could not

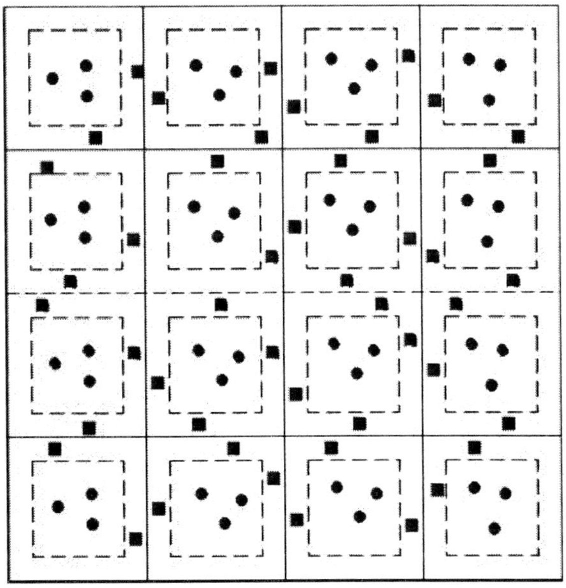

Figure 9. The topology of the network (GW: square; CM: round).

be finished, where the average time for finishing SARC simulation is 90 minutes. Note that the simulation of ASR finished when a small network (80 nodes) is simulated. This demonstrates that the proposed cluster based anonymous routing protocol will have better scalability than flat routing protocols, because of the greatly reduced routing overhead due to constrained local flooding.

7. Related Works

Anonymous communication protocols have been studied intensively in wired networks. The concept of mix was proposed in [12], and was employed in various anonymous communications proposals for the Internet, such as P^5 [13]. A similar but different concept, crowd, was introduced in [16] for Internet web transactions. However, these methods can not be directly applied in MANET due to the lack of fixed infrastructure.

Secure anonymous routing for MANET attracts a lot of attentions lately. In [5], a protocol was proposed to allow trustworthy intermediate nodes to participate in the path construction protocol without jeopardizing the anonymity of the communicating nodes. The basic idea is that each node will classify its neighbors into different security levels, and a common key is shared between them for each level. Only nodes at certain level can receive and transfer data during routing. ANODR [7], an anonymous on-demand routing protocol for mobile ad hoc networks, is based on "broadcast with trapdoor information", in which a cryptographic onion [15] is used for route pseudonym establishment. Since in ANODR each node en route only knows the next node with a fake identity, it can prevent strong attackers from tracing a packet flow back to its source or destination and ensure that attackers cannot discover the real identities of local transmitters. In routing response of ANODR, the node ID is transmitted to next node. Although it is a pseudonym, the attacker can launch

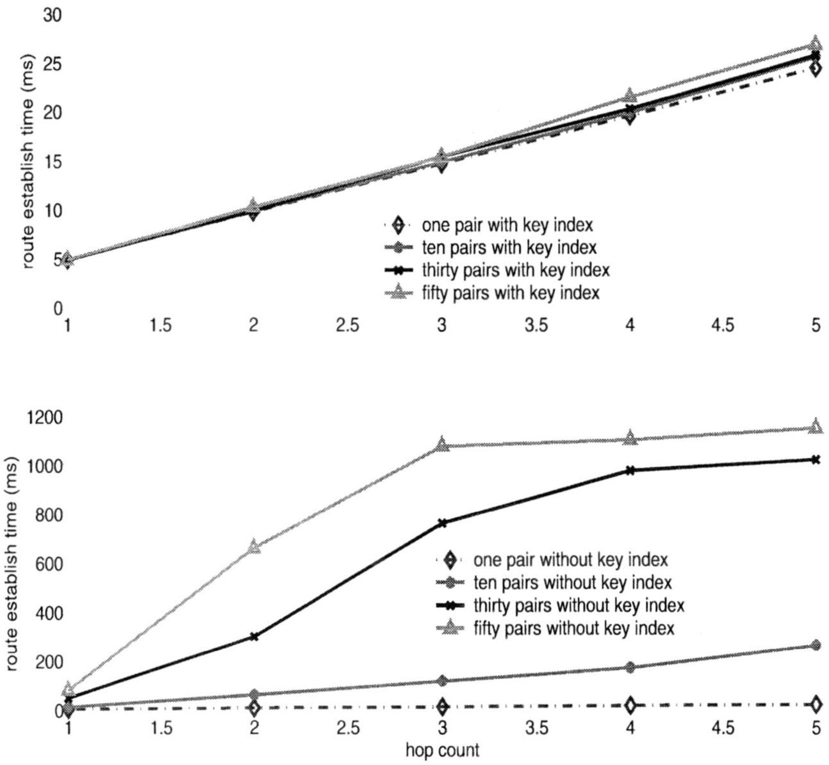

Figure 10. Inter-cluster route establish time (with and without key index).

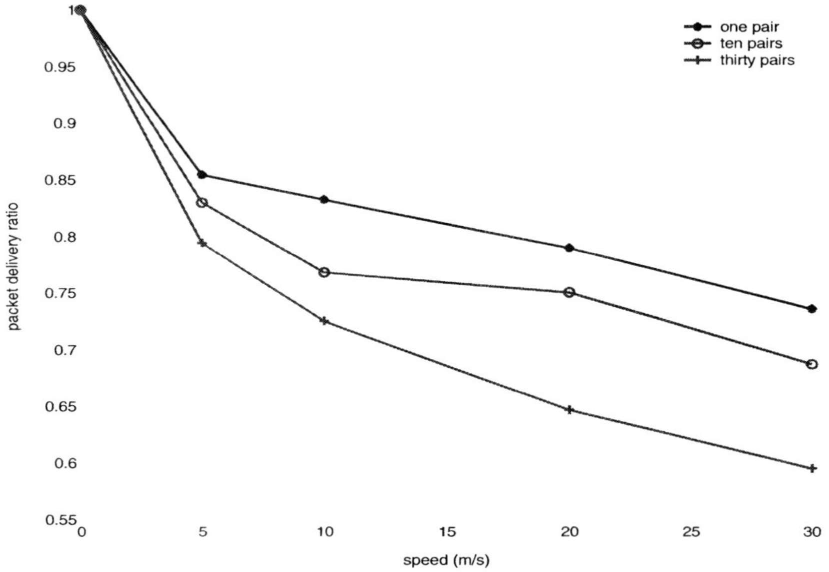

Figure 11. Packet delivery ratio under different node speeds.

a Denial-of-Service (DoS) attack based on that ID. ASR (Anonymous Secure Routing) [6] solves this problem by a scheme in which node one sends a temporary public key to node two in routing request, and in routing response node two transmits its pseudonym encrypted by the key to node one. Besides location privacy and route anonymity, ASR also supports limited hop count and destination verification by intermediate node so as to provide more security properties. A similar scheme to ASR is proposed by Cheng and Agrawal [30]. An on-demand anonymous routing scheme using bloom filters (ODAR) is proposed in [29]. ODAR guarantees source, destination anonymity, but not linkability anonymity. A recent work [9] proposed an anonymous on-demand routing protocol, termed MASK, based on a new cryptographic concept called pairing. An anonymous neighborhood authentication protocol is used and MASK fulfills the routing and packet forwarding tasks without disclosing the identities of participating nodes under a rather strong adversarial model. Although the above works addressed secure anonymous routing for MANET, none of them considered cluster based architecture. An extensive simulation study of SDAR, AnonDSR, ANODR, ASR, and MASK is given in [31]. A summary comparing SDAR [5], AnonDSR [21], ANODR [7], ASR [6] and the proposed scheme is given in Table III.

An on-demand position-based private routing algorithm, called AO2P, is proposed by Wu and Bhargava [32]. Anonymity for a destination relies on the difficulty of matching a geographic position to a real node ID. AO2P may be appropriate for MANET with high mobility nodes because node mobility enhances anonymity by making the match of a node ID with a position momentary. Other anonymous routing schemes using nodes' position information include [33] and [34].

Security in cluster based MANET was considered in several recent studies. The authors in [10] proposed a security architecture based on public key schemes and distributed certification. The CHs perform admistrative functions and hold shares of a network key for node authentication using proactive digital signatures. A similar design is proposed in [11] for cluster-based Near-Term Digital Radio (NTDR) ad hoc networks. A security infrastructure is provided for secure intra-cluster and inter-cluster communications. However, anonymity is not considered in these studies. [18] presents a scheme to ensure secure communication and to provide anonymity and location privacy in hybrid ad hoc networks. This proposal can only be effectively used for networks with fixed and powerful access points. A key management and secure routing protocol for cluster based MANET is proposed in [19]. This scheme uses asymmetric cryptography completely, which requires very high computational capacity. A fairly recent work [33] addressed the anonymous communication problem in MANET using positioning information of the nodes. Although it is based on the fact that broadcasting in a certain zone will hide the destination node, a similar idea as in our previous work [36], the scheme developed in [33] assumes that each node knows its own position and can learn other nodes' positions as well, thus it does not provide location anonymity. A hierarchical anonymous on-demand routing protocol (HANOR) has been proposed in [35]. HANOR extends ANODR to cluster based MANET. The focus of [35] is to show good scalability of anonymous routing protocols in a cluster based architecture, which is true for any hierarchical routing protocols in general. However, critical network elements such as the CHs are not hidden in HANOR, thus they may be located and attacked by adversaries. In addition, since the inter-cluster routing packets will be broadcast and processed by every node rather than just the GWs, the overhead is still high for HANOR.

Table 3. Comparison of secure anonymous routing protocols

	SDAR [5]	AnonDSR [21]	ANODR [7]	ASR [6]	SARC
trapdoor	Public key	Symmetric key with pre-established key index	Symmetric key	Symmetric key	Symmetric key with key hashing as index
RREQ	Onion using public key	Onion using symmetric key; symmetric key encrypted by source provided public key	Onion using symmetric or public key;	A temporary public key is issued by each node on the path; onion is not used	Intra-SARC: simple broadcast; Inter-SARC: A temporary public key is issued by each node on the path; onion is not used
RREP	onion	onion	Pseudonyms are attached and modified at each intermediate node	Pseudonyms are attached and modified at each intermediate node	Intra-SARC: simple broadcast; Inter-SARC: Pseudonyms are attached and modified at each intermediate node
Data Comm	onion	onion	In the clear	Use tag; and data shuffle (detail is not given)	Use hash value of the pseudonyms and the XOR of the next-hop pseudonym and data
Identity privacy (Sender)	Yes	Yes	Yes	Yes	Yes
Identity privacy (Receiver)	Yes	Yes	No	Yes	Yes
Identity privacy (Intermediate nodes)	No	No	Yes	Yes	Yes
Identity privacy (Sender-receiver)	Yes	Yes	Yes	Yes	Yes
Weak location privacy	Yes	Yes	Yes	Yes	Yes
Strong location privacy	No	No	No	Yes	Yes

Note that the current manuscript is a drastically improved version of our previous work on secure anonymous routing for cluster based MANET [36]. The improvements include: (1). A cluster based security architecture is employed and described in detail, including how the keys are managed, how the CH is chosen, and how to handle the nodes that join or leave a cluster, etc. (2). The efficiency of SARC is considered by introducing key indexing and the tradeoff between anonymity and efficiency is discussed. (3). A new technique for data forwarding is proposed using simple XOR operations to prevent data traffic from traffic analysis and packet analysis attacks. (4). Analytical results of anonymity analysis of SARC is given using information theoretic measures of anonymity. (5). Extensive simulations are carried out for a large MANET to evaluate the performance of SARC, where the node mobility is taken into account by using the Random Waypoint mobility model [27].

8. Conclusions

In this study, a Secure Anonymous Routing scheme for Cluster based MANET termed SARC is proposed. In SARC, intra-cluster routing uses a common broadcast channel to provide anonymity, while inter-cluster routing uses a sequence of temporary public keys as the trapdoor information. One of the unique features of SARC is that critical network elements, such as the cluster heads, are hidden from adversaries during the routing process. Key indexing is used to maximize routing efficiency and a technique based on XOR operations for data forwarding is applied to provide anonymity and maximize efficiency during data transmissions. The proposed routing scheme satisfies the principles of efficient anonymous routing in mobile networks, i.e., the proposed routing scheme are both identity-free and on-demand [8]. Analytical results are derived using information theoretic measure for anonymity analysis of SARC. Both anonymity analysis and attack analysis show the effectiveness of the proposed scheme. The overhead of both intra-cluster and inter-cluster anonymous routing is low. In addition, the computational complexity of data forwarding is also low due to the use of symmetric ciphers rather than public key schemes. In addition, SARC is demonstrated by simulation to scale well in large scale MANET.

References

[1] C.R. Lin and M. Gerla, "Adaptive Clustering for Mobile Wireless Networks", *IEEE Journal on Selected Areas of Communications*, vol.15, no.9, pp.1265-1275, Sep 1997.

[2] L. Ritchie, H. Yang, A. Richa, and M. Reisslein, "Cluster overlay broadcast (COB): MANET routing with complexity polynomial in source-destination distance", *IEEE Transactions on Mobile Computing*, vol.5, no.6, pp.653-667, Jun 2006.

[3] R. Ramanathan and M. Steenstrup, "Hierarchically-organized, Multihop Mobile Wireless Networks for Quality-of-Service Support", *Mobile Networks and Applications*, vol.3, no.1, pp.101-119, 1998.

[4] B. Schneier, *Applied Cryptography: Protocols, Algorithms, and Source Code in C*, 2nd Ed, John Wiley & Sons, 1996.

[5] A. Boukerche, K. El-Khatib, L. Xu, L. Korba, "SDAR: A Secure Distributed Anonymous Routing Protocol for Wireless and Mobile Ad Hoc Networks", *In 29th Annual IEEE International Conference on Local Computer Networks (LCN'04)*, pp. 618-624, Tampa, Florida, USA, November 16 - 18, 2004.

[6] Bo Zhu, Zhiguo Wan, Mohan S. Kankanhalli, Feng Bao, Robert H. Deng, "Anonymous Secure Routing in Mobile Ad-Hoc Networks", *The 29th Annual IEEE Conference on Local Computer Networks (LCN) 2004*, Tampa, Florida, U.S.A., 2004.

[7] Jiejun Kong, Xiaoyan Hong, and Mario Gerla, "ANODR: ANonymous On Demand Routing with Untraceable Routes for Mobile Ad-hoc Networks", *In Proceedings of ACM International Symposium on Mobile Ad Hoc Networking & Computing (MobiHoc 2003)*, Annapolis, MD, June 2003.

[8] Jiejun Kong, Xiaoyan Hong, M.Y. Sanadidi, and Mario Gerla, "Mobility Changes Anonymity: Mobile Ad Hoc Networks Need Efficient Anonymous Routing", *In Proceedings of IEEE ISCC*, 2005.

[9] Yanchao Zhang, Wei Liu, and Wenjing Lou, "Anonymous communications in mobile ad hoc networks", *IEEE INFOCOM 2005*, Miami, FL, March 2005.

[10] M. Bechler, H.J. Hof, D. Kraft, F. Pahlke and L. Wolf, "A Cluster-Based Security Architecture for Ad Hoc Networks", *IEEE INFOCOM 2004*, Hong Kong, March 2004.

[11] V. Varadharajan, R. Shankaran and M. Hitchens, "Security for cluster based ad hoc networks", *Computer Communications*, Vol.27, pp.488-501, 2004.

[12] David Chaum, "Untraceable Electronic Mail, Return Addresses, and Digital Pseudonyms", *Communications of the ACM*, vol.24, no.2, 1981.

[13] Rob Sherwood, Bobby Bhattacharjee, and Aravind Srinivasan, "P5: A Protocol for Scalable Anonymous Communication", *In the Proceedings of the 2002 IEEE Symposium on Security and Privacy*, May 2002.

[14] Nathalie Weiler, "Secure Anonymous Group Infrastructure for Common and Future Internet Applications", *In 17th Annual Computer Security Applications Conference (ACSAC'01)*, pp.401-410, December 10 - 14, 2001.

[15] M. Reed, P. Syverson, D. Goldschlag, "Anonymous Connections and Onion Routing", *IEEE Journal on Selected Areas in Communication Special Issue on Copyright and Privacy Protection*, 1998.

[16] Michael K. Reiter and Aviel D. Rubin, "Crowds: anonymity for Web transactions", *ACM Transactions on Information and System Security (TISSEC)*, Vol.1, Issue 1, pp.66 - 92, 1998.

[17] M. Jiang, J. Li, and Y.C. Tay, "Cluster Based Routing Protocol (CBRP) Function Specifications", *IETF Draft draft-ietf-manet-cbrp-spec-00.txt*, Aug. 1999.

[18] S. Capkun, J. Hubaux, and M. Jakobsson, "Secure and Privacy-Preserving Communication in Hybrid Ad Hoc Networks", *EPFL Tech. Report*, 2004.

[19] H. Lin, Y. Huang, and T. Wang, "Resilient Cluster-Organizing Key Management and Secure Routing Protocol for Mobile Ad Hoc Networks", *IEICE Trans. Commun.*, vol.E88-B, no.9, pp.3598-3613, Sep 2005.

[20] Rajani Poosarla, Hongmei Deng, A. Ojha, and D.P. Agrawal, "A cluster based secure routing scheme for wireless ad hoc networks", *23rd IEEE International Performance, Computing, and Communications Conference (IPCCC 2004)*, pp. 171-175, 2004.

[21] R. Song, L. Korba, and G. Yee, "AnonDSR: Efficient Anonymous Dynamic Source Routing for Mobile Ad-Hoc Networks", *SASN05*, November 7, 2005, Alexandria, Virginia, USA.

[22] Y. Hu, A. Perrig and D.B. Johnson, "Ariadne: A secure On-Demand Routing Protocol for Ad hoc Networks", *MobiCom 2002*, pp. 12-23, Atlanta, GA, Sep 2002.

[23] L. Korba and R. Song, "Scalability, Security Technologies and Mobile Applications", *In Proceeding of the first International Workshop on Mobility Aware Technologies and Applications (MATA'04)*, 2004.

[24] A. Serjantov and G. Danezis, "Towards an information theoretic metric for anonymity", *In Privacy Enhancing Technologies (PET)*, 2002.

[25] C. Diaz, "Anonymity Metrics Revisited", *DAGSTUHL seminar on anonymous communication and its applications*, 2005.

[26] http://www.eskimo.com/~weidai/benchmarks.html

[27] D. B. Johnson and D. A. Maltz, "Dynamic Source Routing in Ad Hoc Wireless Networks", *In Mobile Computing*, T. Imielinski and H. Korth (ed.), Vol. 353, pp. 153-181, Kluwer Academic Publishers, 1996.

[28] S. Seys and B. Preneel, "ARM: Anonymous Routing Protocol for Mobile Ad Hoc Networks", *Proc. International Conference on Advanced Information Networking and Applications (AINA)*, 2006.

[29] D. Sy, R. Chen and L. Bao, "ODAR: On-Demand Anonymous Routing in Ad Hoc Networks", *Proc. of The Third IEEE International Conference on Mobile Ad-hoc and Sensor Systems (MASS)*, Vancouver, Canada, Oct 2006.

[30] Y. Cheng and D. P. Agrawal, "Distributed Anonymous Secure Routing Protocol in Wireless Mobile Ad Hoc Networks", *In OPNETWORK 2005*, Aug 2005.

[31] J. Liu, J. Kong, X. Hong, and M. Gerla, "Performance Evaluation of Anonymous Routing Protocols in Mobile Ad-hoc Networks", *In IEEE Wireless Communications and Networking Conference (WCNC)*, Las Vegas, Nevada, USA, Apr 2006.

[32] X. Wu, and B. Bhargava, "AO2P: Ad Hoc On-Demand Position-Based Private Routing Protocol", *IEEE Transactions on Mobile Computing*, vol. 04, no. 4, pp. 335-348, July/August 2005.

[33] X. Wu and E. Bertino, "An Analysis Study on Zone-based Anonymous Communication in Mobile Ad Hoc Networks", *to appear in IEEE Transactions on Secure and Dependable Computing*, available at http://www.cs.purdue.edu/homes/wu/HTML/paper_purdue/zap_2clum.pdf.

[34] M. Rahman, M. Mambo, A. Inomata, and E. Okamoto, "An anonymous on-demand position-based routing in mobile ad hoc networks", *International Symposium on Applications and the Internet (SAINT 2006)*, Jan 2006.

[35] J. Liu, X. Hong, J. Kong, Q. Zheng, N. Hu, and P. Bradford, "A Hierarchical Anonymous Routing Scheme for Mobile Ad-Hoc Networks", *In IEEE MILCOM,* Washington D.C., Oct 23-25, 2006.

[36] L. Qian, N. Song, and X. Li, "Secure Anonymous Routing in Clustered Multihop Wireless Ad Hoc Networks", *IEEE CISS 2006*, Princeton, NJ, March 22-24, 2006.

In: Wireless Communications Research Trends
Editor: Tong S. Lee, pp. 83-122
ISBN 978-1-60021-674-9

Chapter 3

TRENDS AND CHALLENGES FOR MOBILITY MANAGEMENT IN IP-BASED NEXT GENERATION WIRELESS NETWORKS

Christian Makaya* and Samuel Pierre†
Mobile Computing and Networking Research Laboratory (LARIM)
Department of Computer Engineering, École Polytechnique de Montréal

Abstract

A major trend in next generation wireless networks (NGWN) or fourth generation wireless networks (4G) is the coexistence of diverse but complementary architectures and wireless access technologies. This heterogeneity brings several design and deployment challenges. Among them, integration of existing heterogeneous wireless networks requires the design of efficient mobility management schemes, at IP layer as well as lower layer, to enable seamless roaming of users. IP technology is the best choice for interworking and integration of various radio access technologies and is the main drive of networks evolution towards all-IP core networks. IP mobility management is a crucial issue in heterogeneous mobile environments. Several IPv6-based mobility management schemes have been proposed and the most known of them is Mobile IPv6 (MIPv6). Despite some advantages, MIPv6 and its extensions are hindered by several shortcomings, such as handoff latency, signaling overhead, packet loss. Thus, they fail to fulfil requirements of real-time applications. Moreover, with growing demand for real-time and multimedia applications, quality of service (QoS) support in heterogeneous wireless networks is of primary importance in order to improve system performance and users' satisfaction. However, QoS provision mechanisms like IntServ/RSVP and DiffServ have been developed for wired networks and are not optimized for mobile and wireless environments. This chapter addresses mobility management issues and QoS provision in IP-based NGWN. We present recent techniques addressing these problems and discuss their limitations as well as outstanding challenges that still need to be addressed to motivate research activities for the de-

*Correspondence to: Christian Makaya, Department of Computer Engineering, École Polytechnique de Montréal, P.O. Box 6079, Station Centre-ville, Montréal, Québec, Canada H3C 3A7Phone: +1-514-340-4711 ext. 2126, Email: christian.makaya@polymtl.ca.

†E-mail address: samuel.pierre@polymtl.ca.

sign of efficient protocols and mechanisms in NGWN. Finally, performance analysis of IPv6-based mobility management protocols is provided to show their pros and cons.

1. Introduction

Current trends in mobile wireless networks evolution are directed towards all-IP networks and to be independent of access technologies. In other words, all signaling between various entities in networks is exchanged at IP-layer, in order to achieve convergence and interworking of different access technologies. The initial evolution towards the all-IP mobile networks is the addition of IP multimedia subsystem (IMS) to the international mobile telecommunication 2000 (IMT-2000) network with the aim of offering IP-based real-time multimedia services. Next generation or 4G wireless networks (NGWN/4G) are expected to exhibit heterogeneity in terms of wireless access technologies, personalized and user-oriented services, high usability and more capacity [1]. In fact, to attract benefits and more business opportunities, value-added services with lower cost are necessary for success and survival of service providers in NGWN. Also, users will have more demands for seamless roaming across different types of wireless networks, support of various services (e.g., voice, data, game) and QoS guarantees. To enable legacy and value-added services across heterogeneous networks, usage of gateways may be required. Although, wireless and mobile technologies aim to provide ubiquitous information access to users on move, there is no single one that is able to simultaneously provide high bandwidth, low latency, low power consumption, and wide area data services to a large number of mobile users.

Conceptually, an NGWN architecture can be viewed as many overlapping wireless Internet access domains as shown in Figure 1 and is so-called wireless overlay networks [2]. NGWN will support a large variety of applications, with different QoS requirements, running on different types of mobile terminals and connected to various types of networks. This means, an NGWN architecture must be flexible and open, capable of supporting all these (different) type of networks, terminals and applications. A hierarchical network structure is recommended for NGWN to ensure scalability and facilitate management. Technological advances in evolution of portable devices have made possible support of multiple access technologies.

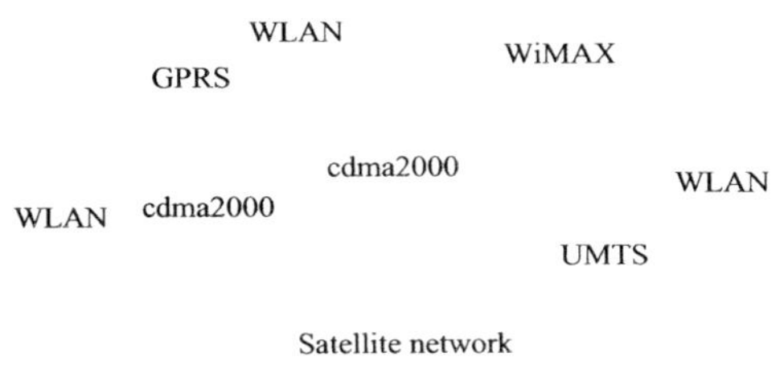

Figure 1. Overview of 4G/NGWN network architecture.

At present, no solution exists which comprehensively addresses the entire scale of heterogeneity and no wireless technology meets all QoS requirements such as low packet loss rate, signaling overhead, delay and jitter, all time. Existing wireless networks, like

3G/UMTS, 3G/CDMA2000 and WLAN/IEEE 802.11, have been extensively studied on an individual basis. Integration of these systems seems unavoidable due to potential benefits of their complementarity. Two major architectures (*loose* and *tight coupling*) for 3G/WLAN interworking have been proposed by both 3G wireless network initiative projects, 3GPP and 3GPP2, for their respective system [3, 4]. However, this integration will bring new challenges such as architecture and protocols design, interworking, mobility management, QoS guarantees and security issues.

Mobility management is one of the key topics in order to support global roaming of mobile node (MN) across different networks in an efficient way. In current homogeneous wireless networks such as WLAN and 3G cellular, mobility management focuses mainly on physical mobile object such as user and terminal. However, in NGWN, mobility is also a logical concept rather than only a physical one. In fact, mobility could just means the change of the logical location of network point of attachment instead of user's geographic position. In other words, handoff may occur between two different cells without necessarily moving out of the coverage area. Then, it is critical to provide seamless roaming support based on intelligent and efficient mobility management schemes.

Mobility management with QoS provision in NGWN remains a challenging and complex task. In homogeneous networks, mobility management has been widely and comprehensively studied in literature and surveyed in [5]. However, usage of mobility management schemes proposed for homogeneous wireless networks in NGWN leads to several drawbacks. Two generic QoS models, Integrated Services (IntServ) [6] with it signaling protocol Resource ReSerVation Protocol (RSVP) [7] and Differentiated Services (DiffServ) [8], have been proposed by the Internet Engineering Task Force (IETF) for wired networks. However, due to characteristics of wireless networks such as scarcity of resources, unpredictable available bandwidth, variable error rates, users mobility rate, they could be impractical. Hence, new and efficient mobility management techniques with QoS guarantee are needed for heterogeneous wireless networks in order to achieve global seamless roaming and service continuity.

There is no straightforward solution that takes into account the entire mobility management requirements and heterogeneity of NGWN. Instead, literature offers several proposals, each with their pros and cons, many challenges remain to be addressed. In this chapter, we present challenges of 4G/NGWN networks in terms of protocols design, architecture and services. We focus on handoff management and overview recent handoff schemes that aim to provide mobility over various access network technologies. Furthermore, we discuss limitations of these handoff techniques and challenges that still need to be addressed to achieve seamless roaming across heterogeneous wireless networks. Recall that, handoff is a process by which an MN changes it current attachment towards a new subnet. In contrast to related literature, we performed a detailed study of IPv6-based mobility management protocols and several others mobility management protocols which handle vertical handoff. IPv6 technology is considered as the basis of future all-IP networks. For example, 3GPP decision is to adopt IPv6 as the only IP version for the IMS. Hence, we focus mainly on IPv6 as convergence technology of various access networks (e.g., WLAN, 3G cellular networks, mobile ad hoc networks, WiMAX, etc.).

The remainder of this chapter is organized as follows. Next section describes characteristics pertaining to mobility management while in Section 3., some mobility management schemes are presented. QoS challenges in mobile environments are investigated in Section 4.. Performance analysis of IPv6-based mobility management protocols, with respect to various network parameters, is carried out to point out their limitations and advantages in Section 5.. Concluding remarks are presented in the last section.

2. Mobility Management

Mobility management is essential for roaming of users with mobile terminals to access services in progress through wireless networks. In other words, mobility management enables system to locate roaming terminals in order to deliver data packets and maintain connections with them when moving into new area or subnet.

2.1. Basic Principles and Concepts

In NGWN, there is four types of high-level mobility: *terminal, personal, service* and *session mobility* [5]. They basically dealt with network and application layer and it seems very logical to attempt their combination. However, a naïve simultaneous implementation of mobility management schemes of these different mobility types would lead to performance inefficiency and even system disorder. Basic mobility management schemes are domain-based and splits mobility into micro-mobility and macro-mobility. Macro-mobility occurs when an MN crosses two different wireless domains during its roaming, while micro-mobility refers to movement within a specific wireless domain. Many performance and scalability requirements should be taken into account when trying to select or design mobility management schemes in NGWN, including:

- signaling traffic overhead: control data load should be lowered to an acceptable range;
- routing efficiency: to exclude redundant transfer, the routing paths between correspondent nodes (CNs) and MNs should be optimized;
- QoS provision: establishment of new QoS parameters in order to deliver data traffic should be supported by a mobility management scheme; disruption of service must be kept to a minimum level during establishment or renegotiation of new connection;
- seamless handoff: handoff algorithm should minimize packet loss and be fast enough to ensure that an MN can receive IP packets at its new location within a reasonable time interval in order to reduce packet delay;
- security: different levels of security must be supported by mobility management schemes such as data encryption and user authentication while limiting the traffic and time of security process;
- authentication, authorization and accounting (AAA) services: issues for AAA services are especially important in heterogeneous wireless environments as different radio access networks are likely be managed by different administrative domains.

Mobility management is based on two components, namely *location* and *handoff management*. Location management enables the system to locate users and their terminals between call arrivals. In other words, it allows the network to track MN movement and update the information pertaining to the location between consecutive communications. In NGWN, MNs can be reached through multiple access networks, then new challenges arise pertaining intelligent location management techniques. Locating MN in NGWN requires interoperability between several wireless systems that do not necessarily use the same technology. Thus; signaling traffic load may increase and network performance could decrease. Various location management schemes have been proposed in the literature to reduce wastage of network resources such as bandwidth. Several survey papers address location management in wireless networks. Among them, in [9], authors provide a comprehensive survey of location management approaches and suggest future research avenues. Location management in NGWN bring challenges for the design of mobility management. First, when coverage areas fully overlap, through which networks should an MN perform location registration? Within which one? How should the up-to-date user location information be stored? [10]. Another issue concerns the manner in which the exact MN location would be determined within a specific time constraint. In addition, an efficient use of network resources and the enhancement of scalability, reliability and robustness must be taken into account.

Handoff management enables a network to maintain a connection with MN during its movement and change its network point of attachment. Whenever an MN changes its attachment point, it sends a request for handoff initiation to current access point (AP) or base station (BS) towards target AP. Control is handed over to target AP by the current one after initiation step. Change of attachment point may also lead to IP address changes then leading to IP-layer handoff. In fact, after connection to new AP is established, an MN is able to receive router advertisement message sent by new access router. If the prefix of router advertisement message is the same as the prefix of previous IP address, this means that previous and new APs belong to same subnet. Then, an MN can switch to the new AP without changing the IP address. Otherwise, an MN must perform network layer handoff, which results in IP address changes. After acquisition of new IP address, data may now be sent to that address. Handoff management is a major issue in mobility management since an MN can trigger several handoffs during a session as it will be in NGWN. Three main approaches could be used to control handoff process: mobile-controlled handoff (MCHO), network-controlled handoff (NCHO) and mobile-assisted handoff (MAHO). Most proposals based on these approaches were designed for homogeneous networks.

In NGWN, more tradeoffs are required for these approaches to allow best selection of target access network (AN) where an MN should hand over. In fact, handoff triggering is performed either in AN and/or by MN while handoff process execution requires several entities located in both home network (HN) and AN, particularly for vertical handoffs. Managing AN or technology selection in mobile device and/or by access router (AR) to trigger handoff could only be conducted with locally available information such as link quality, signal strength, and AR capabilities and/or load. However, other information recorded in HN such as operator policy, AN global load and user preferences, can be relevant to make such selection. Consequently, when selection decision is made according to AN information which is restricted to MN and AR knowledge, it could lead to a suboptimal access network selection. In fact, managing computation of handoff decision on MN side may be limited to its processing capabilities and other factors cited above. It seems that efficient mobility management schemes is a crucial need.

In IPv6-based wireless networks, QoS may be defined by different metrics such as packet loss, handoff latency and signaling overhead cost. Handoff latency at MN side is the time interval during which an MN cannot send or receive any packets during handoff and it is composed of L2 handoff latency and L3 handoff latency. The time interval when an MN disconnects with the air-link of current AP/AR and connects to the air-link of new AP/AR refers to L2 (link switch) handoff latency. On the other hand, L3 handoff latency is the sum of movement detection latency, IP addresses configuration delay and binding update (BU) procedure latency. The overall handoff latency may be high and leads to packet loss, which is intolerable for real-time applications. The packet loss is defined as the number of lost packets during handoffs and is proportional to handoff latency. An efficient mobility management mechanisms must ensure that handoffs are fast (i.e., minimal packet delay), smooth (i.e., minimal packet loss) and scalable (i.e., lower signaling overhead). An handoff with these combined characteristics is known as *seamless handoff*. In other words, a seamless handoff should be performed with minimal packet loss and low packet delivery delay. This means that, seamless handoff schemes should have following features: minimum handoff latency, low packet loss and limited handoff failure.

2.2. Vertical Handoff Management

In NGWN, mobile users will encounter both intra- and inter-system handoffs. Then, it is essential that applications running on mobile terminals remain unaffected by users' movement. With coexistence of various wireless access technologies, there is two kinds of handoffs in NGWN: *horizontal* and *vertical handoffs*. Horizontal handoff occurs when an MN is moving between APs of the same network technology. When APs belong to different networks (e.g., IEEE 802.11 and UMTS), this movement refers to vertical handoff. Characteristics of NGWN make implementation of vertical handoff more challenging as compared to horizontal handoff. Various schemes have been proposed in literature for horizontal handoff [5]. However, much less effort was deployed for vertical handoff research. Although some studies are now available in the literature, the proposed approaches are still hindered by several shortcomings [11]. Vertical handoff is usually asymmetric and can be classified into two types namely, *upward* and *downward handoff* [2]. Hence, a handoff to a wireless overlay with a larger cell size and lower bandwidth per unit area is an upward vertical handoff, for example from WLAN to 3G cellular network. On the other hand, a downward vertical handoff is a handoff to a wireless overlay with a smaller cell size and usually higher bandwidth per unit area, for example from 3G cellular network to WLAN. The aim of vertical handoff techniques is to achieve efficient interface management that gives better power balance and to perform handoff to the most appropriate access network at the right time. That means the mobile user is being always best connected [12].

Handoff process in NGWN can be divided in three phases: network discovery, handoff decision and handoff execution. Network discovery is the process that allows an MN to know which wireless networks are reachable. The simplest way for an MN with multiple radio interfaces to discover reachable wireless networks is to keep all radio interfaces always on. However, keeping an interface active all the time consumes battery power even without sending/receiving any packets and also bandwidth usage. Radio interface may be periodically activated to receive service advertisements. However, the activating frequency will directly affect network discovery process. In fact, an MN that activates radio interfaces with high frequency may discover reachable networks quickly but its battery may run out very soon. On the other hand, an MN that activates radio interfaces with low frequency may increase power efficiency, but it may discover reachable wireless networks slowly. Hence, there is a trade-off between network discovery time and power efficiency [13]. Efficient radio interfaces management is then crucial. Although, passive connectivity may be valuable, it leads also to battery power consumption. Paging seems as a solution to this problem. However, the design of efficient paging scheme is very hard. Often, vertical handoff schemes do not support paging nor does it differentiates between an active and idle MN. Absence of paging support can lead to significant power wastage, especially in NGWN where a single device can have multiple wireless interfaces and may need to maintain multiple simultaneous bindings.

Handoff decision is the process of deciding when to perform handoff. In homogeneous networks, handoff decision is typically driven by metrics which are strictly related to received signal strength (RSS) level, channel and resources availability. However, in NGWN, RSS from different networks can not be compared directly and each network has specific characteristics. Then, handoff decision based on signal strength as alone criterion may be inefficient in NGWN. Mobility in NGWN is either logical or physical, then user profile and preferences seem to be important when performing vertical handoff. For instance, user may have access preferences based on price, power consumption, speed, security and other requirements. Hence, more complex metrics combining a higher number of parameters such as price, bandwidth, priority, power consumption, service type, system performance and user preferences, network and MN conditions, have to be defined for handoff in NGWN [14]. Design of handoff decision function which evaluates simultaneously these various metrics is crucial in NGWN and remains a research challenge. After decision to handoff is taken, handoff execution process comes into play for association with the new wireless network. Note that an MN should observe if the new wireless is consistently better than the current one before performing handoff, to avoid ping-pong effect.

3. Mobility Management Protocols

For users roaming across NGWN, complete mobility management scenario may be subdivided into two types of roaming aspects: *intra-domain* or *micro-mobility*, and *inter-domain* or *macro-mobility*. A wireless domain or simply domain is a large wireless access network managed by a single administrative authority. Macro-mobility occurs when an MN crosses two different wireless domains while roaming, while micro-mobility refers to movement within a specific wireless domain. According to technology used in each domain, these two roaming aspects could be subdivided into intra-system and inter-system mobility. The former refers to movement between different domains of the same system (similar network interfaces and protocols) while the latter focuses on movement between different backbones, protocols, technologies, or service providers. Several protocols have been proposed for mobility support in NGWN at different layers of TCP/IP protocols stack.

3.1. Macro-mobility Protocols

Mobile IPv6 (MIPv6) [15] was proposed for mobility management at IP layer and allows MNs to remain reachable despite its movement within IP environments by the Internet Engineering Task Force (IETF). Each MN is always identified by its home address (HoA), regardless of its current point of attachment to the network. While away from its home network, an MN is also associated with a care-of address (CoA), which provides information about the MN's current location. In fact, when the MN crosses the boundary of its current serving subnet, movement detection and router discovery (router solicitation/advertisement -RS/RA- messages exchange) are performed for identification of new point of attachment and new access router (NAR). Further the MN acquires new CoA through stateless or stateful IPv6 address autoconfiguration mode [16]. Duplicate address detection (DAD) procedure is performed, through neighbor solicitation/advertisement (NS/NA) messages exchange, to ensure that the configured CoA is likely to be unique on the new link. DAD is a time consuming process; in fact, according to [16], DAD execution takes at least 1000 milliseconds to detect there is no duplicate address in the link. Based on a premise that DAD is far more likely to succeed than to fail, in [17] a modification of stateless address autoconfiguration and IPv6 neighbor discovery processes, termed optimistic DAD (oDAD), is proposed in order to minimize address configuration delays by eliminating DAD completion time.

After acquisition of CoA, an MN sends a binding update (BU) message to home agent (HA), informing it about the new address and also to the correspondent nodes (CNs) to enable route optimization. Note that, CN does not need to implement the specific MIPv6 functionalities if route optimization usage is not required. In this case, CN will send all packets destined to MN at its home address. In response to BU message, the binding acknowledgment (BAck) will be returned. Note that, before BU process is performed at CN, return routability procedure (RRP) is required to insure that BU message is authentic and not from malicious MN. Return routability procedure is based on home address test and care-of address test. In home address test, MN sends a Home Test Init (HoTI) message to HA, which forwards them to CN. The CN responds with Home Test (HoT) message addressed to MN's HoA and contains a secret Home Key Token. The HoT is forwarded by HA to the MN's current address. On the other hand, during the care-of address test, MN sends a Care-of Test Init (CoTI) message directly through optimized path to CN, and the latter responds with the Care-of Test (CoT) message containing a secret Care-of Keygen Token. HoTI and CoTI message are sent at the same time and the procedure requires very little processing at CN. Figure 2 presents the sequence of message flows used in MIPv6 based on stateless address autoconfiguration.

To enable service continuity and QoS provision, seamless handoff is of great importance, which means low packet loss, low latency and minimum service disruption during handoff. Handoff latency is the time interval during which an MN cannot send or receive any application traffic during handoffs. The handoff latency is composed of L2 handoff latency and L3 handoff latency. The time interval when an MN disconnects with the air-link of current access point (AP) and connects to the air-link of new AP refers to L2 handoff latency. On the other hand, L3 handoff latency is the sum of movement detection latency, CoA configuration latency, DAD process delay and BU process latency. The overall hand-

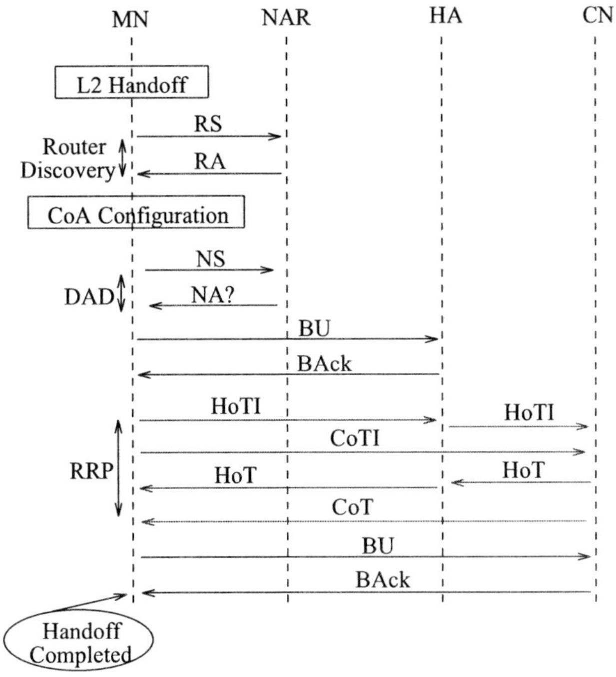

Figure 2. Signaling messages sequence for MIPv6.

off latency may be long enough and leads to packet loss, which is intolerable for real-time applications, i.e., delay-sensitive and loss-sensitive applications such as VoIP. Thus, efficient mobility management techniques are needed for heterogeneous wireless networks in order to achieve global seamless roaming and service continuity.

MIPv6 has some well-known disadvantages such as overhead of signaling traffic (especially for MN with high mobility rate or if MN is located far away of HA or CNs), high packet loss rate and handoff latency, thereby causing a user-perceptible deterioration of real-time traffic [18]. Signaling overhead generated by MNs increases when their number increases and will result in scalability problem with MIPv6. Moreover, BU process at CN requires, at minimum, two round-trip times between MN and its CN. This could be unsuitable for real-time applications, for example when round-trip times is higher than 200 milliseconds. Also, MIPv6 handles local mobility of MNs in the same way as it handles global mobility. Simultaneous mobility is another problem MIPv6 faces due to route optimization, which can occur when two communicating MNs have ongoing session and they both move simultaneously [19].

These weaknesses have led to investigations of other solutions that enhance MIPv6 and aim to handle local mobility of MNs within the boundaries of an access network to alleviate core network from the burden of increased signaling traffic. Anticipation of handoff approaches are also used to minimize service disruption.

Session Initiation Protocol (SIP) [20] is a signaling protocol defined by IETF and offers numerous benefits including provision of session/call control and scalability. SIP provides a framework for multimedia applications management and has the potential capability to support advanced high-level mobility management in heterogeneous networks. In order to handle mid-session mobility, SIP uses re-INVITE message. In fact, when one of the two parties moves, its sends a re-INVITE message to the other party, informing it about its new location, e.g., its new IP address. In addition to the re-INVITE message sent directly to correspondent node (CN), the mobile node (MN) also registers its new location in the foreign network to the SIP server in its home network. Figure 3 shows the signaling messages for SIP mobility over IPv6. SIP does not support inherent return routability procedure. However, the new CoA of an MN can be verified by using cryptographic technique such as SIP identity [21].

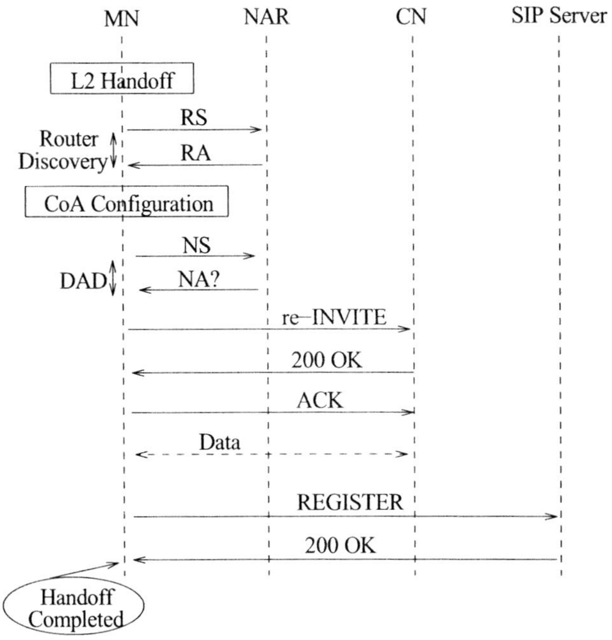

Figure 3. Signaling messages sequence of SIP over IPv6.

Although SIP provides its own mechanisms for reliable transmission, it runs over TCP or UDP to carry its signaling messages and is thus limited by performance of these protocols over wireless links. In fact, it is hard to keep TCP session alive while an MN changes its IP address in SIP-based mobility management. Moreover, processing of SIP messages in intermediate and destination servers may take considerable amount of time and introduces unacceptable handoff latency. Also, each time an MN acquires a new IP address, it must register this address with the SIP server in home network. Hence, this could create a high load on the home server and introduce long handoff delay when an MN is far away from its home network. Furthermore, SIP is more generous in message sizes since SIP's messages are text based. Hence, a pure SIP mobility management approach would generate higher signaling load compared to MIPv6. On the other hand, it is very difficult to extend SIP for tackling TCP mobility since SIP is featured by mobility awareness to applications. SIP and MIPv6 can be integrated to allow support of terminal and personal mobility and to complement each other in order to support different applications that are transparent, mobility aware and UDP/TCP-based. However, this integration should be done carefully to avoid unnecessary redundancy of some functionality.

The transport layer is considerably affected by mobility of an user. In fact, it must be able to quickly adapt its flow and congestion control parameters to the new network situations during and after handoffs. Hence, transport layer seems a candidate for mobility support in IP networks. Mobility management schemes at transport layer are often referred as cross-layer approaches because they operate by exploiting network an data-link layer information for network detection and IP address management [22]. Stream Control Transmission Protocol (SCTP) is a new reliable transport layer protocol proposed in [23] and is featured by multi-streaming and multi-homing contrarily to TCP and UDP. Multi-homing feature allows an association (i.e., connection between two SCTP endpoints) to maintain multiple IP addresses. The multi-homing feature of SCTP enables it to be used for Internet mobility support, without support of network routers or special agents. SCTP includes the so-called ADDIP extension to enable an SCTP endpoint to dynamically add a new IP address or delete an unnecessary IP address. Moreover, it enables to change the primary IP address used for the association during an active session. When an event such as add, delete or change occurs, the MN will notify the corresponding event to the CN by sending an ASCONF (Association Configuration Change) chunk with Add IP Address parameter to inform the CN of the new IP address. The CN will add the new IP address to the list of association addresses and will reply with an ASCONF ACK chunk to MN.

While MN is moving, it may change the primary path to the new IP address by path management function. The MN can also informs CN to delete the IP address of previous network from the address list by sending ASCONF chunk with Delete IP Address parameter when MN confirms that the link of previous network has failed permanent. This extension of SCTP, is known as Mobile SCTP (mSCTP) [24, 25] is under study and can be used to provide soft handoff for MNs that are moving into different IP network domains during the active session. Mobility management implementation at transport layer has some benefits, such as simplified network infrastructure, location privacy and smooth handoff. However, one of the shortcomings of transport layer mobility management approach is their dependence on other layers for location management. Also, an authentication scheme is required to prevent spoofing when each transport layer protocol implement binding update. On the other hand, the key issue of mSCTP is that SCTP is not used by applications, hence to have mSCTP used in a widespread fashion would involve modifying lot of applications. Furthermore, if the MN keeps moving back and forth between two subnet, as signaling is required for every CN, then mSCTP can be inefficient. In fact, if there are many CNs, this could result into a signaling overhead could. mSCTP does not handle location management support by itself; thus a proposal is to reuse IP-based mobility management protocols, such as MIPv6, for location management in mSCTP.

Higher layer-based mobility protocols (e.g., SIP, SCTP, mSCTP) are focused primarily on mobility management on the end-to-end basis. Furthermore, they do not have the potential to achieve low handoff latency. Mobility support at transport layer makes the network architecture simple by working without adding new entities within the network. Simultaneous mobility or handoff problem happen when the MN and CN are both mobile, and they both move at around the same time. SIP extension for mobility management, mSCTP and MIPv6 use direct binding updates between an MN and a CN. Then, MIPv6, mSCTP and SIP are vulnerable to the simultaneous mobility problem.

3.2. Micro-mobility Protocols

Two main micro-mobility protocols proposed by IETF in order to improve the performance of MIPv6 are HMIPv6 (Hierarchical Mobile IPv6) [26] and FMIPv6 (Fast Handovers for Mobile IPv6) [27].

3.2.1. Hierarchical Mobile IPv6 (HMIPv6)

HMIPv6 handles handoff locally through a special node called MAP (Mobility Anchor Point). The MAP, acting as a local HA in the network visited by the MN, will limit the amount of MIPv6 signaling outside its domain and reduce delays associated with location update process. An MN residing in MAP's domain is configured with two temporary IP addresses: a regional care-of address (RCoA) on the MAP's subnet and an on-link care-of address (LCoA) that corresponds to the current location of the MN. The RCoA is communicated to HA and all active CNs for packets routing. The MAP, will receive all packets addressed to the RCoA of the MN and will encapsulate and forward them to MN's current LCoA. As long as an MN moves within MAP's domain, it does not need to transmit BU messages to its HA and CNs. However, an MN must notify the MAP of its movements, resulting in a change of its LCoA. The RCoA does not change during MN roaming within MAP domain. This makes the MN's mobility transparent to HA and active CNs.

When an MN crosses a new MAP's domain, moreover from registering with new MAP, the BU messages need to be sent by the MN to its HA/CNs to notify them of its new virtual location. In fact, an MN performs two types of binding update with HMIPv6: local and global. Global binding update occurs when an MN moves out of its MAP domain while local binding update is performed when an MN changes its current IP address within a MAP domain. Hence, for global binding update, the MN first registers with a local MAP and thereby obtains a regional care-of address (RCoA) on the MAP's link, then registers this RCoA to HA and CNs. Figure 4 presents the generic sequence of message flows used in HMIPv6 for intra-MAP roaming. Handoff performance is improved with HMIPv6 by reducing handoff latency and signaling overhead. However, HMIPv6 cannot meet requirements for traffic that is delay sensitive such as voice over IP (VoIP), due to packets loss and services disruption [18, 29]. Moreover, MAP has a central position and this exposes mobility management to possible MAP failures. In other words, MAP could be a single point of failure. This issue could be resolved by deploying a hierarchy with multi-level MAP domains.

3.2.2. Fast Handovers for Mobile IPv6

FMIPv6 was proposed to reduce handoff latency and to minimize services disruption during handoffs pertaining to MIPv6 operations such as movement detection, binding update and addresses configuration. Link layer information (L2 trigger) is used either to predict or rapidly respond to handoff events. Depending whether an AR or the MN initiates the handoff, FMIPv6 can be either network-initiated or mobile-initiated. The triggers for the handoff decision are beyond the scope of [27] and are left optional. However, the two mains possibilities are a link-specific event (L2 trigger) occurring in the MN or in the network and

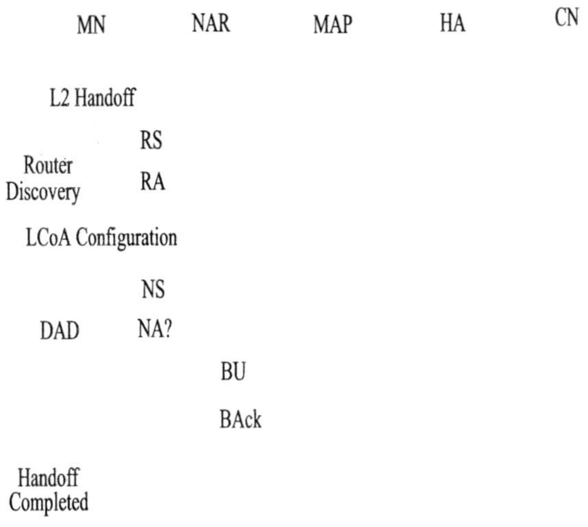

Figure 4. Signaling messages sequence for HMIPv6.

router discovery performed by an MN. A preliminary list of link layer triggers is proposed in [28].

Consider the mobile-initiated case, when an MN detects its movement toward NAR by using L2 trigger, it sends a router solicitation for proxy (RtSolPr) message, which contains the AP identifier, to its current AR (PAR in Figure 5) in order to obtain information about NAR and neighbor networks. On receiving this request, PAR replies with a proxy router advertisement (PrRtAdv) message that contains the link layer address, the IP address and the subnet prefix of the NAR. Note that, when PAR receives RtSolPr, it must resolve the identifier to subnet specific information before to generate PrRtAdv message. If the new AP is unknown to PAR, its has to respond indicating that the latter is unknown. In this case, the MN must stop fast handoff protocol operations on the current link, but may use a reactive handoff mode from the new link as explained below. Upon receipt of PrRtAdv, the MN configures a prospective new CoA (NCoA) by using the information about the NAR and sends fast binding update (FBU) to PAR with the NCoA. On reception of FBU, PAR starts fast handoff procedure by sending handoff initiate (HI) message to NAR, which includes the request for verification of NCoA and for establishment of bi-directional tunnel between PAR and NAR. In response to HI message, NAR performs DAD procedure and then responds with handoff acknowledgment (HAck) message.

After receiving HAck message, PAR sends the result to MN by using a fast binding update acknowledgment (FBAck) message and establishes a binding between previous CoA (PCoA) and NCoA and tunnels any packets addressed to PCoA to NCoA in NAR's subnet. Since the precise time that the MN will switch networks (link layer handoff) is unpredictable, FBAck message is sent to both previous link and new link. This ensure that the MN will receives the FBAck message either via the PAR or NAR indicating the successful binding. According to the implementation, the NAR can start buffering these forwarding packets until the MN arrives on its link. The MN announces its presence on the new link by sending router solicitation (RS) message including fast neighbor advertisement (FNA) option to NAR. Then, NAR will start delivering the buffered and new incoming packets to MN. Bi-directional tunnel remains active till the MN complete the MIPv6's BU procedure which occurs after FMIPv6 process. Sequence of message flows used in FMIPv6 is illustrated in Figure 5 for MN-initiated handoff of predictive mode. A counterpart to predictive mode of FMIPv6 is reactive mode. Reactive mode refers to scenario in which an MN sends FBU message from NAR's link but FBAck message has not been received yet. In reactive mode, the registration of the MN with NAR is delayed until L2 handoff is completed. Reactive mode can be done either intentional or serve as a fall-back when a predictive mode could not complete successfully, for example if the L2 handoff was completed before FBAck message has been received at the MN.

Although FMIPv6 paves the way on how to improve MIPv6 performances in terms of handoff latency, it is still hindered by several problems such as QoS support and scalability. In fact, FMIPv6 does not effectively reduce global signaling and packet losses, which may lead to unacceptable service disruption for real-time applications. Moreover, bi-directional tunnel between PAR and NAR may bear a non-optimal routing path if there is no direct link between PAR and NAR. Hence, this non-optimal routing path increases end-to-end packet delivery delay and wastes bandwidth. In original FMIPv6, NAR consumes storage space to buffer packets forwarded by PAR before to deliver them to MN, which occurs once MN establishes an IP connectivity with NAR. Moreover, these transferred packets have no QoS guarantee before the new QoS path is set up and no security association protection over the wireless link. Also, FMIPv6 requires assumption that the mutual security associations are already established between ARs and they need to share their information. This assumption may be difficult to satisfy if the access network is managed by different service providers. Hence, it is crucial to manage buffer efficiently, security association establishment and to minimize the registration latency.

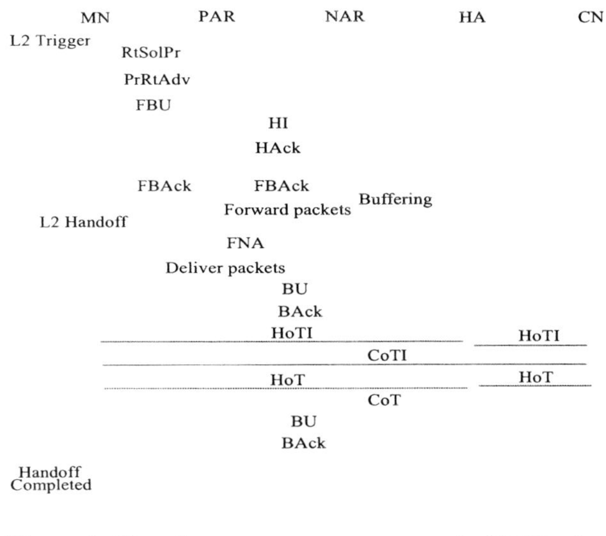

Figure 5. Signaling messages sequence for FMIPv6.

3.2.3. Fast Handover for HMIPv6 (F-HMIPv6)

Combination of HMIPv6 and FMIPv6 motivates the design of Fast Handoff for HMIPv6 (F-HMIPv6) [30, 31] to allow more network bandwidth usage efficiency. F-HMIPv6 can be either network-initiated or mobile-initiated like FMIPv6. In F-HMIPv6, the bi-directional tunnel is established between MAP and NAR, rather than between PAR and NAR as it is the case in FMIPv6. Hence, the MN exchanges signaling message for handoff with the MAP, rather than with PAR. After signaling message exchanges based on FMIPv6 messages, MN follows normal HMIPv6 operations by sending local BU (LBU) to MAP. When MAP receives the LBU with new LCoA (NLCoA) from MN, it will stop packets forwarding to NAR and then clear tunnel established for fast handoff. In response to LBU, MAP sends local BAck (LBAck) to MN and the remaining procedure will follow HMIPv6's operations. Sequence of message flows used in F-HMIPv6 is illustrated in Figure 6 when MN moves from PAR to NAR within MAP's domain, and the MAP already has adequate information on the link-layer address and network prefix of each AR.

Note that F-HMIPv6 may inherit some drawbacks of both FMIPv6 and HMIPv6, for example synchronization issues and signaling overhead. In fact, in F-HMIPv6, if the MN would perform the handoff right after sending the FBU to the MAP, all the packets transferred to the previous LCoA, during the period that the FBU requires to arrive to the MAP, would be lost. Additionally, if the MN would perform the handoff right after sending the FBU, it would not immediately receive any redirected packet for the same reason, increasing the handoff latency and packet losses [18]. Moreover, with elimination of inter-AR signaling message exchange (e.g., HI and HAck) in F-HMIPv6, context transfer between PAR and NAR as in FMIPv6 is not possible. Although buffering and tunneling of packets from PAR or MAP to NAR reduces packet loss, it places burden on routers and entails extra signaling and retention of state in routers. With current IPv6-based handoff protocols, seamless mobility is not guaranteed. A comparative analysis of IP-based mobility protocols was conducted in [18, 29] and design issues for seamless mobility amongst heterogeneous NGWN/4G were identified.

3.2.4. Context Transfer and Router Discovery

FMIPv6 typically requires some a priori knowledge of target network where an MN will handoff and the next access router, including the IP address of the router. However, this issue was out of the scope of FMIPv6 protocol. To achieve seamless mobility across various access technologies and networks, MN needs to have information about wireless network to which it could attach. Also, it is necessary to transfer information (context transfer) related to the MN from the current access router to the next one in order to re-establish service. To enable these procedures, Candidate Access Router Discovery (CARD) protocol [32] and Context Transfer Protocol (CXTP) [33] have been proposed. They avoid usage of limited wireless resources and provide fast mobility and secure transfer.

Their key objectives are to reduce latency and packet losses, and to avoid re-initiation of signaling to and from MN from the beginning. However, context transfer is not always possible, for example when MN moves across different administrative domains. The new network may require the MN to re-authenticate and to perform signaling from beginning rather than to accept its transferred context. Moreover, the entities exchanging context or

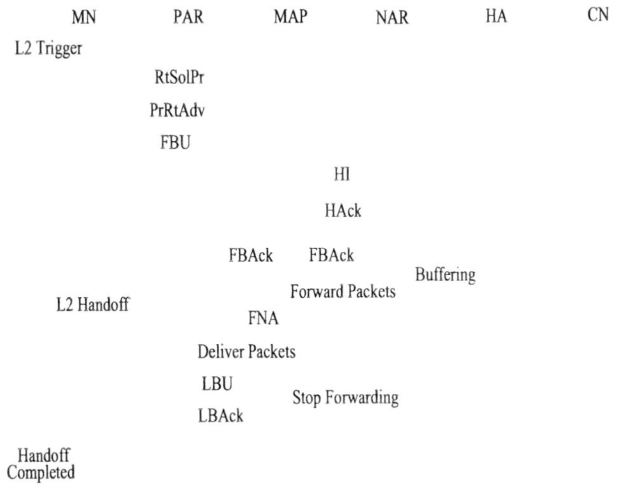

Figure 6. Signaling messages sequence for F-HMIPv6.

router identity must authenticate each other. This could be a tedious process in NGWN due to coexistence of various networks. Furthermore, the requesting AR sends a CARD request message to its peer only when the candidate access router (CAR) table entries timeout. Hence, CARD cannot enable dynamic collection of information about neighbor ARs.

3.3. Cross-layer Protocols

Protocol design and implementation under TCP/IP stack usually follow layered framework to enable simplicity an scalability. However, this layered strategy does not favor an optimal solution, especially with wireless networks. In fact, higher layer protocols are often affected by unreliable lower layer protocols. For example, data rate supported by a wireless link depends on interference, which depends on traffic at neighbors subnet. Network performance improvement can be obtained through collaboration among protocols of different layers. This collaboration or interaction is known as a cross-layer design approach.

Cross-layer design allows integration of different layers of TCP/IP stack and jointly optimize functions of each layer protocol. Figures 7 and 7 show traditional interaction and cross-layer interaction under TCP/IP stack.

Cross-layer approaches [34] have gained more attention in research community for the design of efficient mobility management protocols. However, interactions between layers and the degree of synchronization between radio handoff and IP-based mobile registration process are still not well defined. Further investigations are required.

In spite of the advantages of cross-layer design approach, certain issues must be considered carefully. Cross-layer approach remains risky due to protocol layer abstraction loss, an unexpected impact on the future design of the network, difficult management and maintenance problems and incompatibility with existing protocols [35]. Then, those factors and aspects must be taken into account and any proposed solution must be modular to allow

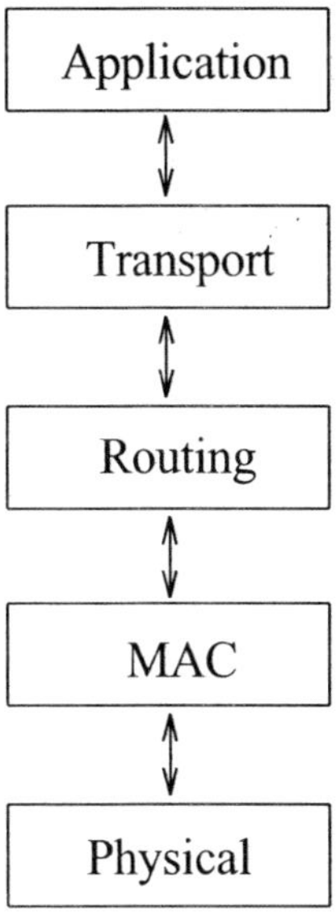

Figure 7. Traditional interactions.

future changes which do not require new design. Two main open issues arise with cross-layer design approach: what are the consequences of cross-layer integration? What is the longevity and the reusability of the solutions?

3.4. Vertical Handoff Protocols

Several IP-based handoff protocols proposed in the literature for supporting horizontal and vertical handoff may appear suitable [10, 11]. But there is still a lot of work to do for further improvements in NGWN/4G. Handoffs in IP-based NGWN involve changes of access point at link layer and routing changes at IP layer. Many approaches have been proposed in order to perform vertical handoff with their strength and drawbacks. In [36], integrated architecture and radio interface selection schemes have been proposed based on signal strength and priority of radio interfaces. As aforementioned, these parameters are not appropriate for handoff decision in NGWN. Moreover, MN must evaluate passively handoff conditions even when applications run just fine under current network. This introduces unnecessary power consumption and network resources usage.

To reduce energy consumption of MNs, without degrading throughput level, an approach named WISE (Wise Interface SElection) [37] for vertical handoff between WLAN and 3G networks has been proposed. WISE introduces a new entity called VDC (Virtual Domain Controller) in a 3G-core network, which acts as a central point for controlling 3G network and WLAN. With WISE, handoff decision is performed based on network load and energy consumption of radio interfaces. However, requirements such as security of services and applications are not considered. Also, VDC node may be a potential single point of failures. HOPOVER (HandOff Protocol for OVERlay networks) [38] is a mobile IP-based approach that handles vertical and horizontal handoffs. Although, HOPOVER enables low signaling overhead, it requires APs to maintain excessive information about MNs.

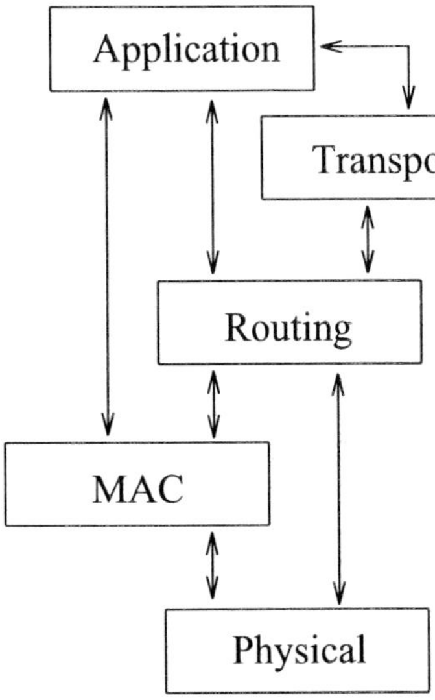

Figure 8. Cross-layer interactions.

Furthermore, to allow various entities to exchange signaling messages among themselves, handoff process using HOPOVER calls for complete standardization among different network service providers.

Policy-based architecture has been proposed by IETF in order to implement a set of rules to manage and control access to network resources and is particularly useful for QoS management [39]. Two main logical entities for policy control-based architecture are the Policy Decision Point (PDP) and the Policy Enforcement Point (PEP). To enable better choice for vertical handoff, several papers have proposed an utility or cost function as a measurement of network quality. A policy-enabled handoff decision algorithm is proposed in [40] based on cost function that considers several factors like bandwidth, power consumption and monetary cost. Cost function presented in [40] is very preliminary and it cannot handle more sophisticated scenarios. In order to maximize user QoS, handoff decision algorithms have been proposed in [14] for vertical handoff and metrics that characterize NGWN are identified. However, proposed cost function could lead to singularity problems if connections are free of charge. Furthermore, handoff instability problems and mobility management at IP layer were not considered.

The common drawbacks of available vertical handoff schemes in the literature are related to inefficiency of radio interface management. Air-interface always on or periodic activation are often used. Both approaches lead to high power consumption and resources usage. Hence, there is a need for more efficient radio interface management scheme which could be based on adaptive approach. In heterogeneous multi-access networks, like 4G/NGWN, MN must be able to detect easily and efficiently the presence and availability of new networks. The issue that arise is how and when to activate an idle interface to detect the coverage of other technologies. An important feature of heterogeneous network is that an MN with multiple interfaces can select the most appropriate radio access network (RAN) from a number of available RANs. More attention must be payed for the trade-off between power consumption efficiency and network discovery time.

Traditionally, received signal strength (RSS) and channel availability are used to assign MNs to specific attachment points and to manage handoffs. A direct comparison of RSS values or other related metrics (e.g., signal-to-noise ratio) may cause performance results to be misinterpreted. Hence, RSS comparison is not sufficient for handoff management in NGWN environments. Other factors such as monetary cost, mobile node and network conditions, data transmission rate, battery lifetime, service level, peer agreements between access and service providers, user's preferences and network availability must be considered. Design of handoff decision function which evaluates simultaneously these various metrics and parameters is crucial in NGWN and remains a research challenge. Support of seamless handoff is needed to ensure minimal disruption in data delivery to MNs. Such issues should be taken into account for NGWN by minimizing location update/registration and paging signaling and by handling fast and smooth handoffs. Also, an efficient way to reduce the costs related to both location update and paging remains an open issue. The location information storage area, be it a distributed or centralized architecture, is an issue which affects NGWN efficiency, robustness and scalability.

4. QoS in Wireless IP Networks

Host mobility is notorious for its significant impact on QoS parameters of real-time applications. QoS defines a system's nonfunctional characteristics which affect the perceived quality of results and could be measured with throughput, reliability, delay, jitter, packet loss, and so on. QoS goals at IP layer are basically focused on packet loss and packet delay. Reducing both parameters may be overambitious where multimedia applications are considered. Hence, there is a trade-off between packet loss and packet delivery delay. With the growing demand for multimedia and real-time applications, QoS support in heterogeneous networks is of primary importance and necessary to improve system performance and users' satisfaction. Internet QoS provisioning techniques like IntServ/RSVP [6] and DiffServ [8] have been developed in context of wired networks. In IntServ/RSVP model, network resources are explicitly identified and reserved. Network nodes classify incoming packets and use resource reservation to provide a high level QoS. IntServ model has three types of services: guaranteed service, controlled load service and best effort service. It works well for small scale networks. RSVP [7] is a resource reservation set up protocol designed for IntServ model by IETF. In fact, RSVP is not a routing protocol but a control or signaling protocol, which allows resource reservation for real-time applications before transmission of data.

In DiffServ model, resources are not explicitly reserved. Instead, traffic is differentiated into a set of classes and network nodes provide priority-based treatment according to these classes. DiffServ involves less processing and offers better scalability for large networks even with heavy traffic. DiffServ Code Point (DSCP) field of the IP packet header is used by DiffServ in order to determine the service types of the data traffic, by specifying a per-hop behavior (PHB) for that packet. PHBs are implemented by means of packet scheduling and buffer management mechanisms in the core nodes. Expedited forwarding (EF) and assured forwarding (AF) are considered as PHBs to allow delay and bandwidth differentiation. Bandwidth broker (BB) is also required for DiffServ model to allow monitoring and management of available bandwidth in Diffserv domain.

IntServ/RSVP and DiffServ are not optimized for mobile IP environments where they could be impractical. In fact, if an IP-based handoff protocol is used for mobility management, each time an MN crosses another subnet, it obtains a new CoA. Hence, it is more difficult to reserve network resources between CN and MN in case of high mobility rate. For example, if RSVP is used for resource reservation, new reservations over the entire data path must be set up whenever the CoA changes. That may lead to large delays, signaling load and service degradation. Need of efficient QoS mechanisms in mobile IP environments is greater due to the scarcity of resources, unpredictable available bandwidth and variable error rates. In mobile IP environments, IntServ model is best applied to access networks due to fine grained classification, whereas core networks can scale better when DiffServ model is applied. Coexistence of IntServ and DiffServ can offer end-to-end QoS. However, border nodes between DiffServ and IntServ domain must establish the correspondence of class of service in both models. 3GPP and 3GPP2 have adopted DiffServ as QoS model for their IP packet domains, although IntServ/RSVP could be supported optionally as well [41, 42].

4.1. IntServ-based Wireless QoS Models

Many mobile extensions of RSVP have been proposed and are surveyed in [43]. Basic idea of these extensions is to restrict resource reservation only to the modified signaling path. However, they have several shortcomings and are not able to provide seamless mobility with guaranteed QoS. In fact, RSVP PATH message must be reinitiated at least locally and these approaches incur latency in end-to-end resource reallocation and increasing signaling load over wireless interface. Mobile RSVP (MRSVP) protocol [44] has been proposed for provision of guaranteed QoS. MRSVP allows resource reservation from locations an MN may visit (called MSPEC or Mobile Specification set) in order to obtain a level of QoS which is not affected by mobility. It uses proxy agents to make passive reservations from unvisited access point (AP) on behalf of MN. However, MRSVP is hindered by several shortcomings. First, RSVP must be enhanced in order to support passive and active reservations and introduction of several proxy agents together with their communication protocol increases network complexity. Finally, a lot of information regarding active and passive reservations during handoff must be maintained at APs.

In order to improve performance upon excessive resource reservations of MRVSP, Hierarchical MRSVP (HMRSVP) [45] has been proposed. According to HMRSVP, resources are only reserved when an MN resides in the overlapping area of the boundary cells. Although this scheme outperforms MRSVP in terms of forced termination, session completion and reservation probabilities while achieving the same QoS, it does not cater to the other drawbacks introduced by MRSVP. These two protocols use the fact that coverage areas of different networks partially overlap. Hence, they cannot be easily extended to support QoS in NGWN based on a hierarchical heterogeneous environment, where coverage area of one network could be completely covered by another network. Moreover, the required number of proxy agents must match the number of cells or APs. This could result in excessive financial deployment, especially when the number of cells increase as it will be the case for NGWN.

Another approach to associate IPv6-based handoff protocols and RSVP is presented in

[46] and called HPMRSVP (Hierarchical Proxy MRSVP) to support QoS of real-time applications in hierarchical mobile environments. HPMRSVP uses fast handoff (FMIPv6) principles to make anticipated resource reservation. This protocol guarantees resource reservation in IPv6-based mobile environments during an ongoing session which is initiated with SIP [20] protocol. Resource reservation between MN and its CNs is limited within the access network rather than in end-to-end manner. In other words, the access network is responsible for upholding session, which optimizes the radio link usage. A performance comparison is made with MRSVP and shows that HPMRSVP outperforms it in terms of setup time for resource reservation paths, throughput and packet delay. Also, HPMRSVP yields better results than MRSVP in terms of reservation blocking, forced termination and session completion probabilities.

4.2. DiffServ-based Wireless QoS Models

Third generation wireless initiatives, 3GPP and 3GPP2, have mandated dynamic service negotiation capability [47, 48]. Although advantages of dynamic service negotiation have been well accepted, there is no universal standardized protocol for this purpose in an end-to-end fashion. For instance, 3GPP has proposed a protocol for negotiation of Radio Access Bearer (RAB) [49, 50]. However, this protocol is specific to radio link and is limited only to 3GPP networks, i.e., UMTS networks. Furthermore, Packet Data Protocol (PDP) Context Modification [51] can be used for QoS profile of an active session at IP layer in 3GPP networks. However, PDP Context is limited between MN and Gateway GPRS Support Node (GGSN) and cannot be used for end-to-end service negotiation in heterogeneous networks.

Several protocols have been proposed in the literature for dynamic service negotiation with their pros and cons. Comparison of some service negotiation protocols is provided in [52]. Among them, we can mention QoS-NSLP (QoS NSIS[1] Signaling Layer Protocol) [53] proposed by IETF NSIS working group. QoS-NSLP uses soft state and peer-to-peer refresh messages as a primary state management mechanism similarly to RSVP. Such periodic refresh messages consume battery power of mobile devices as well as wireless bandwidth, which are premium in wireless networks. Hence, QoS-NSLP may not be well suited for wireless environments. Dynamic Service Negotiation Protocol (DSNP) based on DiffServ is proposed in [54] for Service Level Specification and Agreements (SLS/SLA) at IP layer in a network with heterogeneous link layer. When using DSNP for service negotiation, mobile devices do not required to send periodic refresh messages nor to maintain TCP connection alive. DSNP's QoS architecture introduces two major components called QoS Global Server (QGS), which acts as a centralized node in each domain and QoS Local Node (QLN), which is an edge router of DiffServ domain.

Although DSNP is independent of link layer, interworking between DSNP and radio QoS protocols is not defined. In DSNP, an MN needs to communicate with QGS to obtain SLS/SLA. The drawbacks of DSNP is that QGS's burden is very heavy, because it must communicate with each MN and manages its QoS parameters. Moreover, QGS should know information of each QLN in order to allocate proper resources within QLN. Scalability problem occurs, because with the increasing number of MNs, all requests from MNs within a large-scale network will be sent to QGS. Hence, QGS will become a bottleneck of

[1]NSIS stands for Next Step In Signaling.

the system. Handoff delay and QGS response time increase significantly when the number of MNs increases. Another negotiation protocol for real-time applications over wireless networks is proposed in [55] and it follows almost similar ideas as in DSNP by using distributed QoS architecture rather than centralized one used in DSNP. Performance evaluation conducted shows that high network resources utilization (e.g., higher bandwidth efficiency and shorter delay) may be achieved comparatively to DSNP.

4.3. QoS Requirements in NGWN

In NGWN, available resources are scarce; therefore, efficient resource reservation mechanisms and radio resource management should be applied. Efficient resource reservation mechanisms should reserve resources only when it is certain that they will be used. In IPv6-based mobile environments, it is necessary to have a flexible mechanism to adjust dynamically resource reservation in various network entities. In general, old and new path have many common intermediate routers. Then, it is important that the new resource reservation path can not be fully re-established, but only the portion where path changes. This path portion resides usually in the access network. The support of easily manageable end-to-end QoS is a very challenging task.

In NGWN, entities and mechanisms must be defined for the scalable allocation and control of the resources in various access networks. They must be able to offer and guarantee end-to-end QoS, maintaining user connectivity and QoS level, while users move across different networks. There is a need for dynamic QoS negotiation method to dynamically adjust QoS requirements of MNs because contracted QoS level may not be honored. In fact, if originally requested QoS level cannot be authorized, MN could negotiate a lower QoS level rather than relinquish the session. Moreover, dynamic QoS negotiation mechanisms could allow networks to use their resources more efficiently. One of the promising solutions for QoS provision is a hybrid model based on combination of IntServ for the access network and DiffServ for the core network. However, at the transition nodes of networks, parameters of IntServ and those of DiffServ will need to map with each other properly.

It is essential to establish context transfer, if applicable, for MN to new AR (NAR), which is kept in previous AR (PAR) before handoff. Information carried may relate to AAA (authentication, authorization and accounting) process, header compression and QoS requirements. It seems an attractive procedure for NGWN when it maybe used between different administrative domains having service level agreements (SLAs). However, coordination with handoff control is required for its timely execution. Moreover, context transfer maybe not applicable if PAR and NAR belong to different administrative domains. In fact, the complete AAA procedure re-initiation may be required. New challenges relating to mobility management are introduced by QoS provision in NGWN. Significant discussion and ongoing standardization efforts within IETF on low-latency handoff, context transfer, QoS guarantee and IP paging will help bring it a step closer to the design and deployment of NGWN.

Despite attractive results of current IPv6-based mobility management protocols and other mobility management schemes presented in Section 3., seamless roaming and service continuity is not guaranteed. In other words, these protocols still have packet loss, signaling overhead, considerable handoff latency and jitter. Then, QoS guarantee still stays an open issue in wireless and mobile IP networks. As seen in above paragraphs, QoS is linked to treatment of data traffic. However, an emerging area of QoS is level of security available in the network. In this case, security includes authenticating and authorizing mobile users, securing communication channels and equipments. For instance, with the increasing distributed and heterogeneous nature of wireless networks defining NGWN, authentication problem becomes crucial and significant.

5. Performance Analysis

In this section, performance analysis of IPv6-based mobility management schemes proposed in the literature is done from several QoS metrics such as packet loss, handoff latency and signaling overhead cost.

5.1. Analytical Model

We define the following parameters to compute handoff latency and total packet loss: T_{L2} the L2 handoff latency or link switching delay, T_{RD} the round-trip time for router discovery process, T_{DAD} the time for DAD process execution, T_{RR} the delay taken for the MN to perform return routability procedure and $T_{X,Y}$ one-way transmission delay of a message of size s between nodes X and Y. If the processing delay at each node is ignored, $T_{X,Y}$, when one endpoint is an MN, is computed as follows:

$$T_{X,Y}(s) = \frac{1+q}{1-q}\left(\frac{s}{B_{wl}} + L_{wl}\right) + (d_{X,Y} - 1)\left(\frac{s}{B_w} + L_w + \varpi_Q\right) \tag{1}$$

where $d_{X,Y}$ is the average number of hops between node X and Y, B_{wl} (resp. B_w) the bandwidth of the wireless (resp. wired) link and L_{wl} (resp. L_w) the wireless (resp. wired) link delay, q the probability of wireless link failure and ϖ_Q the queueing delay at each router in the Internet [56]. The following assumptions are made for the performance analysis:

- router advertisement messages always include subnet prefix information and only one is enough to determine the movement of MN;
- router discovery and DAD are performed in sequence;
- just one neighbor solicitation message is enough to confirm an IP address of MN;
- the one-way delays between two nodes in both (downstream and upstream) direction are the same;
- for fast handoff schemes, the MN does not execute L2 handoff immediately after the transmission of FBU, i.e., when FBU is sent from the previous link, all downstream packets sent to MN from PAR prior to the reception of FBU are delivered to MN;
- processing delay and routing table lookup delay are negligible compared to access and transmission delay;
- transmission delay between MN and CN is shorter than the sum of delays between MN and HA, and between HA and CN.

Other parameters introduced for this analysis are: t_s inter-session arrival time random variable, t_c subnet (AR's coverage area) residence time random variable, t_d AN/MAP domain residence time random variable, N_c number of subnet crossing, N_d number of AN/MAP domain crossing, P_c subnet crossing probability and P_d AN/MAP domain crossing probability. Residence time random variables t_c and t_d are assumed i.i.d. (independently and identically distributed) with probability density function f_c and f_d, respectively.

5.1.1. Traffic and Mobility Models

User mobility and traffic models are crucial for efficient system design and performance evaluation. Usually MN mobility is modeled by the cell residence time and various types of random variables are used for this purpose [57]. Two user mobility models commonly used in wireless networks are: random-walk mobility model and fluid-flow mobility model. Random-walk model is more appropriate for pedestrian movements where mobility is confined to a limited geographical area such as business and residential buildings. On the other hand, fluid-flow model is more suitable for MNs with high mobility, infrequent speeds and direction changes [58]. We consider a traffic model having two levels, session and packets. Session duration follows distribution with inter-session rate λ_s while packet generation and arrival rate follow Poisson process. In NGWN, although the incoming calls or sessions follow Poisson process (i.e., inter-arrival time are exponentially distributed), the inter-session arrival times may not be exponentially distributed [57]. Other distribution models, like hyper-Erlang, Gamma and Pareto have been proposed to model various time variables in wireless networks. However, performance evaluation conducted in literature [57] shows that exponential model can be appropriate for cost analysis. In fact, exponential model provides an acceptable tradeoff between complexity and accuracy. However, Gamma distribution is very realistic for mobility model by considering changes in the speed and direction of the MN.

Let μ_c and μ_d be the border crossing rate of an MN out of a subnet (AR) and out of an AN/MAP domain, respectively. When an MN crosses an AN/MAP domain border, it also crosses an AR border. If we assume that all subnets are of circular shape and form together a contiguous area and each AN/MAP domain is composed of M equally subnets, we have $\mu_d = \frac{\mu_c}{\sqrt{M}}$. Note that the roaming probability depends on an MN's movement pattern in its original network but not in its destination network. We assume that subnet residence time follows the Gamma distribution with probability density function f_c given by:

$$f_c(t) = \frac{\rho^{\omega} t^{\omega-1}}{\Gamma(\omega)} e^{-\rho t} \tag{2}$$

where ω is the shape parameter, ρ is the scale parameter with $\rho = \omega\mu_c$ and $\Gamma(\omega)$ is the Gamma function, which is defined as $\Gamma(\omega) = \int_0^{\infty} t^{\omega-1} e^{-t} dt$. The mean and variance of subnet residence time are $1/\mu_c$ and $\frac{1}{\omega\mu_c^2}$, respectively. Similarly, AN/MAP domain residence time follows Gamma distribution and the probability density function f_d is obtained by replacing μ_c with μ_d.

Modeling the probability distribution of the number of boundary crossing during a call plays a huge role in cost analysis for cellular networks. This will be the case again for IP-based wireless networks. Figures 9 and 10 show timing diagram for typical mobile user crossing subnet and access network boundary, respectively. For Figure 9, suppose that the MN is in coverage area of access router i (AR_i), when the previous call arrives and is accepted by the MN. It then moves to AR_j during inter-session/service time. Interval time between the time instant the new call is initiated at the MN and the instant the MN moves out of the subnet (AR_i) if the new call is not completed is called a residual subnet residence time (t_{rs}). In case of inter-AN/MAP movement illustrated in Figure 10, the residual access network residence time is denoted t_{ra}.

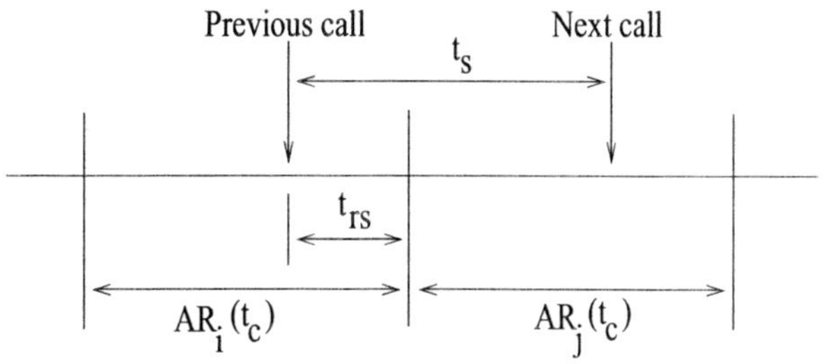

Figure 9. Timing diagram for subnet boundary crossing.

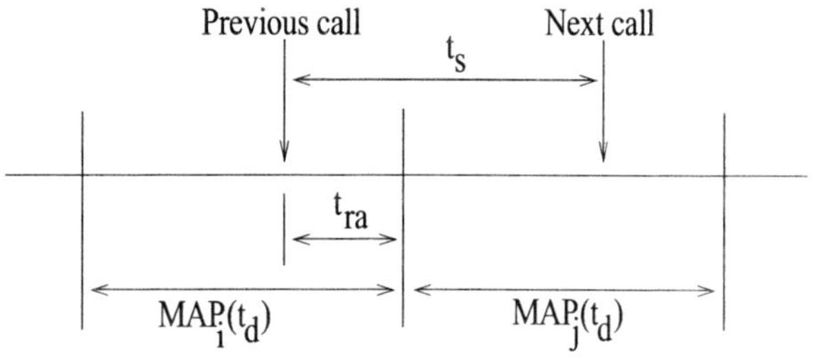

Figure 10. Timing diagram for access network boundary crossing.

For simplicity and easy derivation of analytical expressions, we will use exponential distribution. Note that exponential distribution is a particular case of Gamma distribution. Hence, under exponential distribution assumption, the probability (crossing probability) that there is at least one local (respectively global) binding update between two consecutive sessions to the MN, P_c (respectively P_d) is given by:

$$\begin{aligned} P_c = Pr(t_s > t_c) &= \int_0^{\infty} Pr(t_s > t_c) f_c(u) du = \frac{\mu_c}{\mu_c + \lambda_s} \\ P_d = Pr(t_s > t_d) &= \int_0^{\infty} Pr(t_s > t_d) f_d(u) du = \frac{\mu_d}{\mu_d + \lambda_s}. \end{aligned} \quad (3)$$

Probabilities that an MN experiences k subnets boundary crossing and m access network boundary crossing during its session lifetime correspond to probabilities mass function (PMF) of random variables N_c and N_d, respectively and are expressed by [59]:

$$\begin{aligned} Pr(N_c = k) &= P_c^k(1 - P_c) \\ Pr(N_d = m) &= P_d^m(1 - P_d). \end{aligned} \tag{4}$$

Then, the average number of location binding updates during an inter-session time interval under subnet crossings ($E(N_c)$) and AN/MAP domain crossings ($E(N_d)$) are given by:

$$\begin{aligned} E(N_c) &= \sum_{k=0}^{\infty} kPr(N_c = k) = \sum_{k=0}^{\infty} kP_c^k(1 - P_c) = \frac{\mu_c}{\lambda_s} \\ E(N_d) &= \sum_{m=0}^{\infty} mPr(N_d = m) = \sum_{m=0}^{\infty} mP_d^m(1 - P_d) = \frac{\mu_d}{\lambda_s}. \end{aligned} \tag{5}$$

5.1.2. Signaling Overhead Cost

With MIPv6, MN performs binding update to HA/CNs each time it moves to a new point of attachment in another subnet; then local binding update signaling cost is equal to global binding update signaling cost. However, fast handoff schemes are based on anticipation using L2 trigger, then signaling cost depends on the probability that L3 handoff does or does not occur after L2 trigger. Location binding update signaling cost depends heavily on the average number of location updates during inter-session time. Depending on movement type and mobility management protocol, two kinds of location binding update could be performed: local and global binding update. Global binding update procedure refers to registration of CoA to HA/CNs for MIPv6 and FMIPv6 or RCoA to HA/CNs for HMIPv6, while local binding update refers to registration of LCoA to MAP for HMIPv6. For MIPv6 and FMIPv6, local binding update signaling cost is equal to global binding update signaling cost.

If an MN has ongoing session when it roams between subnets, it is more critical to reduce the signaling cost after entering the new subnet, for the QoS of ongoing session. In contrast, if the roaming MN does not have an active session when it crosses subnet boundary, additional registration in the new subnet will result in unnecessary signaling overhead. Since IPv6-based mobility management protocols proposed by IETF do not support paging functionalities, total signaling cost, C_T, may be divided into location binding update signaling cost (C_{BU}) and packet delivery cost (C_{PD}). Then, total signaling cost (control and data packets traffic) is expressed as follows:

$$C_T = C_{BU} + C_{PD}. \tag{6}$$

Note that signaling cost required for authentication and for L2 handoff are the same for all protocols; then, they are omitted.

To perform signaling overhead analysis, a performance factor called session-to-mobility ratio (SMR) is introduced and it is similar to call-to-mobility ratio (CMR) defined in cellular networks [60]. SMR represents the relative ratio of session (or packet) arrival rate to the user mobility rate: $SMR = \lambda_s/\mu_c$. Let C^g the global binding update cost to HA/CNs and C^l the local binding update cost to MAP, respectively. Average signaling cost (C_{BU}) for location binding update with HMIPv6 is given by:

$$C_{BU} = E(N_c)C^l + E(N_d)C^g = \frac{1}{\lambda_s}\left(\mu_c C^l + \mu_d C^g\right) = \frac{1}{SMR}\left(C^l + \frac{1}{\sqrt{M}}C^g\right). \quad (7)$$

Let $d_{X,Y}$ be the average number of hops between nodes X and Y, $C_{X,Y}$ be the transmission cost of control packets between nodes X and Y and PC_X be the processing cost of control packet at node X. Packet transmission cost in IP networks is proportional to the distance in hops between source and destination nodes. Furthermore, the transmission cost in a wireless link is generally larger than the transmission cost in a wired link [60]. Thus, the transmission cost of a control packet between nodes X and Y belonging to the wired part of a network can be expressed as $C_{X,Y} = \tau d_{X,Y}$ while $C_{MN,AR} = \tau\varphi$, where τ is the unit transmission cost over wired link and φ the weighting factor for the wireless link. Global and local binding update signaling costs for MIPv6 and its extension are given in Table 1 and Table 2. P_s is the probability of anticipated handoff signaling success for fast handoff protocols.

Table 1. Expression of signaling costs.

$C^g_{MIPv6} = C^l_{MIPv6}$	$= 4C_{MN,AR} + 2PC_{AR} + C_{hc}$
C^l_{HMIPv6}	$= 2(2C_{MN,AR} + PC_{AR} + C_{MN,MAP}) + PC_{MAP}$
C^l_{FMIPv6}	$= P_s S_s + (1 - P_s)(S_f + S_r) + C_{hc}$
$C^l_{FHMIPv6}$	$= P_s S^l_s + (1 - P_s)S^l_f + S^l_h$

Table 2. Expression of partial signaling costs.

C_{hc}	$= 2(C_{MN,HA} + N_{CN}C_{MN,CN}) + PC_{HA} + N_{CN}PC_{CN} + C_{rr}$
C_{rr}	$= 2(C_{MN,HA} + N_{CN}C_{HA,CN} + N_{CN}C_{MN,CN}) + PC_{HA} + N_{CN}PC_{CN}$
S_f	$= 3C_{MN,PAR} + 2C_{PAR,NAR} + 3PC_{AR}$
S_s	$= 4C_{MN,PAR} + 3C_{PAR,NAR} + 2C_{MN,NAR} + 5PC_{AR}$
S_r	$= 2C_{MN,PAR} + 2C_{PAR,NAR} + 2C_{MN,NAR} + 3PC_{AR}$
S^l_f	$= 3C_{MN,MAP} + 2(C_{MAP,NAR} + PC_{MAP}) + PC_{AR}$
S^l_s	$= 4C_{MN,MAP} + 3C_{MAP,NAR} + 2C_{MN,NAR} + 3PC_{MAP} + 2PC_{AR}$
S^l_h	$= P_s[2(C_{MN,NAR} + C_{NAR,MAP}) + PC_{NAR} + PC_{MAP}] + (1 - P_s)C^l_{HMIPv6}$

Packet delivery cost is composed of packet transmission cost and packet processing cost. In IP-based networks, the transmission cost between two entities is proportional to the number of hops between these entities while processing cost is associated to packet processing such as mapping table lookup and routing table lookup. Furthermore, the transmission cost in a wireless link is generally larger than the transmission cost in a wired link. Let s_c and s_d the average size of control packets and data packets, respectively and $\kappa = s_d/s_c$. The cost of transferring data packets is κ greater than the cost of transferring control packets. IP routing table lookup is based on the longest prefix matching. Hence, complexity of IP address lookup is proportional to the logarithm of the length of routing table, while the size of mapping table is proportional to the number of MNs located in the coverage area of MAP domain, in case of HMIPv6.

In IPv6-based handoff protocol, two modes of routing are available: triangular routing and optimized routing. Let D_o and D_t denote packet delivery cost for optimized routing and triangular routing, respectively; $E(S)$ average session size in number of packets and ϕ ratio of packets transiting to HA before the completion of binding update process. Then, packet delivery cost per session, C_{PD}, is given as follows:

$$C_{PD} = \phi E(S) D_t + (1 - \phi) E(S) D_o. \tag{8}$$

Packet delivery costs associated to both routing modes for HMIPv6 protocol are given by:

$$\begin{aligned} D_o &= \eta U_o + P_{MAP} \\ D_t &= \eta U_t + P_{MAP} + P_{HA} \end{aligned} \tag{9}$$

where P_{MAP} and P_{HA} denote processing cost at MAP and HA for packet delivery, respectively. Moreover, η, U_o and U_t are unit transmission cost, the hop distance of optimized path and the hop distance of non-optimized (triangular) path, respectively. Processing costs at HA and MAP are expressed as follows:

$$\begin{aligned} P_{HA} &= \phi \Psi_{HA} \\ P_{MAP} &= \lambda_s E(S)[\theta N_{MN} + \delta \log(M)] \end{aligned} \tag{10}$$

where θ and δ are weighting factors, Ψ_{HA} the unit packet processing cost at HA and N_{MN} the total number of users located in AN/MAP domain. For MIPv6 and FMIPv6 schemes, processing cost at MAP, P_{MAP}, is not considered. Moreover, for fast handoff schemes, due to bi-directional tunnel between PAR and NAR for FMIPv6 or between MAP and NAR for F-HMIPv6, packet forwarding cost should be taken into account.

5.1.3. Handoff Latency and Packet Loss

Table 3 shows comparison of IPv6-based mobility management protocols with respect to handoff latency during message update[2]. For fast handoff schemes, handoff latency depends on information availability, and on which link fast handoff messages are exchanged. P_s is the probability that the anticipated fast handoff signaling succeeds. D^p_{FMIPv6} and D^r_{FMIPv6} means respectively handoff latency of predictive mode and reactive mode of FMIPv6, similarly for $D^p_{FHMIPv6}$ and $D^r_{FHMIPv6}$ for F-HMIPv6. Since average delay needed for an MN authentication is the same for all protocols; then, it is omitted.

Predictive mode of FMIPv6 cannot perform anticipated IP-handoff for inter-AN [29]; then handoff latency of FMIPv6 becomes same as for MIPv6. Similar remarks apply to HMIPv6 and F-HMIPv6 for inter-AN/MAP movements. With MIPv6 and HMIPv6, packet

[2] pBR and nBR mean previous and new border router (BR), respectively.

loss occurs during handoff latency or service disruption latency. In fact, the number of packet losses is proportional to handoff latency. This is also the case for FMIPv6 and F-HMIPv6 if there is no efficient buffer management. In fact, for fast handoff schemes, there is no packet loss in theory, unless buffer overflows happen.

Table 3. Handoff latency for intra-AN/MAP movement.

Protocol	Handoff latency
D_{MIPv6}	$T_{L2} + T_{RD} + T_{DAD} + 2T_{MN,HA} + T_{RR} + 2T_{MN,CN}$
D_{HMIPv6}	$T_{L2} + T_{RD} + T_{DAD} + 2T_{MN,MAP}$
D^{p}_{FMIPv6}	$T_{L2} + 2T_{MN,NAR}$
D^{r}_{FMIPv6}	$T_{L2} + 2T_{MN,NAR} + 3T_{NAR,PAR}$
D_{FMIPv6}	$P_s D^{p}_{FMIPv6} + (1 - P_s) D^{r}_{FMIPv6}$
$D^{p}_{FHMIPv6}$	$T_{L2} + 2T_{MN,NAR}$
$D^{r}_{FHMIPv6}$	$T_{L2} + T_{DAD} + 2T_{MN,MAP}$
$D_{FHMIPv6}$	$P_s D^{p}_{FHMIPv6} + (1 - P_s) D^{r}_{FHMIPv6}$

5.2. Numerical Results

The network topology considered for performance analysis is illustrated in Figure 11. For protocols which dot not involve hierarchical mobility management the MAPs act as a normal intermediate (border) router. All links are supposed to be full-duplex in terms of capacity and delay.

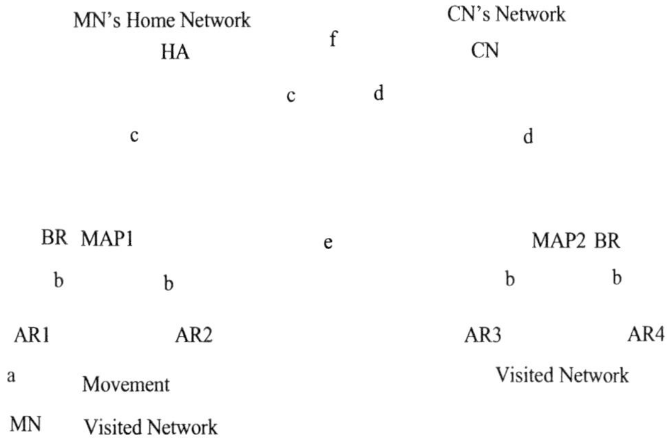

Figure 11. Network topology used for analysis.

Parameters and values used in performance evaluation are listed in Table 4. Except when DAD time, wireless link delay and packet service rate are considered as variable pa-

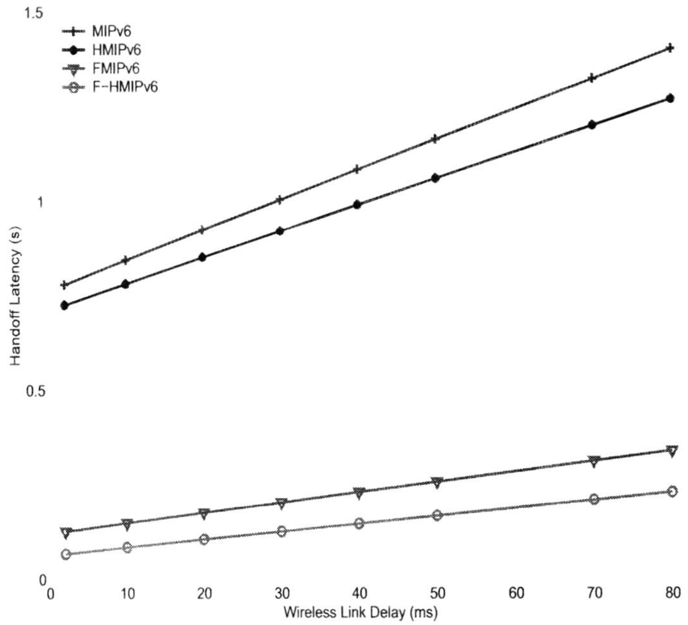

Figure 12. Impact of wireless link delay on handoff latency.

rameters, their values are fixed and given in Table 4. Other parameters for cost computation are: $\tau = 1$, $\varphi = 10$, $\theta = 0.1$, $\delta = 0.2$, $PC_{AR} = 8$, $PC_{HA} = 24$, $PC_{CN} = 4$ and $PC_{MAP} = 12$ and $\Psi_{HA} = 20$. We assume that distance between different domains are equal, i.e., $c = d = e = f = 10$ and we set $a = 1, b = 2$. Without lost of generality, CN is set up as a traffic source and the MN acts as a sink receiving packets from CN.

In Figure 12, we see that, handoff latency increases linearly with the wireless link delay. Wireless link delay between MN and AR depends on their distance and the channel condition. MIPv6 has the highest handoff latency, while F-HMIPv6 performs better than all other schemes. With FMIPv6, packet forwarding between PAR and NAR may be non-optimal, then lead to additional handoff delays. It is well known that the maximum tolerable delay for interactive conversation is approximately 200 milliseconds. Hence, F-HMIPv6 and FMIPv6 can meet this requirement when the wireless link delay is set up below to 50 milliseconds.

Figure 13 shows handoff latency for various DAD delay. If we compare each point of handoff latency with DAD delay, we observe that DAD process counts for a large portion of handoff delay. Then, it is important to decrease DAD delay in order to decrease handoff latency. With FMIPv6 and F-HMIPv6, all time consuming procedure like DAD and addresses configuration are performed by anticipation. Then, handoff latency does not depend on DAD delay as we can observe in the figure showing handoff latency versus DAD delay.

Figure 14 shows total packet loss versus packet arrival rate to MN and we observe that packet losses are far less for FMIPv6 and F-HMIPv6 than for MIPv6 and HMIPv6. The effect of handoff in IPv6-based mobile environments is dominated by packet loss, which is due to L2 handoff and MIPv6 operations. In fact, due to the lack of any buffering mecha-

Table 4. Simulation parameters set up.

Parameters	Symbols	Values
DAD time	T_{DAD}	500 ms
Router discovery delay	T_{RD}	100 ms
L2 handoff latency	T_{L2}	50 ms
Wired link bandwidth	B_w	100 Mbps
Wireless link bandwidth	B_{wl}	11 Mbps
Wired link delay	L_w	2 ms
Wireless link delay	L_{wl}	10 ms
Wireless link failure probability	q	0.5
Number of ARs per AN	M	2
Control packet size	s_c	96 bytes
Data packet size	s_d	200 bytes
Session arrival rate	λ_s	0.1
Average session size	$E(S)$	100

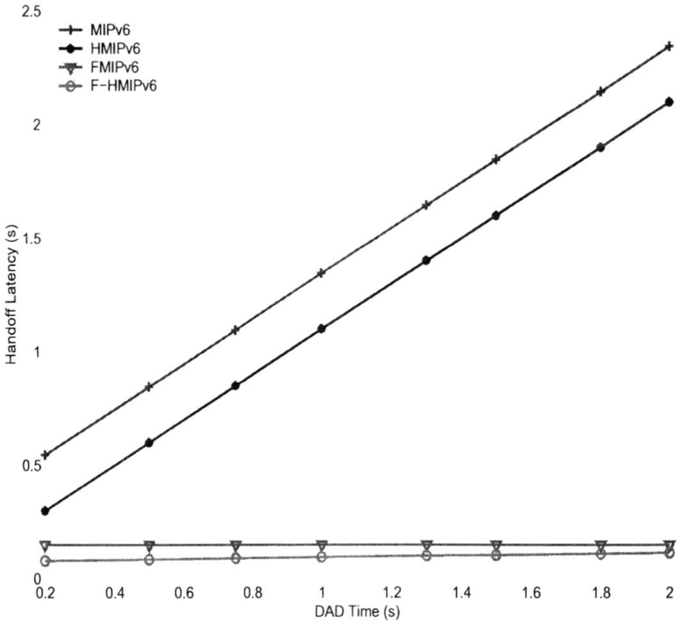

Figure 13. Impact of DAD delay on handoff latency.

nism, all in-flight packets will be lost during the handoff in MIPv6 and HMIPv6. However, in FMIPv6 and F-HMIPv6, packet loss begins when the L2 handoff is detected till the buffering mechanism is initiated. Similar results are shown in Figure 15 for packet loss versus the wireless link delays.

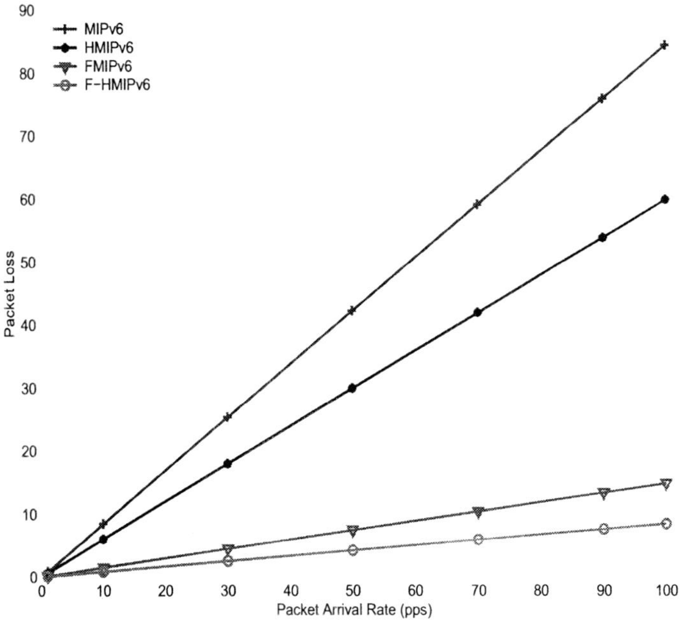

Figure 14. Impact of packet arrival rate on packet loss.

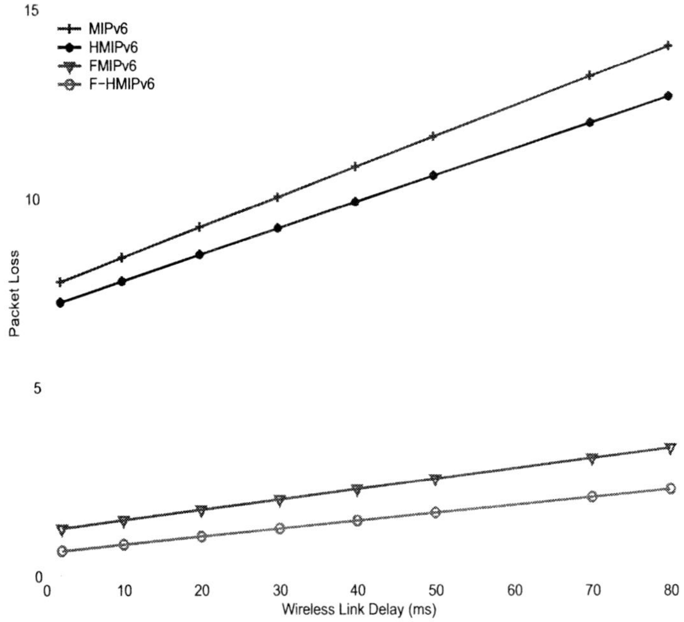

Figure 15. Impact of wireless link delay on packet loss.

Figure 16 illustrates signaling cost during handoff as a function of SMR. When SMR is small, the mobility rate is larger than the session arrival rate; then, the MN changes its point of attachment frequently due to its mobility, there is several handoffs. These handoffs will cause exchange of several messages among different network entities and MN and will increase signaling overhead. However, when the session arrival rate is larger than mobility rate (i.e., SMR is larger than 1), the binding update is less often performed. In this case, packet delivery cost is more dominant factor. In other words, signaling overhead decreases as the frequency of subnet change decreases and for high mobility (i.e., low subnet residence time), the signaling overhead is considerably reduced for HMIPv6. Additional messages introduced in fast handoff schemes lead to signaling overheads. As shown in Figure 16, signaling cost with FMIPv6 and F-HMIPv6 are greater than MIPv6 and HMIPv6, respectively due to these additional messages. However, these signaling overheads are compensated by better handoff latencies and packet loss as we saw in results reported above. Local mobility management under HMIPv6 enables better signaling cost performance than other all protocols. In fact, when SMR is low, a significant cost saving is obtained with the binding update in HMIPv6.

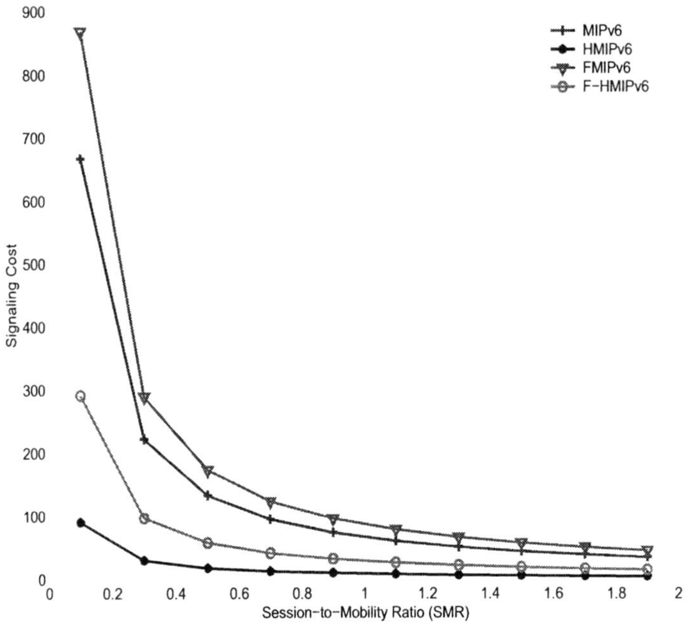

Figure 16. Impact of session-to-mobility ratio on signaling overhead cost.

Due to additional packet processing at MAP for HMIPv6, there is an extra cost for packet delivery. Packet delivery cost is depicted in Figure 17 for all of the concerned protocols as function of packet arrival rate (λ_p). Fast handoff based schemes (FMIPv6 and F-HMIPv6) outperform MIPv6 and HMIPv6, and they are more efficient when λ_p increases. This means that FMIPv6 and F-HMIPv6 are more adequate for real-time applications where periodic packets are sent at high rates. This suggests that it is important to reduce packet delivery cost for scalable services. In order to reduce packet delivery cost, it is possible to minimize the lookup latency in a mapping table by using efficient search algorithm.

For varying prediction probability, P_s, Fig. 18 shows the packet delivery cost. Packet delivery cost decreases when the accuracy of P_s increases for fast handoff schemes. Due to

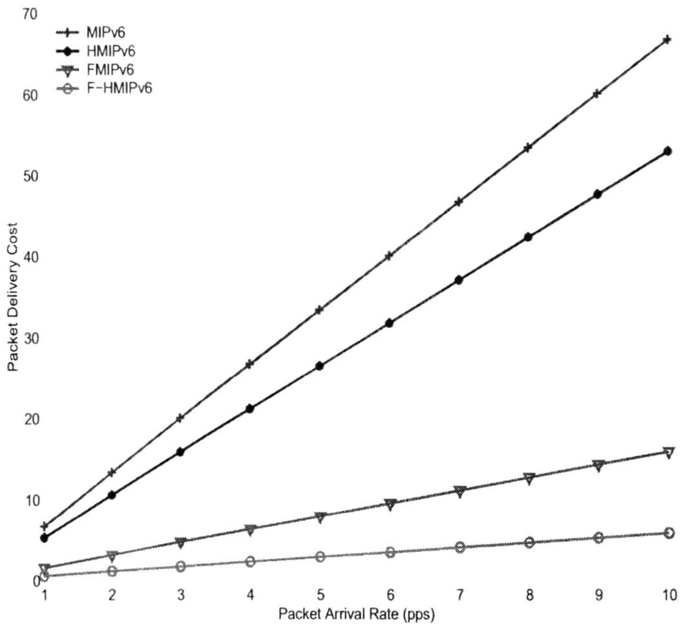

Figure 17. Impact of packet arrival rate on packet delivery cost.

additional packet processing at MAP for F-HMIPv6, there is an extra cost for packet delivery with inaccuracy prediction. In fact, in this case, F-HMIPv6 turns to HMIPv6, as we can see when $P_s = 0$. HMIPv6 and MIPv6 are not affected by prediction probability. For high values of P_s, F-HMIPv6 performs better than FMIPv6. Since there is a relation between handoff latency and packet delivery cost, similar behavior will be observed when comparing handoff latency with prediction probability. Hence, it is necessary to have good prediction mechanisms to allow better performance for F-HMIPv6. Since IPv6-based handoff protocols support route optimization, CNs must maintain binding cache of its communicating MNs. Hence, when MN performs BU process with entities located outside its access network, it executes BU to all of its communicating CNs. Then, the BU cost increases linearly with the number of CNs. For $SMR = 0.3$, Figure 19 shows the impact of the number of CNs on the signaling cost.

6. Conclusion

This chapter provided a survey of some mobility management protocols and QoS guarantee approaches available in the literature and addressed their limitations. We observed that the available mobility management schemes are not appropriate to support seamless roaming and services continuity in future heterogeneous IP wireless networks. NGWN's challenges were also presented as being a guideline that must be taken into consideration for further research activities. After analyzing results of performance evaluation we can state that F-HMIPv6 enables improvement in terms of handoff latency and packet loss rather than all other IPv6-based handoff protocols (i.e., MIPv6, HMIPv6 and FMIPv6). But, this performance is traded off by its signaling traffic overhead an required buffer space when compared

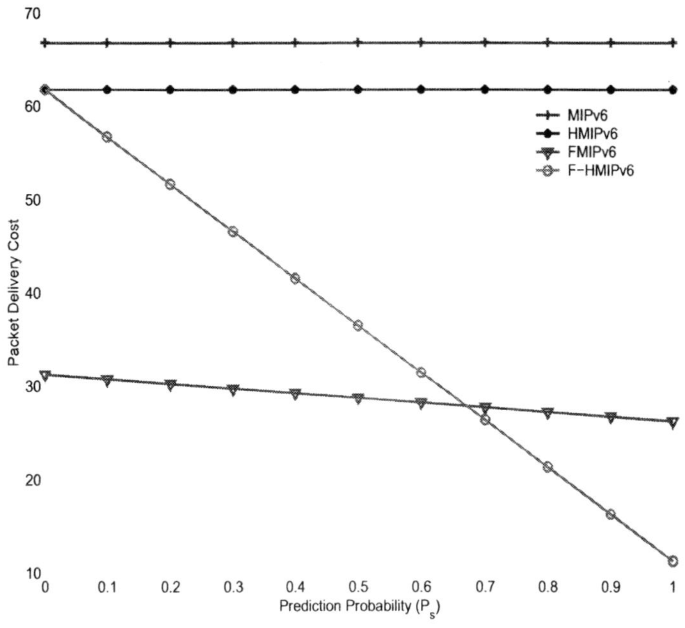

Figure 18. Impact of prediction probability on packet delivery cost.

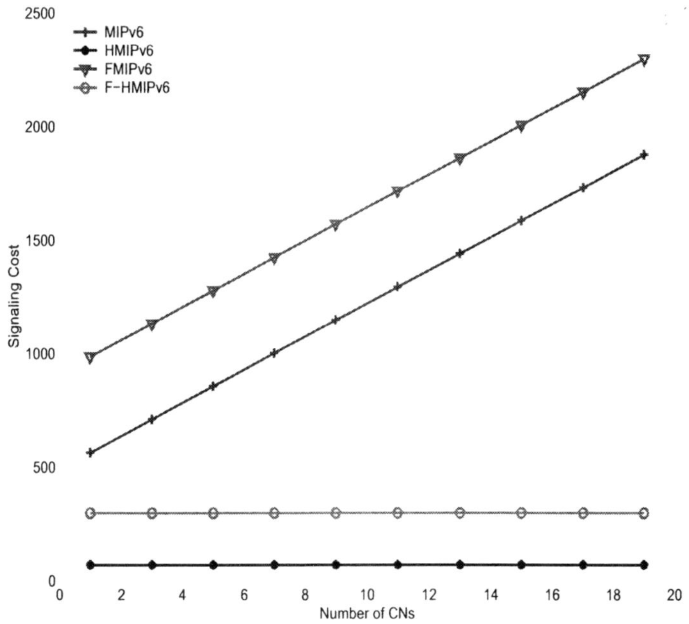

Figure 19. Impact of number of active CNs on signaling overhead cost.

to HMIPv6.

However, it is necessary to have good handoff prediction mechanisms to allow better performance for F-HMIPv6 compared to FMIPv6. The important features that put fast handoff schemes(i.e., FMIPv6 and F-HMIPv6) on top of the list are the fact that the time consuming operations are performed before link layer handoff and the tunneling functionality between previous and new subnet. In spite of several proposals for L2 and L3 handoff latency reduction, none of them can provide optimal mobility management and lot of them must be further improved.

With coexistence of a various wireless technologies in NGWN, no single specific mobility management protocol is expected to work in all situations. Hence, the design of a unified mobility management protocol that will suit all of different networks requires tremendous efforts. Furthermore, QoS provision remains an open issue in mobile environments. One of the promising solutions is a hybrid model based on IntServ for the access network and DiffServ for the core network. Open research issues in NGWN are also discussed in this chapter, to motivate research for efficient protocols and architecture design. These issues should be solved to enable deployment of IPv6-based next generation wireless networks and to achieve convergence of wired and wireless networks.

References

[1] Hui, SY; Yeung, KH. Challenges in the migration to 4G mobile systems. *IEEE Communications Magazine*, Dec. 2003, vol. 41, no. 12, 54-59.

[2] Stemm, M; Katz, RH. Vertical handoffs in wireless overlay networks. *ACM Mobile Networking and Applications* (MONET), Dec. 1998, vol. 3, no. 4, pp. 335-350.

[3] 3GPP TS. 3GPP System to WLAN Interworking; System Description (Release 6). 3GPP TS 23.234 v6.3.0, Mar. 2004.

[4] 3GPP2 TS. 3GPP2-WLAN Interworking; Stage 1 Requirements. 3GPP2 S.R0087-0 v1.0, July 2004.

[5] Akyildiz, IF; McNair, J; Ho, JSM; Uzunalioglu, H; Wang, W. Mobility management in next generation wireless systems. *Proceedings of the IEEE*, Aug. 1999, vol. 87, no. 8, pp. 1347-1384.

[6] Braden, R; Clark, D; Shenker, S. Integrated services in the Internet architecture: an overview. IETF RFC 1633, June 1994.

[7] Braden, R; Zhang, L; Berson, S; Herzog, S; Jamin, S. Resource ReSerVation Protocol (RSVP), version 1 Functional Specification. IETF RFC 2205, Sept. 1997.

[8] Blake, S; Black, D; Carlson, M; Davies, E; Wang, Z; Weiss, W. An architecture for differentiated services. IETF RFC 2475, Dec. 1998.

[9] Wong, VWS; Leung, VCM. Location management for next generation personal communication networks. *IEEE Network*, Sept./Oct. 2000, vol. 14, no. 5, pp. 18-24.

[10] Akyildiz, IF; Xie, J; Mohanty, S. A survey of mobility management in next-generation all-IP-based wireless systems. *IEEE Wireless Communications*, Aug. 2004, vol. 11, no. 4, pp. 16-28.

[11] Zhu, F; McNair, J. Multiservice vertical handoff decision algorithmss. *EURASIP Journal on Wireless Communications and Networking*, 2006, vol. 2006, 13 pages.

[12] Gustafsson, E; Jonsson, A. Always best connected. *IEEE Wireless Communications*, Feb. 2003, vol. 10, no. 1, pp. 49-55.

[13] Chen, WT; Liu, JC; Huang, HK. An adpative scheme for vertical handoff in wireless overlay networks. In *Proceedings of the* 10th *International Conference on Parallel and Distributed Systems* (ICPADS), July 2004, vol. 10, pp. 541-548.

[14] McNair, J; Zhu, F. Vertical handoffs in fourth-generation multinetworks environments. *IEEE Wireless Communications*, June 2004, vol. 11, no. 3, pp. 8-15.

[15] Johnson, DB; Perkins, CE; Arkko, J. Mobility support in IPv6. IEFT RFC 3775, June 2004.

[16] Thomson, S; Narten, T. IPv6 stateless address autoconfiguration. IETF RFC 2462, Dec. 1998.

[17] Moore, N. Optimistic duplicate address detection. IETF RFC 4429, April 2006.

[18] Pérez-Costa, X; Torrent-Moreno, M; Hartenstein, H. A performance comparison of mobile IPv6, hierarchical mobile IPv6, fast handovers for mobile IPv6 and their combination. *ACM Mobile Computing and Communications Review*, Oct. 2003, vol. 7, no. 4, pp. 5-19.

[19] Wong, KD; Dutta, A; Schulzrinne, H; Young, K. "Simultaneous mobility: analytical framework, theorems and solutions," *Wireless Commun. and Mobile Computing*, accepted for publication, Sept. 2006.

[20] Rosenberg, J; Schulzrinne, H; Camarillo, G; Johnston, A; Peterson, J; Sparks, R; Handley, M; Schooler, E. SIP: session initiation protocol. IETF RFC 3261, June 2002.

[21] Peterson, J; Jennings, C. Enhancements for authenticated identity management in the session initiation protocol (SIP). IEFT RFC 4474, August 2006.

[22] Eddy, W. At what layer does mobility belong? *IEEE Communications Magazine*, Oct. 2004, vol. 42, no. 10, pp. 155-159.

[23] Ong, L; Yoakum, J. An introduction to the stream control transmission protocol (SCTP). IETF RFC 3286, May 2002.

[24] Koh, SJ; Chang, MJ; Lee, M. mSCTP for soft handover in transport layer. *IEEE Communications Letters*, March 2004, vol. 8, no. 3, pp. 189-191.

[25] Koh, SJ, Xie, Q, Park, SD. Mobile SCTP (mSCTP) for IP handoff support. IETF draft, draft-sjkoh-msctp-01.txt, Oct. 2005, work in progress.

[26] Soliman, H; Castelluccia, C; Malki, KE; Bellier, L. Hierarchical mobile IPv6 mobility management (HMIPv6). IETF RFC 4140, Aug. 2005.

[27] Koodli, G. Fast handovers for mobile IPv6. IETF RFC 4068, July 2005.

[28] Gupta, VG; Johnston, D. A generalized model for link layer triggers. *IEEE 802.21 Media Independent Handoff Working Group*, http://www.ieee802.org/handoff/march04_meeting_docs/Generalized_triggers-02.pdf, March 2004.

[29] Gwon, Y; Kempf, J; Yegin, A. Scalability and robustness analysis of mobile IPv6, fast mobile IPv6, hierarchical mobile IPv6, and hybrid IPv6 mobility protocols using a large-scale simulation. In *Proceedings of IEEE International Conference on Communications* (ICC), June 2004, vol. 7, pp. 4087-4091.

[30] Jung, HY; Soliman, H; Koh, S. Fast handover for hierarchical MIPv6 (F-HMIPv6). IETF draft, draft-jung-mipshop-fhmipv6-00.txt, Oct. 2005, work in progress.

[31] Jung, HY; Kim, EA; Yi, JW; Lee, HH. A scheme for supporting fast handover in hierarchical mobile IPv6 networks. ETRI Journal, Dec. 2005, vol. 27, no. 6, pp. 798-801.

[32] Leibsch, M; Singh, A; Chaskar, H; Funato, D; Shim, E. Candidate access router discovery (CARD). IETF RFC 4066, July 2005.

[33] Loughney, J; Nakhjiri, M; Perkins, C; Koodli, R. Context transfer protocol (CXTP). IETF RFC 4067, July 2005.

[34] Kota, SL; Hossain, E; Fantacci, R; Karmouch, A. Cross-layer protocol engineering for wireless mobile networks: Part 1. *IEEE Communications Magazine*, Dec. 2005, vol. 43, no. 12, pp. 110-111.

[35] Kawadia, V; Kumar, PR. A cautionary perspective on cross layer design. *IEEE Wireless Communications*, Feb. 2005, vol. 12, no. 1, pp. 3-11.

[36] Buddhikot, M; Chandranmenon, G; Han, S; Lee, YW; Miller, S; Salgarelli, L. Integration of 802.11 and third-generation wireless data networks. In *Proceedings of IEEE International Conference on Computer Communications* (INFOCOM), April 2003, vol. 1, pp. 503-512.

[37] Minji, N; Nakjung, C; Yongho, S; Yanghee, C. WISE: Energy-efficient interface selection on vertical handoff between 3G networks and WLANs. In *Proceedings of IEEE International Symposium on Personal, Indoor, and Mobile Radio Communications* (PIMRC), Sept. 2004; pp. 692-698.

[38] Du, F; Ni, LM; Esfahanian, AH. HOPOVER: A new handoff protocol for overlay networks. In *Proceedings of IEEE International Conference on Communications* (ICC), May 2002, vol. 5, pp. 3234-3239.

[39] Yavatkar, R; Pendarakis, D; Guerin, R. A framework for policy-Bbased admission control. IETF RFC 2753, Jan. 2000.

[40] Wang, HJ; Katz, RH; Giese, J. Policy-enabled handoffs across heterogeneous wireless networks. In *Proceedings of* 2nd *IEEE Workshop on Mobile Computing Systems and Applications* (WMCSA), Feb. 1999, pp. 51-60.

[41] 3GPP TS. End-to-end quality of service (QoS) concept and architecture (Release 6). 3GPP TS 23.207 v6.3.0, June 2004.

[42] 3GPP2 TS. Support for end-to-end QoS - Stage 1 requirements. 3GPP2 S.R0079 v1.0, May 2004.

[43] Moon, B; Aghvami, H. RSVP extensions for real-time services in wireless mobile networks. *IEEE Communications Magazine*, Dec. 2001, vol. 39, no. 12, pp. 52-59.

[44] Talukdar, AK; Badrinath, BR; Acharya, A. MRSVP: A resource reservation protocol for an integrated services network with mobile hosts. *Wireless Networks*, Jan. 2001, vol. 7, no. **1**, pp. 5-19.

[45] Tseng, C; Lee, G; Liu, R. HMRSVP: A hierarchical mobile RSVP protocol. *Wireless Networks*, Mar. 2003, vol. 9, no. **2**, pp. 95-102.

[46] Abondo, C; Pierre, S. Hierarchical proxy mobile resource reservation protocol for mobile IP networks. In *Proceedings of IEEE International Conference on Wireless and Mobile Computing, Networking and Communications* (WiMob), Aug. 2005, vol. 2, pp. 228-234.

[47] 3GPP TS. Quality of service (QoS) concept and architecture (Release 6). 3GPP TS 23.107 v6.1.0, March 2004.

[48] 3GPP2 TS. All-IP core network multimedia domain: IP multimedia subsystem - stage 2. 3GPP2 X.S0013-002-0 v1.0, Dec. 2003.

[49] 3GPP TS. RAB quality of service negotiation over I_u (Release 4). 3GPP TR 25.946 v4.0.0, March 2001.

[50] 3GPP TS. RAB quality of service renegotiation over I_u (Release 4). 3GPP TR 25.851 v4.0.0, March 2001.

[51] 3GPP TS. General packet radio service (GPRS); Service description; Stage 2 (Release 6). 3GPP TS 23.060 v6.5.0, June 2004.

[52] Sarangan, V; Chen, JC. Comparative study of protocols for dynamic service negotiation in next generation wireless Internet. *IEEE Communications Magazine*, March 2006, vol. 44, no. 3, pp. 151-156.

[53] Manner, J; Karagiannis, G; McDonald, A. NSLP for quality-of-service signaling. IETF draft, draft-ietf-nsis-qos-nslp-11.txt, June 2006, work in progress.

[54] Chen, JC; McAuley, A; Sarangan, V; Baba, S; Ohba, Y. Dynamic service negotiation protocol (DSNP) and wireless DiffServ. In *Proceedings of IEEE International Conference on Communications* (ICC), May 2002, vol. 2, pp. 1033-1038.

[55] Wang, X; Wang, W. An efficient negotiation protocol for real-time multimedia applications over wireless networks. In *Proceedings of IEEE Wireless Communications and Networking Conference* (WCNC), March 2004, vol. 4, pp. 2533-2538.

[56] McNair, J; Akyildiz, IF; Bender, MD. Handoffs for real-time traffic in mobile IP version 6 networks. In *Proceedings of IEEE Global Telecommunication Conference* (GLOBECOM), Nov. 2001, vol. 6, pp. 3463-3467.

[57] Fang, Y; Ma, W. Mobility management for wireless networks: modeling and analysis. In: Guizani M, editor. *Wireless Communication Systems and Networks*. New York: Kluwer Academic Publishers; June 2004, pp. 473-512.

[58] Wang, W; Akyildiz, IF. Intersystem location update and paging schemes for multitier wireless networks. In *Proceedings of ACM International Conference on Mobile Computing and Networking* (MOBICOM), Aug. 2000, pp. 99-109.

[59] Xiao, Y; Pan, Y; Lie, J. Design and analysis of location management for 3G cellular networks. *IEEE Transactions on Parallel and Distributed Systems*, April 2004, vol. 15, no. 4, pp. 339-349.

[60] Xie, J; Akyildiz, IF. A novel distributed dynamic location management scheme for minimizing signaling costs in mobile IP. *IEEE Transactions on Mobile Computing*, July/Sept. 2002, vol. 1, no. 3, pp. 163-175.

In: Wireless Communications Research Trends
Editor: Tong S. Lee, pp. 123-163 ISBN 978-1-60021-674-9

Chapter 4

ENERGY AWARE MEDIUM ACCESS CONTROL PROTOCOLS

Jagoba Arias, Itziar Marín and Aitzol Zuloaga
Universidad del País Vasco
Bilbao, Spain

Abstract

In the last years, a great research effort has been done to reduce the power consumption of integrated circuits, specially microprocessors. The main advantage of this trend is the fact that battery operated devices may live longer with the same small size cells. However, the use of wireless technologies to transmit information is an energy consuming activity that may exhaust the batteries: the use of high frequency carriers and the need of emitting at least some milliwatts of RF signal consume a relatively large amount of energy that must be invested carefully to ensure that it is not wasted. This article describes which are the main sources of energy waste in medium access control protocols, analyzes the pros and cons of the wireless MAC protocols proposed so far and gives a series of directives to design new and more efficient protocols in the future. The result is an article that any researcher or engineer working with wireless battery operated devices must read in order to ensure that his or her devices are as long-lived as possible.

1. Introduction

Lowering energy consumption is one of the major issues in wireless sensor design. As many nodes in this kind of networks are battery powered or rely on unpredictable power sources, the node designer must bear in mind that *all* operations performed in a node require energy consumption and, therefore, limit the life time of the node. Analyzing the most energy-avid peripherals in a typical sensor node, we realize that the radio interface consumes even more energy than the microprocessor. Therefore, special care must be taken when designing communication protocols among nodes, so that the radio peripheral is only turned on when necessary. As the first non-physical layer in the communication stack is the Medium Access Control (MAC), we will carefully examine which strategies must be followed to design an energy efficient protocol that enables nodes to exchange information among them.

Most low cost radio interfaces can work only in half-duplex fashion, i.e. they can emit or receive, but cannot do both simultaneously. Energy consumption is different in these modes and, depending on the output power and carrier frequency, the radio interface will require different amounts of energy to work. Changing the status of the radio module (mainly switching it on and off) cannot be done for free, either: it will require some time with the local oscillator running, which is also translated into energy consumption.

The most important sources of energy loss in MAC protocols are the following [1]:

- Errors in incoming frames: whenever a frame must be discarded and retransmitted, energy is wasted. Of course, random error bits due to unpredictable interference cannot be avoided but there are other sources of interference that can (and should) be avoided. For example, if two nodes transmit a frame or a fraction of a frame at the same time, a third node that is listening to both of them will observe a collision and will not be able to decode any of the frames correctly. In this situation, the three nodes have wasted energy (either by transmitting or by listening and decoding an erroneous frame).

- Idle listening: in many MAC protocols, there are some lapses of time when the nodes simply wait for a frame to arrive, with their radio receivers on. When this happens, the nodes are consuming energy while doing nothing and this situation may be avoided by correctly synchronizing the nodes, so that they know when their neighbors will start transmitting information.

- Overhearing: whenever a node is listening to some information that is addressed to some other node, is wasting energy.

- Communication overhead: the presence of headers in frames, acknowledge packets and all control schemes that require transmitting information from one node to the other will require energy and will not transmit information. Therefore, overhead should be kept to a minimum to guarantee that the actual information travels from the source to the sink using the minimum amount of energy.

The base for all MAC protocols in wireless sensor networks is to reduce energy consumption by only switching the radio interface on when it is necessary. The rest of the time, nodes will keep their RF modules off, reducing power consumption. The percentage of time each node has its radio on is called *duty cycle*. To keep battery usage as low as possible, the duty cycle must also be low. However, when the communication interface is off, nodes cannot exchange information, so they will require some kind of synchronization to decide when they are going to be able to communicate with their neighbors. The organization of all nodes to make this possible is the purpose of MAC protocols.

Although energy consumption reduction is the main objective for this type of MAC protocols, there are also secondary objectives that should not be neglected. For instance, a MAC scheme that keeps the radio off for ever does not consume too much energy but it fails to complete the main objective of any communication protocol: exchanging information. Other merit figures used to compare MAC protocols for wireless sensor networks are:

- *Latency*: this is the time elapsed since a frame is emitted until it arrives to the sink. As all intermediate nodes in the network keep switching their communication interfaces on and off, the relaying of packets is not performed in a fluent way. On the contrary, frames may travel at different bursts and they may sometimes cross several hopes one after the other and sometimes the may get stuck at some node for a relatively long lapse of time.

- *Throughput*: this is the effective bitrate that may be achieved by a node. If nodes must switch their RF interfaces on and off, the rate at the link layer will be much lower than the physical bitrate in the radio link.

The rest of this chapter is organized as follows: in section 2. we will analyze the most commonly used approaches for the design of MAC protocols for wireless sensor networks. Thus, depending on the scheme used, we propose different kinds of classification parameters for the algorithms. Section 3. describes the most significative MAC protocols, their major contributions and their pros and cons. And last but not least, in section 4. we describe the conclusions of this chapter, with the main points to bear in mind while designing a MAC protocol for wireless sensor networks.

2. Protocol Classification

There are many aspects to be taken into account when designing MAC protocols. Therefore, there are different ways of classifying these protocols, depending on different characteristics of the protocols.

2.1. Medium Access Scheme

In classical wireless networks, there are different schemes to provide communication channels to nodes. However, not all of them are practical for wireless sensor networks, mainly due to the per node cost and power consumption restrictions. For instance, FDMA (Frequency Division Multiple Access) or CDMA (Code Division Multiple Access) require rather complex hardware resources that make nodes more expensive and more energy-greedy. As a matter of fact, the most widely used Medium Access Scheme is TDMA (Time Division Multiple Access), because it uses one single channel, which is shared by all the nodes in the network. However, as the wireless channel is a shared resource several aspects must be taken into account when designing a TDMA based MAC protocol:

- When a node is transmitting a frame, it will be heard in a certain area around the emitter. If a second node close to the first one starts emitting some other message, both packets will collide and the nodes that could receive both messages will not receive none of them.

- If the distance between two emitters is large enough, both of them can emit their messages simultaneously, because the attenuation in the medium will avoid the collision.

TDMA based MAC protocols must provide procedures to avoid or minimize collisions among nodes in the network. These procedures tell nodes when they are allowed to emit packets and when they should remain silent, even if they have information to forward. This organization may be performed in two different flavors:

- Nodes may be allowed to transmit at any time, provided that they check if the channel is being used by any other node. This task is performed at the physical level, by sensing the presence of a carrier signal in the medium. This approach is called CSMA (Carrier Sense Medium Access).
- Time is divided into time slots. Each node is assigned one of these slots, so that it is only allowed to transmit in a certain lapse of time. This slot assignment must be performed carefully to avoid possible collisions. In WSN MAC literature, these protocols are usually known as TDMA protocols (although, strictly talking, CSMA protocols are also time division protocols).

The survival of these two different approaches implies that both of them have advantages and disadvantages. The CSMA approach is not very efficient in terms of energy consumption, because it allows frames to collide. However, it is easily scalable for networks with a large number of nodes – which is usually the case for sensor networks – because almost no scheduling information must be shared among the nodes. If one of them dies (e.g. because its batteries have been exhausted) or a new node is added to the network, no major changes in network organization must be performed. On the other hand, time-slot based algorithms are much more efficient when energy must be conserved, because every node not only knows when it must emit a message, but also when other nodes will emitting theirs. This really helps avoiding problems such as overhearing or collisions. Nevertheless, any change in network topology will trigger a rearrangement of the slot distribution and, therefore, managing large and dynamic networks with this scheme will require a penalty in exchange for the efficiency in the static mode.

Derived from the intrinsic characteristics of the medium access approach, different problematic scenarios may appear. One of them is the *hidden terminal* problem. Let us suppose that there are three nodes within a wireless network (A, B and C), as shown in figure 1(a). B can exchange information with the other two nodes, but there is no wireless link between A and C. If A sends a frame to B, C will not be able to hear this frame and, in a CSMA based algorithm, it could start transmitting a frame for node B, although it will collide with the frame sent by A.

A second problem is the *exposed node* problem. Let us consider a scenario with four nodes (A, B, C and D), as shown in figure 1(b). The nodes are deployed in a line, so that each node can communicate with its two neighbors, but it cannot receive messages from the rest. The two nodes in the ends of the line (A and D) only have one neighbor. Whenever node B sends a message to A, node C will detect that the channel is busy and, in order to avoid collisions, it will remain silent until B finishes its transmission. However, if the destination node for C's frames is D, this situation is unnecessary, because D cannot hear the nodes emitted by B and, therefore, both nodes B and C can transmit frames simultaneously, provided that the destination for these frames are A and D.

Another problem that may appear when TDMA schemes are used is the *masked terminal* problem. This scenario is quite similar to the hidden terminal problem and it may appear when trying to avoid it. Let us suppose that there are five nodes in the same network (A, B, C, D and E), as in figure 1(c). As in the previous cases, we will also suppose that each node can only exchange information with its immediate neighbors. If nodes B and D simultaneously send packets to nodes A and E respectively, C will not hear any of them. Of

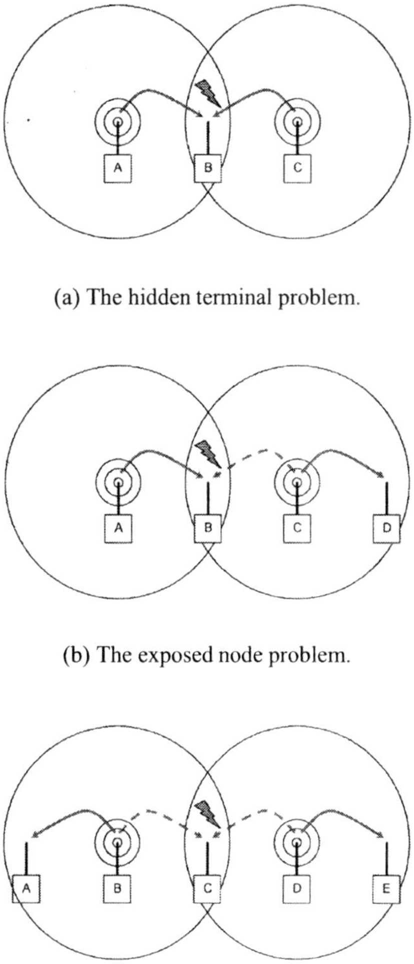

(a) The hidden terminal problem.

(b) The exposed node problem.

(c) The masked terminal problem.

Figure 1. Medium access problems in wireless networks.

course, if the packets only contain data this is not really a problem (C was not the destination for them) but if the frames were control packets used for arbitrating the access to the communication channel, this collision would make that node C missed some information that would be important for avoiding further data collisions.

2.2. Placement of the Network Organizer

Scheduling when each node is allowed to transmit information is a complex task when there is a large number of nodes in the network. Even when this task is performed in a reduced environment, ordering the access to a shared communication channel can be done in two different ways:

- Some nodes are elected as arbiters of the communication channel and they decide which of their neighbors is allowed to use the channel at each moment. Thus, the rest of the nodes in that area must simply follow the schedule imposed by the arbiter. This way of organizing the medium access is called *centralized access*.
- There is no hierarchy and all nodes have the same right to establish their own schedule. The organization of the medium access is performed following a series of rules that assure that collisions and other forms of energy wastage are minimized. This way of organizing the medium access is called *distributed access*.

Of course, both approaches have their advantages and disadvantages. Centralized access schemes usually allow a better time slot usage and avoids possible collisions but it requires the presence of an arbiter that will have to perform a heavier task than the rest of nodes. Therefore, this node will exhaust its battery at a higher rate than the rest of its neighbors. Besides, whenever there is a change in the network topology (e.g. a node dies or a new one is added to the network), it has to communicate with the arbiter and major changes in channel organization will appear. On the other hand, distributed access schemes allow a flat hierarchy of nodes which eases the addition and removal of nodes to the network structure, but collisions cannot be completely avoided, which is translated in a poorer energy efficiency of the protocol.

2.3. Topology Dependence

Wireless Sensor Networks are not formed by individual nodes that may want to communicate with each other. On the contrary, data flow in WSNs usually has a certain pattern: sensors will report the events they detect to a sink, which is in charge of interfacing the network with the final user and the final user might send some configuration data to the sensors. Although the direction of the data flow is more related to the network layer than to the link layer, paying attention to this issue may be useful for improving the quality of service of the network, e.g. by reducing latency.

In this sense, we distinguish two different type of algorithms:

- *Topology dependent protocols* take network topology in consideration and try to optimize the access to the medium in such a way, that data frames do not stop in a certain node simply because the node must go to the sleep mode as soon as it has received the frames. This restriction in medium access forces all nodes to take into account the network topology, which corresponds to higher layers in the protocol (i.e. the network layer).
- *Topology transparent protocols*, on the contrary, provide a medium access organization that minimizes the latency and maximizes the throughput of the network, without having to take into account the topology of the network. Therefore, these algorithms are easily scalable and provide a better inter-node communication scheme. These nodes are usually based on TSMA (Time-Spread Multiple Access), an organization scheme where time is divided in frames, which are subdivided into q subframes and these into q slots. The transmission slots, when each node is allowed to transmit is defined by a polynomial of order k, which belongs to a Galois Field $GF(q)$. In this

schemes, collisions are not avoided, but they are minimized attending to the properties of Galois fields.

Topology transparent protocols provide a general-purpose medium access procedure that may be useful in some application level protocols which require inter-sensor communication. However, topology dependent protocols are more adapted to the specific characteristics of WSNs and provide a better network behavior, reducing energy consumption.

3. Protocol Description

In this section we describe the evolution of MAC protocols, which are their main contributions, their advantages and disadvantages.

3.1. Overview of Existing Protocols

In the 1990s a great research activity was performed in the field of wireless computer networks. The availability of laptops with radio network interfaces had made it possible to produce wireless networks without any kind of infrastructure: the nodes simply passed the packets from one to the other, until the packets arrived to their destination. Ad Hoc networks were at this point a reality. Although all nodes in these networks were battery powered, the energy restrictions were not too hard, because the nodes could be connected to the electric network at any time, which in fact did not represent any applicability restriction to the network. However, in the late 90s the first steps were given towards power aware wireless networks [2–4]. At the beginning, all these studies were focused on power reduction in typical laptop peripherals (such as displays, hard drives or even I/O devices) but then, mainly thanks to the advances made in cellular telephony, the radio interface was identified as one of the greediest energy consuming peripherals in Ad Hoc network nodes. Therefore, some researchers focused their interests on saving energy by turning on and off the radios. At the same time, low cost radio integrated circuits and low power microcontrollers made it possible to build large networks that monitor wide geographical areas. This new application did require that the nodes worked unattended for a long time, without a stable source of energy and new MAC protocols that used energy as efficiently as possible were needed.

The first MAC protocol explicitly designed for WSNs was S-MAC. Based on previous protocols for Ad Hoc networks, its authors analyzed the main sources of energy waste and they proposed exchanging quality of service for energy savings: nodes will follow a schedule that told them when they should turn their radios on and off. However, this scheme did not bear in mind the needs of the network in terms of data flow and latency. In order to improve the deficiencies shown by S-MAC, some other protocols followed: both ER-MAC and TRAMA proposed TDMA mechanisms that improved the energy consumption efficiency, avoiding collisions. While TRAMA followed a schedule sharing protocol for node synchronization, similar to that used in S-MAC, ER-MAC proposed a slot management algorithm that did not treat all nodes equally, to overcome the excess of tasks (and therefore, the excess of energy consumption) that had to be performed by some specific nodes in the network. On the other hand, Sift evolved the 802.11 protocol for Ad Hoc networks and mixed it with the ideas expressed in S-MAC. Thus, the resulting algorithm was a contention window based protocol that reduced the collisions in S-MAC, but that treated the traffic created in sensor networks in the same way as computer network traffic was treated, not allowing the nodes to save as much energy as possible. However, almost simultaneously appeared T-MAC. This was an evolution of the S-MAC protocol that tried to resolve the lack of adaptability to the network flow shown by this algorithm. To do so, instead of using schedules with a fixed duty cycle, it proposed a variable duty cycle scheme, where the idle period had a fixed length and the active period could have a variable duration. Thus, when the network detected an event, it was quickly reported to the sink, investing energy in reducing latency, but the rest of the time, the nodes would save power by lowering their duty cycle as much as possible.

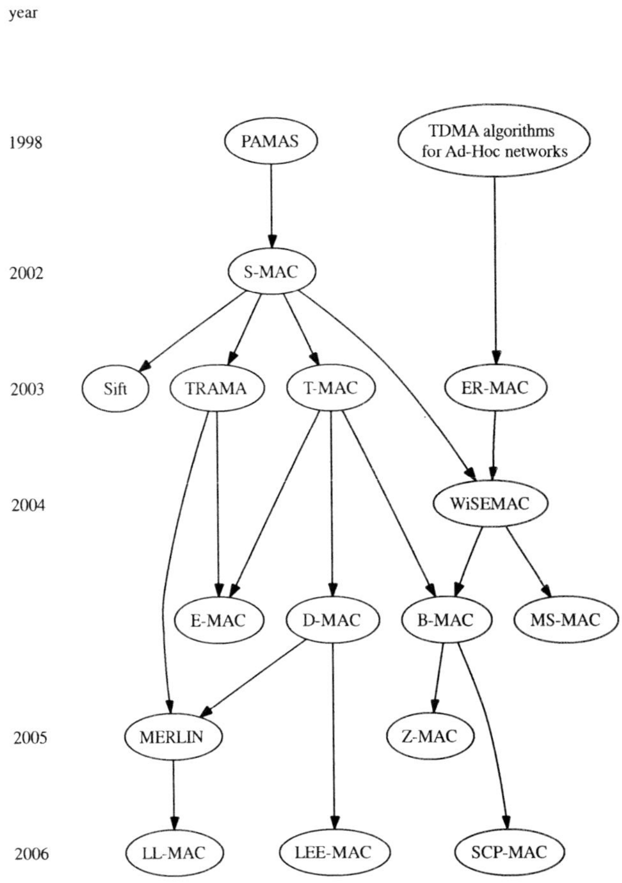

Figure 2. Evolution of MAC algorithms for wireless sensor networks.

With the development of the different protocols, new challenges appear. For example, if nodes in the network move, schedules will change because not all nodes in the network have to be synchronized. Mobility in this case will increase the number of frame exchanges to keep nodes in a certain location within the same schedules. The MS-MAC protocol addresses this problem, providing an algorithm which is half way between the hard schedules of S-MAC and the mobility features of 802.11. Apart from mobility issues, there are other points of improvement in MAC protocols, such as quality of service or second order energy consumption reduction. Other minor points where energy may be saved are the way in which the node tests the channel to detect if someone else is sending a message (introduced in WiseMAC), or the amount of energy needed to perform the transitions among the node modes: listen, sleep and transmit. Sometimes, it is much more expensive, speaking both in terms of energy and time, to put a node to sleep if it must become active a few milliseconds afterwards. Algorithms such as E-MAC try to resolve this problem by analyzing the amount of energy needed to perform the whole transition and how much energy would be required if the node simply stayed awake. In order to improve latency, protocols such as D-MAC, LL-MAC or LEE-MAC organize schedules all throughout the network so that messages do not get stuck at a certain node, waiting for the next active slot.

In the last years, the main effort in MAC design for WSNs has been focused on simplifying the protocols, to pursue different objectives: reduction of the number of schedules in the network, reduce the number of hardware resources required for implemented the MAC protocols and enabling the scalability of the algorithms. These are the objectives followed by Z-MAC, B-MAC, MERLIN, etc.

In the rest of the section, we will describe all these protocols chronologically to understand what are their main contribution and which are their main drawbacks.

3.2. PAMAS

The PAMAS (Power Aware Multi-Access protocol with Signaling) [5] is one of the first efforts made in power consumption reduction in wireless networks. This protocol had not been designed keeping in mind the restrictions present in WSN, but it is a first step, on which many other MAC protocols for wireless sensor networks are based. The main objective of this protocol is to reduce the power consumption due to overhearing: when in a wireless network a node sends information to another node, its neighbors cannot exchange information because, otherwise, the frames would collide and no communication would be possible. Thus, if a node wants to transmit and it detects that the channel is busy, then it can simply turn its radio off to save energy, because this interface is unusable until the communication between the two other nodes is over. Based on this simple idea, the PAMAS protocol defines a TDMA MAC scheme, based in two different channels: a data channel, used for actual data exchange, and a signalling channel, for channel control. In the data channel, only data frames are exchanged and in the signalling channel, there are three possible frames: RTS (Ready to Send), CTS(Clear to Send) and Busy Tone (to show that the node is receiving a frame).

Thus, this protocol allows nodes to be in six different states:

- *Idle*: a node enters this state when it has nothing to send or to receive.
- *Await CTS*: if a node in the idle state decides that it should transmit a frame, it sends a RTS packet and enters this state, to wait for the confirmation. The node will remain in this state until a CTS or RTS packet is received or until a timeout period expires.
- *BEB (Binary Exponential Backoff)*: in this state, the node waits for a random time without sending anything, to wait for the channel to get free. This state avoids successive collisions between neighbors. A node enters this state if in the *Await CTS* state the timer expires.

- *Transmit Packet*: if a node in the *Active CTS* state does receive a CTS packet, the node is ready for transmitting the frame, which is performed in this state. Of course, in this state signalling packets are ignored, so that they do not interfere with the communication. When the frame transmission is over, the node returns to the *Idle* state.

- *Await Packet*: if a node in the *Idle* state receives a RTS packet, it responds with a CTS packet and enters this state. When the nodes starts receiving the data packet, it sends a Busy Tone packet and enters the *Receive Packet* state. However, if the data frame does not arrive in a certain lapse of time, the node returns to the *Idle* state.

- *Receive Packet*: in this state, a node receives an incoming packet. After the transmission is over, the node returns to the *Idle* state.

This state machine and the control messages exchanged allow a node to know when it should turn its radio off. As a matter of fact, there are only two situations in which the node should do this: when it has no information to transmit and a neighbor is transmitting data packets addressed to some other neighbor and when a neighbor is receiving packets from someone else. These two situations are easily detected by the presence of data packets in the data channel or Busy Tone packets in the control channel. Therefore, all nodes in the network can decide independently when they should turn their interfaces off, but for how long? If the node is able to listen to the beginning of the transmission, it can compute the length of the information exchange process, taking into account the number of bits of the frames and the bitrate, or even wait long enough for the longest feasible frame to complete. However, it is not that easy when the node cannot hear the frames because the emitter is too far away or because it has turned its radio interface on when the communication had already started.

The PAMAS algorithm resolves this situation by adding two new types of packet for the control channel: *t_probe(l_1,l_2)* and *t_probe_response(t)*. The first packet, *t_probe(l)* is used to ask if there is a neighbor, which is finishing its transmission in the interval $[l_1, l_2]$. Thus, all neighbors that meet this condition will answer with a *t_probe_response(t)*, indicating in the t parameter when will their communication end. The presence of a collision in the *t_proble_response(t)* packet means that there are several nodes whose communications will end in the $[l_1, l_2]$ interval, while the absence of response means that there are no nodes that meet this condition. These two packets are then used for performing a binary search for the time when the longest ongoing transmission will end.

Although the PAMAS algorithm is not a MAC algorithm for WSNs, it establishes some guides to design low power MAC algorithms for wireless networks. It addresses the problem of overhearing and tries to provide a procedure to establish when the nodes should turn their radios off and for how long. However, it is still far from what we should consider an appropriate protocol for WSNs:

- This protocol does not address energy waste as a crucial problem for the network. It does not analyze (and eliminate) all sources of waste, but it simply discovers one of them.

- The MAC management procedure is quite complex with separate channels for signalling and data. Of course, this is very typical in Ad-Hoc networks, but not in

WSNs, where nodes should be kept as cheap (and therefore, simple) as possible. [5] even describes the possibility of keeping the signalling channel on, while the data channel is off, to simplify the power management.

- A major concern of the authors of this algorithm is to show that typical figures of merit for computer networks, such as latency and throughput, are not affected substantially by the fact that nodes are switched on and off. Nodes' interfaces are only turned off when they otherwise would be unusable, because the channel was being used by someone else, so turning off the radios will have no effect on how long will a packet need to get to its destination or how much information per second will the network be able to process. As an example of this, the authors propose a binary searching algorithm for determining when a node should switch its data channel on again, which requires a relatively large number of packets: the authors propose – and practically desestimate – a much simpler algorithm with no binary search that depending on the first *t_proble_response(t)* packet would make a decision, arguing that this would increase latency in the network, and neglecting the saved energy in the process.

- No special care has been taken to check how much memory or computing power will be required to implement this algorithm. Although this algorithm *per se* does not require complex mathematical operations, it does need a random number generator for the Binary Exponential Backoff state and a certain amount of memory to keep track of every node's neighbors' state. Of course, this requirements are easy to achieve in typical nodes for Ad-Hoc networks, but hardware resources in WSN nodes are scarce and they should be wisely managed.

Anyway, the PAMAS algorithm was a start point for the development of the different MAC protocols that bear in mind the particularities of WSNs.

3.3. S-MAC

One of the first MAC protocols specifically designed for wireless sensor networks is Sensor-MAC (S-MAC) [1]. It is based on PAMAS and the IEEE 802.11 standard, and offers a power saving mechanism thanks to a *sleep/active* scheme.

Each node wakes up for the first time and keeps on listening until it hears a transmission. If nothing is listened, it starts its own *sleep/active* scheme, that may be synchronized or not with the schemes of the rest of nodes.

Every node in the network has the same *sleep/active* scheme, that is it, every node listens during the same time duration, but not necessary in the same moments. Nodes try to get synchronized to their neighbors and therefore make their awake periods coincide. However, there might be different schemes of performance within the same network. Therefore, nodes that are located in between to different scheme zones, must listen to both schemes in order to interconnect both sub-networks. This behavior leads to high power waste and these nodes may get their batteries depleted very quickly.

Inherent to the control message access management (RTS/CTS), there is the *overhearing* problem. Nodes need to listen to any RTS and/or CTS of any node in order to know that the medium is occupied and other node is going to start a conversation. S-MAC offers a simple solution for energy saving for *idle listening* situations: nodes go to sleep whenever a control message is received.

However, this conflicts with global latency. If a node, next to the current receiver of the information goes to sleep when the transmission starts, it will not listen to the next RTS of the receiver that wants to forward the packet to it. Consequently, the third node will not be ready and that transmission will not succeed. This implies that a forwarding possibility is lost and it will take longer than necessary to reach the sink.

Before each node starts its periodic *listen/sleep* scheme, it needs to choose a schedule. In order to select one, it follows these instructions:

- It keeps on listening during at least the *synchronization period* of ten seconds. If it does not hear any schedule from another node, it starts its own schedule.
- If a node listens a schedule from a neighbor before choosing or announcing its own schedule, it adopts the received schedule.
- If a node receives a schedule different from its own, there are two different actions. If the node had no neighbors before and the schedule it is using was chose on its own, it will follow the received one. If the schedule it was following was received from one or more neighbors, then it adopts both schedules and will be awake in the listen periods. Nodes that adopt two or more different schedules are called *border nodes* and serve as connection between two zones with different schedules. However, as explained before, this kind of nodes waste much more energy than the rest of the nodes of the network.

The frequency with which each node tries to receive a schedule from other nodes depends on whether the node has a neighbor or not. If it has not any neighbor the node will try to find a schedule more often that if it has already received one. This period is very energy consuming and therefore, it should not be executed very often. This periodic synchronization discovery is ten seconds long and it is performed every two minutes if nodes have at least one neighbor.

Hence, S-MAC's repetition period is two minute-long and consists on:

- Ten seconds of synchronization period.
- More than one hundred executions of *listen/sleep* scheme of one second each. Each *listen/sleep* scheme is divided into 100 ms of listening and 900 ms of sleeping.

In order to maintain their clocks synchronized and correct clock drifts, the *listen* period includes a SYNC phase. Therefore, before sending any transmission request, nodes share synchronization information. For that purpose, the *listen* phase is divided into two parts: the *synchronization* part and the *packet* part. Both parts start with a contention window and the latter is also formed by two phases: medium access control (RTS, CTS and ACK), as explained above, and data packet. An overview of the general performance of S-MAC is presented in figure 3.

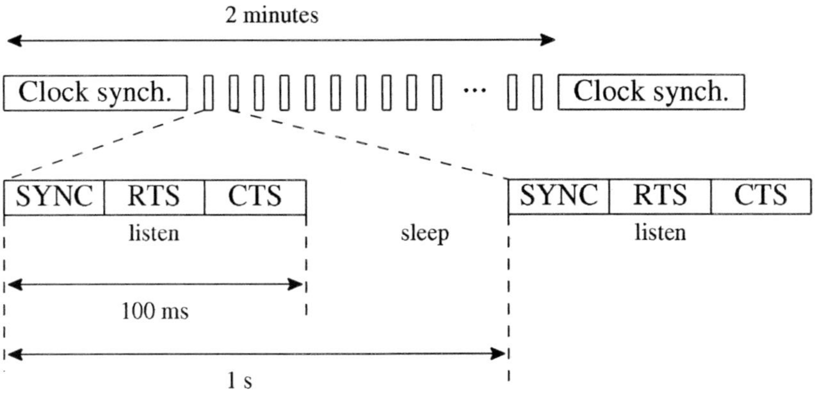

Figure 3. Frame structure and timing in S-MAC algorithm.

Two follow-up papers of the same authors present some performance improvements. The first approach [6], tries to reduce the *overhearing* problem presented in S-MAC with the implementation of another mechanism. The solution proposed in S-MAC incurs in a high latency due to the fact that the nodes went to sleep whenever they heard a control message not related to themselves.

In this first paper, *adaptive listen* is presented. It takes advantage of the content of the CTS message, that indicates the duration of the transmission of the packet that will be transmitted after it. Therefore, when a node listens to a CTS message, it goes to sleep the exact time that the CTS packet says and it wakes up when the transmission has just finished. Consequently, in that moment, the node needs to forward the recently received message and sends a RTS message toward the following node. If the nodes that listened before to the CTS have waken up correctly, then they will listen to this RTS and will allow to forward the message correctly.

However, as the authors analyzed, this approximation offers a solution for a multi-hop performance for a few consecutive hops but no more than 3. This is due to the fact that nodes that could listen to a RTS or CTS of a transmission, have to be in the coverage area of the sender and it is not always possible for a node to overhear the control messages of all next-hop nodes.

Furthermore, if the RTS/CTS packets are generated in an *adaptive listen*, surrounding nodes will not hear them and the improvement of this method would vanish. Finally, if nothing is received during the *adaptive listening*, the node goes back to sleep and follows the normal schedule.

The second approach, [7], faces two different challenges: an end-to-end multi-hop transmission and a global schedule.

As explained before, there could be different performance schemes within the same network. This situation makes the intermediate nodes between two sub-networks with different wake-up moments waste much more energy than the other nodes in the network. A new mechanism is presented, called *Global Schedule Algorithm* to let the whole network converge to a common scheme of performance for all the nodes. Usually, that global schedule is the oldest one in the network.

In order to achieve a global schedule, the GSA algorithm has to face two different problems:

1. How to uniquely identify each schedule

2. How to exchange schedules among nodes

To solve the first problem, GSA identifies each schedule with the identifier of the node that sends it. This way, the owner of the schedule can be determined but there may be a problem if the same node starts different schedules, with the same identifier, every time it reboots.

To exchange the different schedules in the network, in order to converge to a unique one, GSA adds a indicator called *schedule age*. This new value indicates how long that schedule is been working in the network. Schedules that start in the network in different moments will have different *age*. However, it is possible that two schedules start at different moments in the network but with the same wake-up time. Such schedules are the same but with different *schedule age*. Whenever a node receives its own schedule but with older *age*, must update its own *age* in order to unify schedules.

Finally, in trying to solve the *multihop* problem, the authors developed the *Fast Path Algorithm* (FPA) that offers an improved solution for more consecutive hops. This solution is presented in figure 4.

FPA adds more wake-up periods that initially scheduled in order to forward the messages along the different hops. These additional active periods are carefully selected as they occur in the exact moment when the receiver of the packet wants to forward it to other node. This approach offers a low latency channel for data when necessary.

Based upon the GSA algorithm, it is assumed that the network is working with a unique global schedule and therefore all the nodes in the network follow a synchronized scheme.

Once the fast path is established, the data can be sent without additional control messages. If everything works as it is supposed, next hop nodes should be awake and listening to the medium in order to receive the data transmission. This way, packets do not have to wait and can be forwarded without interruption.

As this solution is developed based upon the former paper [6], *adaptive listen* is also active in this approach. If the moment of performing *adaptive listen* overlaps with a fast-path slot, the latter is executed as it offers better performance in terms of latency reduction. If a fast path schedule is established but nodes involved in it, detect that it has not been used for a determined lapse, it is discarded.

Although a single fast-path schedule is studied, the authors state that each node could manage a small number of active paths separately and there could be some fast-paths working simultaneously in the network.

S-MAC and its modifications offer a good approach for a wireless sensor network MAC protocol with a power-aware scheme and some mechanism to perform multi-hop low latency transmission. Nevertheless, it still suffers from *overhearing* and *idle listening*, and the *hidden terminal* problem is not solved yet.

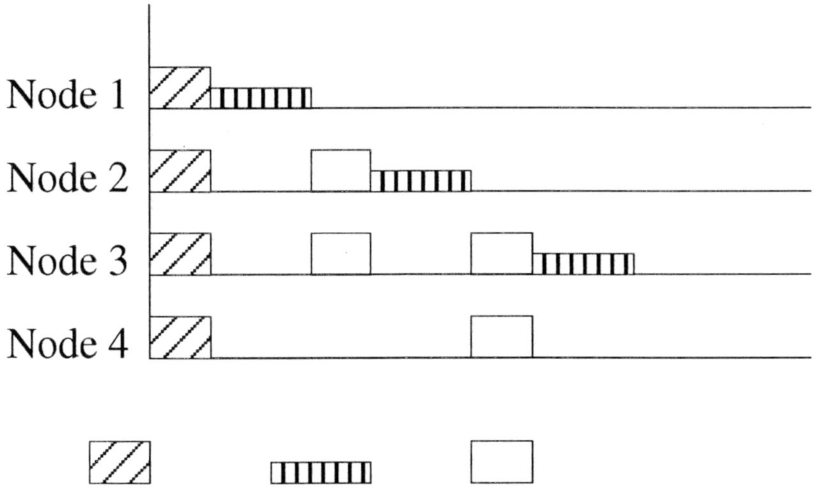

Figure 4. Frame structure and timing in the FPA algorithm.

3.4. ER-MAC

The ER-MAC (Energy and Rate based MAC) [8] protocol proposes a TDMA based protocol, where each node is given some slots when it is allowed to transmit. As we mentioned in section 2., this scheme avoids collisions but it is more difficult to scale. However, the ER-MAC protocol does not decide how the schedules are shared among all the nodes in the network. On the contrary, it establishes a mechanism to decide which nodes should be given a higher priority in the slot allocation process to increase the overall network lifetime.

In a wireless sensor network, the energy consumption rate is not the same for all nodes. For example, peripheral nodes, which are far away from the data sink will not consume as much power as nodes that are closer to the sink, because the latter have to route more packets than the first ones. In the same way, in a more reduced geographical space, some nodes may tend to route more packets than the rest, using more energy in this process. In a wireless sensor network, where nodes do not have a direct link with the sink, if a node runs out of battery, many other nodes may become affected because this link failure may produce changes in routing tables or even leave large groups of sensors without a route to the sink. To make matters worse, the busiest nodes in the network are usually closer to the sink, which means that a failure of one of these nodes will affect a very large amount of peripheral nodes.

Previous TDMA protocols assigned slots to nodes in terms of fairness or available information but remaining energy had not been taken into account. In this case, ER-MAC defines a measurement to know if a node should be given a higher priority within a TDMA group: criticality. This figure is taken from two measures: the amount of remaining energy (E) of the node and the amount of packets originated at the node (F). Criticality for node i, C_i, is computed using energy and traffic data from the rest of neighbors of the TDMA group as:

$$C_i = \frac{E_i}{\max\{E_j\}} + \frac{F_i}{\max\{E_j\}} \tag{1}$$

This value of criticality will be useful to establish which nodes have more remaining energy and more packets to transmit and, therefore, should have a higher priority than nodes that are running out of energy and originate almost no information.

In ER-MAC, all nodes can transmit two different types of packets: data frames, which are used for exchanging information from higher protocol layers, and control frames, used for establishing which is the criticality of each node and to publish the node with the lowest criticality in the group. Initially all nodes are assigned two slots to transmit. They must also keep a table with the slot assignment, so that it can turn their receivers on when the node they are connected to must transmit a frame. Otherwise, they will keep their radios asleep when they do not own the current slot. The energy conservation procedure in this algorithm consists in assigning less slots to those nodes with a lower criticality, so the MAC layer has a state parameter, Radio-power-mode, which will be set to true when the node is able to transmit in only one slot, and it will be false if the node can transmit in two different slots.

In normal operation, all nodes will be transmitting information in their slots and, periodically, they will start a voting phase to detect which is the neighbor with the smallest amount of energy (the "winner"). This voting phase is triggered whenever the energy of a node falls below the energy value of the winner in the last election. In this phase, all nodes emit their energy and packet flow values, so that all members of the TDMA group can compute their neighbors' criticality. The neighbor with the lowest criticality changes its Radio-power-mode to true, and it will be assigned two slots. All looser nodes will set Radio-power-mode to false and, therefore, will be assigned one single slot to transmit. The winner will require less energy to keep being part of the network, because within its slots it can switch the radio completely off if it has no data to transmit. On the other hand, having more slots to transmit than the rest of its neighbors, it is less likely to receive packets in transit to the sink, which will also reduce its outgoing traffic. After the election phase, all nodes in the TDMA group know their neighbors' energy level and criticality so, when one of these neighbors notices that its energy resources are lower than those of the last winner, it will trigger a new election phase.

ER-MAC is not a Medium Access Protocol strictly speaking, as it does not pay attention to the way in which nodes decide when they should transmit information. However, it is an interesting approach when addressing the problem of deciding which nodes in the network – if any – should have special privileges to save energy. In this approach, nodes measure the amount of energy they have left and this algorithm changes the parameters of the underlaying MAC protocol to reduce the energy consumption of those nodes whose batteries are closer to exhaustion. Thus, all nodes in the network consume the same amount of energy (mainly because the most energy-consuming activities are rotated among the nodes), avoiding situations in which the unavailability of a node leaves a large amount of nodes out of service because their data cannot reach the sink.

Nevertheless, the ER-MAC algorithm also has drawbacks. It is based on a slot-like TDMA algorithm, which is difficult to scale and that requires periodical management operations. In fact, each time the ER-MAC algorithm proposes a new winner for energy conservation, the TDMA scheme of the whole TDMA group must be updated, changing slot allocation for all nodes. Furthermore, the mere possibility of a node requesting a new election requires that all nodes in the group must listen to each other, even if they are not expecting information from them, forcing them to overhear information and thus, to waste energy. On the other hand, being ER-MAC a local algorithm, based on local information, it will also have a local effect: the members of the same TDMA group will consume energy at a similar rate, but this cannot be extended to nodes that do not belong to the same group (i.e. that are not close to each other). As a matter of fact, distant nodes are more likely to have different consumption rates, because their environmental situation will be quite different: they will not detect the same events, the amount of information they will have to route to the sink will also be different, etc

In conclusion, the ER-MAC algorithm introduces the possibility of privileging some nodes to prevent them from totally exhausting their batteries, but it fails to provide a scalable global solution to the problem.

3.5. TRAMA

TRaffic-Adaptive Medium Access protocol (TRAMA) [9] proposes a different approach. TRAMA uses a unique communication channel, slotted and time divided for control and data purposes. Control period is formed by *signaling slots* that are accessed in a CSMA fashion whereas TDMA slots are for data transmission and are called *transmission slots*.

Nodes can only join the network during the random access period. Time synchronization is also achieved among nodes in this period as every node must be either transmitting or receiving. This way, nodes exchange signaling packets and can update its neighborhood tables. However, the longer the random access period is, the higher the power consumption is.

Each node implements back-off mechanisms to get access to the medium in the random access period. Back-off timers try to avoid collisions among signaling packets. However random access period does not guarantee a correct delivery of packets and therefore, signaling information could get lost. In order to achieve a high and constant packet delivery rate, the number of signaling slots in the random access period grows proportionally to the number of nodes in the network. However, this figure seems to be too high for WSN requirements. On the other hand, *transmission slots* are collision-free and assures a correct reception of the data.

It tries to manage the slotted medium access by not assigning time slots to those nodes that do not have activity scheduled. If a node is not going to transmit and/or receive any message in a cycle, it keeps on sleeping saving energy. This way, it offers adaptation to different kind of applications:monitoring applications need constant data flow while event-monitoring deployments would generate a great amount of packet traffic when the *observed* event appears and there will be no traffic the rest of the time.

The TRAMA protocol is formed by three different mechanisms: the *Neighbor Protocol* (NP), the *Schedule Exchange Protocol* (SEP) and the *Adaptive Election Algorithm* (AEA).

- The *Neighbor Protocol* collects the neighborhood information included in some small packets that are exchanged during the random access period. This packets, apart from update neighborhood information, are used for knowing if the neighbors are still alive, and therefore checking neighbor connectivity. If a node does not listen to other node's signaling for a certain lapse of time, it will delete it as a neighbor. Thanks to this mechanism, each node knows its one-hop neighborhood and can access to the neighborhood information of each neighbor. This way, each node can know its two-hop neighbors.

- The *Schedule Exchange Protocol* lets nodes exchange their two-hop neighborhood information and their own schedules. Within this information, each node specifies the intended receivers of its traffic in chronological order.

- The *Adaptive Election Algorithm* protocol uses the information collected and exchanged by the previous protocols to decide what to do. For each node and time slot, AEA protocol determines if it has to transmit information to other node, receive a packet from a sender or can go to sleep and save energy. This decision depends on its two-hop neighborhood priority and on one-hop neighborhood schedules. When a node is the owner of a slot it knows that no other node will transmit and collide with its information but when it is not the slot owner, it does not know who the owner is. It could happen that a node needs more than a slot in order to transmit its information. It could also happen that a node, when being the owner of the slot, has nothing to transmit. In that case, there is a list of *Possible Transmitter Set* in order to let those nodes with information to send, use other's transmission slots. This is very useful in networks with few data sources that send information to a small set of receivers.

TRAMA offers higher delays and power consumption than S-MAC but guarantees data packets delivery thanks to a collision free functionality. Furthermore, TRAMA performance can get adapted dynamically to the network traffic conditions thanks to its time slots scheme. However, there are some shortcomings:

- This protocol proposes the use of too many control packets that have to be exchanged among nodes in order to perform correctly. This large amount of transmissions leads to a great power waste.

- Random access periods are too long in order to solve the possible collisions with back-off delays. *Transmission slots* are even longer, as they are fixed in seven times longer than *signaling slots*.

- Moreover, nodes with the TRAMA protocol must keep awake during the whole CSMA period, transmitting or receiving which leads to a working cycle of 12.5%. This percentage is excessive for the kind of networks.

3.6. T-MAC

Timeout-MAC protocol (T-MAC) [10] is based on S-MAC and is focused on saving the energy that S-MAC wastes due to *idle listening*. As each node does not know the exact moment when it will receive any packet, it must keep on waiting awake until it arrives. Moreover, S-MAC keeps on listening even when the packet has already arrived, wasting energy uselessly. T-MAC proposes reducing the *active* time in order to reduce the global consumption of the mote.

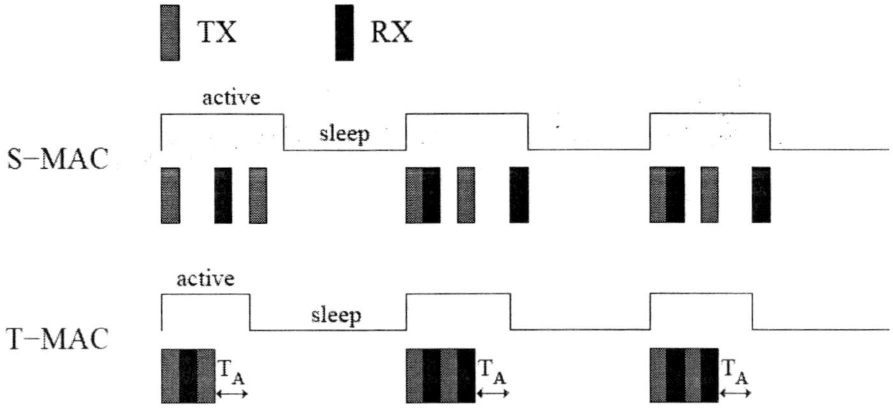

Figure 5. Variable duty-cycle scheme in T-MAC. This approach adapts the MAC layer to the current situation of network traffic.

As presented in figure 3, the *active* time and *sleep* time distribution in the S-MAC working cycles are maintained invariant during its performance. However, this behavior does not fulfill some networks requirements, as data flow could not be constant.

If the active time is longer than necessary, energy is being wasted and therefore the authors propose transmit data in bursts and detect a data transmission end in a *intuitive* way: if nothing is heard for a specified lapse of time, the node switches off the radio.

In the same way as S-MAC, a node can be in the active period or in sleep period. A node will remain in the active period, listening to the medium during a scheduled period of time. If, from the last activation event, it has not received another activation event in a *TA* period (see figure 5), it will finish the active period. Activation events could be:

- The reception of a data packet
- The detection of communication in the medium
- The reception of own's packet transmission end or ACK

However, *TA* duration has to be selected very carefully. A node must not go to sleep while any of its neighbors is still transmitting, because it could be the next destination of the data packet.

If the *TA* value were not correctly chosen, a new problem could appear: the *early sleeping problem*. A node could go to sleep before it can hear a RTS destined to it and therefore miss the opportunity of receiving the packet. Furthermore, this new problem might cause another effect, *overemitting*. This happens when a node transmits a packet thinking that the destination is awake waiting for it. Nevertheless, the receptor has gone to sleep too soon and therefore will not hear the packet. The worst consequence of these effects is the fact that the transmitter has wasted its energy uselessly as it will have to transmit the packet again.

Due to the fact that most traffic in wireless sensor networks flows in one direction, from the nodes to the sink, a solution for this effect is proposed. In the way to the sink, when a node receives a RTS from other node, apart from replying with a CTS, it should send a

Future Request-To-Send message (FRTS) to the next node in the chain in order to inform it that it will be the next. Thank to this mechanism, the future receiver of the packet can go to sleep but knowing that it must wake up later in order to receive the RTS message from the current receiver.

When a node send a RTS, waits for a CTS but if it does not receive it in a *TA* lapse, it goes to sleep. However, the receiving node might not have replied due to:

1. A collision has happened
2. A previous RTS/CTS have been heard from other node
3. The receiving node is asleep

In order to avoid canceling in case 1 and 2, when the receiver is waiting for the packet, RTS retries are implemented. If no reply is obtained after two retries, the node will finally go to sleep.

Apart from the FRTS, authors also offer another original solution: *full-buffer priority*. When a node has its own transmission buffer full, it prefers sending data and emptying the buffer to receiving more packets. For that reason, whenever it receives a RTS message, it immediately sends its own RTS message to its receiver instead of replying a CTS message. A node wins the medium by *stealing* it to another node. This mechanism is very appropriate to *convergecast* networks in which data flows in one direction.

Although T-MAC approach solves the *overhearing* problem that S-MAC suffers, it could incur in other negative effects, such as *overemitting* and *early sleeping*.

3.7. WiseMAC

Enz *et al.* propose in [11, 12] a TDMA/CSMA protocol called WiseMAC. This protocol offers a completely different approach compared to the rest of the solutions presented so far. In the traditional schemes, the receiver must wake up when the transmitter is about to send the information packet. However, in WiseMAC, the transmitter waits for the receiver to wake up and be ready for receiving the data packet.

WiseMAC is quite similar to the *Spatial TDMA and CSMA with Preamble Sampling* protocol presented in [13]. In this approach, two communication channels are used: one TDMA for data transmission and other CSMA for control messages. WiseMAC follows a similar scheme but with a unique communication channel, instead of two.

In WiseMAC, each node has its own *active/sleep* scheme and these schemes do not have to be centralized. Hence, nodes do not need to be synchronized with the rest of the nodes but have to advertise their own scheme to the rest. This way, each node knows about the scheme of the surrounding neighbors and can wake up the moment the intended receiver is waiting for data to arrive. It is considered to be TDMA because each node has a instant to stay listening and therefore to receive data but this moment is not necessarily exclusive to that node.

WiseMAC's mechanism is based on the *preamble sampling* technique. Every T_w, a sensor node samples the medium for activity. If there is no activity, the node goes to sleep and waits T_w until the next medium scan. However, if the node finds the medium busy, it will keep on listening until the data packet arrived or the medium goes idle again.

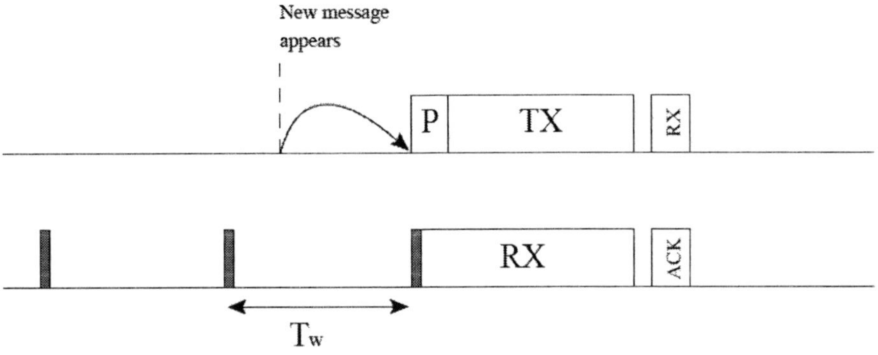

Figure 6. WiseMAC

In order to coincide with the receiver's medium sampling, the transmitter must know the moment the receiver starts listening. The sender wakes up and sends the preamble. The receiver will detect that emission and will determine that the medium is busy. Consequently, it will wait awake for the packet to arrive. On the other side, the transmitter will have started the transmission of the message. At the end of the packet transmission, the receiver sends an updated information about its schedule within the ACK message. This way, WiseMAC's nodes exchange synchronization information. This behavior is depicted in figure xx.

This protocol was originally designed for wireless sensor networks. However, as it has no contention mechanism between the senders, collisions appear very often. Two nodes may sense the absence of a carrier at the same time and afterwards try try to transmit their frames simultaneously. Moreover, if all nodes of the network are not within the coverage range of the rest, the *hidden terminal* problem may also appear. It was only simulated and its implementation in traditional WSN was found very resource consuming.

Finally, it was deployed in an infrastructure wireless sensor networks, formed by a network of access points (AP) and each AP gives service to a group of nodes. APs are continuously powered and do not have to work in a power aware fashion.

In infrastructure networks data flows in two directions: from the APs to the sensor nodes or *vice versa*. Sensor nodes must save their energy as much as possible and therefore, WiseMAC protocol must reduce the idle listening of them in order to receive the packets from the APs. However, since the APs are continuously powered, they can be listening to the medium all the time and sensor nodes can send their data packets as soon as they are formed. WiseMAC only offers a solution for downlink traffic in which the AP must wait until the moment the destination of its data packet is awake.

WiseMAC offers some notable advantages, such as a highly reduced listen period (without *idle listening* and *overhearing*), thanks to the preamble sampling technique, and collision avoidance as the access points are the unique senders of information. It does not require global synchronization as every node is in the cover range of the access point and therefore there is no need of communication among nodes. Moreover, only the access points need to know the scheme of every node in order to send them data packets.

However, it has some shortcomings:

- In the same way as [13], WiseMAC suffers from the *hidden terminal* problem. A node checks if the medium is busy during the preamble and if its free, the node starts sending the packet. This transmission can collide with another one that could be in progress but the sender can not hear.

- It has only been tested in a one hop network.
- It does not offer solution for multihop transmission and therefore, optimization is only achieve when nodes receive information, not when transmitting.

3.8. MS-MAC

MS-MAC [14] was born taking into account that most sensor network MAC protocols had assumed stationary environments of performance for the nodes. Therefore, proposed approaches could not be applied directly to networks with some moving nodes. The authors present a mobility-aware MAC protocol that could work well in both stationary and mobile scenarios. MS-MAC is based upon S-MAC protocol and has as its main objective to adapt this protocol to a network with certain mobility.

When a node is moving within the network, the topology is changing as long as it travels from one point to another. Every node can discover a mobile node within its neighborhood comparing the received signal strength of periodical SYNC messages from the neighbors. If there is a change in the signal level from a neighbor, the node assumes that the neighbor (or itself) are moving. Moreover, analyzing the degree of change in that signal level, mobile's speed value can be predicted. Every node includes this information of estimated speed in its own SYNC message. This way, neighbors of the moving node can create an active zone around it. Active zones are formed two hops away around the mobile node along the trajectory of its movement.

Hence, MS-MAC divides the network into active zones based on mobile speed. Inside active zones, nodes fixes a lower repetition period than the standard in order to allow quicker topology access to the moving nodes. This is achieved modifying the general repetition period of the topology management of S-MAC. This modification is shown in figure 7. S-MAC presents a fixed repetition period of 2 minutes whereas MS-MAC varies its value depending on the requirements of the network at each moment. If some nodes in the network are moving at a low speed, the repetition period is reduced to one minute to refresh the network topology; if the nodes speeds are medium, then the repetition period is settled to 30 seconds and when the velocities are high enough that 30 seconds of refreshment are not enough to update the topology correctly, the nodes keep awake all the time. In this last approximation, the answer of the network to topology changes are immediate but a power-constrained node can not listen to the medium continuously as could get its batteries depleted very quickly.

As it can be seen, MS-MAC offers the possibility of adapting S-MAC to a moving environment thanks to a simple modification. However, stationary protocols do not work well in moving environments and are not efficient when trying to adapt them to topologies with mobility. Moreover, the consumption that this adaptation implies is prohibitive for the kind of sensors we are talking about. Therefore, MAC protocols for sensor networks with moving nodes have to be specifically designed for those environments.

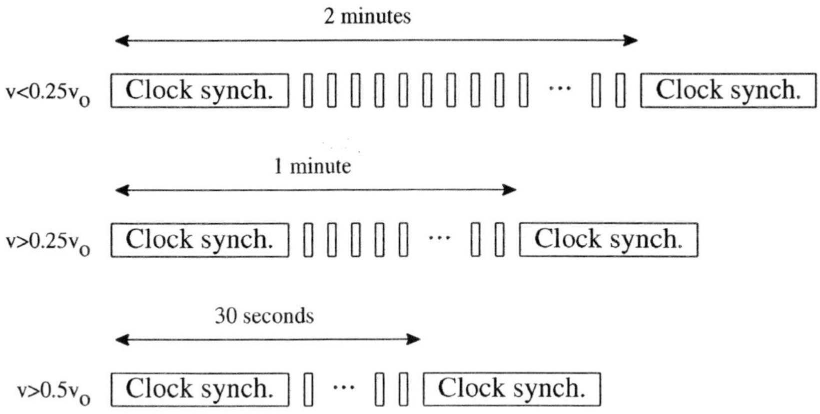

Figure 7. Frame structure and timing for MS-MAC.

3.9. DMAC

DMAC [15] protocol offers a low latency routing mechanism for gathered data in a tree-topology network. It is designed for a *convergecast* performance collecting data from the furthest nodes of the branches of the network to the sink through intermediate routing nodes.

Authors identify the *data forwarding interruption problem* (DFI) that exists in most presented solutions due to coverage limits. As the number of hops between a node and the sink may be large, nodes cannot listen to all messages exchanged throughout this path. It is stated that the former protocols, S-MAC and T-MAC have tried to solve it with additional mechanisms, such *adaptive listening* and *future-request-to-send* (FRTS). DMAC tries to reduce packet collisions and channel contention with a TDMA access scheme and evades forwarding delay due to unsynchronized transmission/listening periods.

The intermediate nodes have a transmission slot immediately after the receiving slot and as long as the node is higher in the tree-topology, its wake-up scheme starts later in order to coincide a transmission slot with a reception slot. In reception, a node listens a packet and answers with an ACK message. In transmission, a node sends a packet and waits for an ACK message from the receiver.

Instead of RTS/CTS control messages, DMAC divides time in a modified slotted ALOHA [16] fashion in which time slots are assigned depending on the position each node occupies inside the tree-topology. This mechanism is presented in figure 8. All nodes in the network wake up in a reaction chain, from the furthest ones to the nearest ones. This way, information can be routed from the sources to the sink in a low latency fashion.

Furthermore, all the nodes belonging to the same hop number, share the same *active/sleep* scheme. Consequently, all of them wake up and start transmission at the same time. Taking into account that each time slot is one-packet-long, all of them must fight for the transmission slot. The probability of collisions increases and the packets might get lost. As only one packet can be sent in each time slot, there is no possibility of retries within the same slot and the retransmissions have to wait until next cycle.

Moreover, all nodes belonging to the same hop number might exchange messages directly, as they could be deployed far away ones from the others. This situation becomes worse as the hop number increases. This leads to the *hidden terminal* problem, similar to the CSMA approaches.

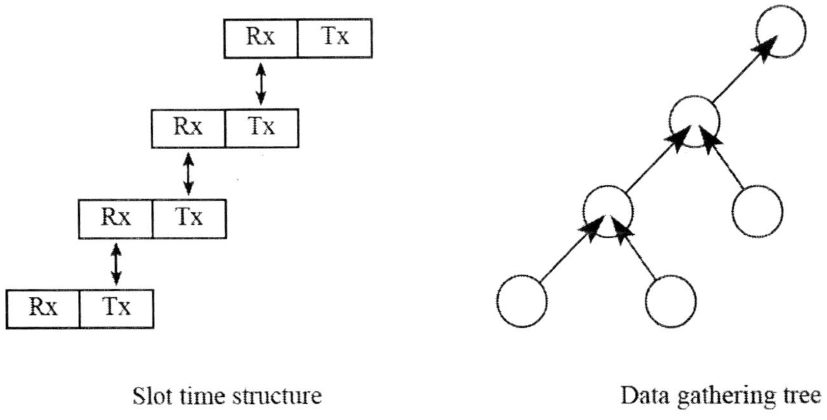

Figure 8. Frame structure and timing for DMAC.

In order to solve this problem, DMAC presents *More-To-Send Packet* (MTS). MTS is an explicit control packet that tries manage the retransmissions of the packets. There are two types of packets: *request MTS* and *clear MTS*. The first one is sent from a child to its parent when (1) the channel is busy and it can not transmit its own data, assuming it has lost the channel or (2) when it has already received another *request MTS* from its children and it has to propagate it upwards. A *request MTS* is sent only once before a *clear MTS* packet is sent. MTS is similar to FRTS of T-MAC, as both serve to ask for future packet transmissions. However, MTS offers the possibility of forwarding the *notification* through the whole path to the sink and T-MAC's solution just let forward the packet one hop further. In [17] a slight variation of this scheme is proposed and, instead o requesting a new slot when the emitter finds that it still has more information to transmit, nodes allocate all the slots they will require at the beginning of the transaction. Thus, no more energy is required to allocate other slots. This is particularly interesting in WSNs if we take into account the traffic patterns in these networks: when a new event appears, there are several nodes that will want to report this event to the sink. Thus, the traffic is generated in bursts and nodes will know exactly how much information they want to transmit *before* they do. Allocating a certain number of nodes will avoid turning the radio on and off, making the communication faster and with fewer radio transitions.

On the other hand, *clear MTS* is sent from a child to its parent when its buffer is empty and there is nothing pending. A node keeps awake almost continuously after receiving a request MTS and returns to its basic duty cycle after sending a *clear MTS* to its parent, that implies there is nothing else to send.

Although DMAC was born for a tree-topology conception, the authors state that just a linear topology solution is been proven so far and the final solution for tree-structures would need of a more complicated management, a mechanism called *data prediction* and the presented *More-To-Send Packet* (MTS), which has not been proven yet.

On the other hand, DMAC does not implement control message in order to manage the high collision probability that appears when the number of nodes that belong to the same hop number increases. It does not even implement a back-off mechanism to contend for the access to the shared medium and it only relays its recovering system in *MTS* messages.

Moreover, it does not take into account the *hidden terminal* problem and finally, it needs an additional mechanism for synchronization among the nodes, called *Reference Broadcast Synchronization* (RBS) [18].

3.10. E-MAC

E-MAC (Energy efficient sensor networks Medium Access Control) [19] is a Time Division Multiple Access (TDMA) based, self-organizing network schema for wireless sensors. The main goal of E-MAC is to minimize the energy consumption in a sensor network preventing the preamble overhead and keeping the number of transceiver state switches to a minimum.

Like other TDMA-based protocols, the time is divided in frames and these are divided again in time slots. A node uses a time slot to transmit its data, but unlike traditional TDMA-based protocols there is no a central manager to assign the time slots to the nodes. Each active node picks its time slot with local network knowledge only. These characteristics make the E-MAC an adaptive network because no base stations are needed and with no energy wasting dealing with transmission collisions.

Searching a more efficient energy use, nodes in E-MAC network have three operation modes: active mode, passive mode and dormant mode. Active nodes are the main elements in the network communication; each of them controls a time slot and can communicate with other active nodes with no collisions, also it accepts data from passive nodes. A passive node keeps track of one active node to receive network messages; it also asks the active node, in a non-collision free communication, for permission to use part of the time slot for data transfer. Nodes in dormant mode are operating in a low power state for an agreed amount of time. This mode is used when node is in a critical situation of energy run out.

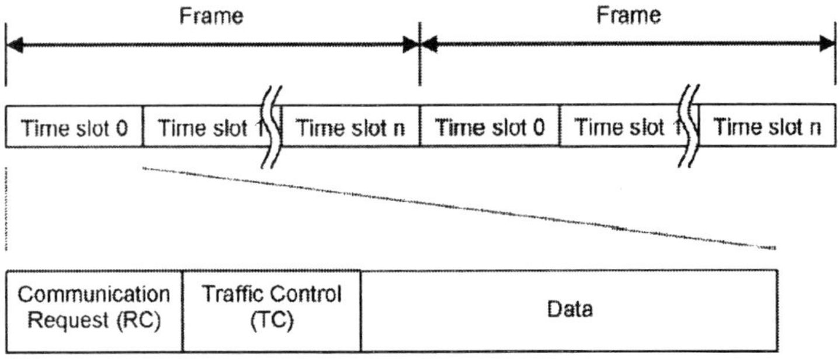

Figure 9. Temporal distribution of frames in E-MAC.

The time slot has three sections, the Communication Request Section (CR), the Traffic Control Section (TC) and the Data Section (see figure 9). The CR section is used by passive nodes to request the use of the data section to transmit a data. Then, the associated active node uses the TC section to identify itself and to acknowledge the requesting passive node. As several passive nodes can be associated to a particular active node, collision of requests can occur. In this case, the active node notifies it in the TC section. But if the data rate is low, and few passive nodes are associated to each active node, the collision probability is low. The active nodes listening the TC sections of neighboring nodes, know the local topology of the network, and they can use the TC section to assist message routing. To

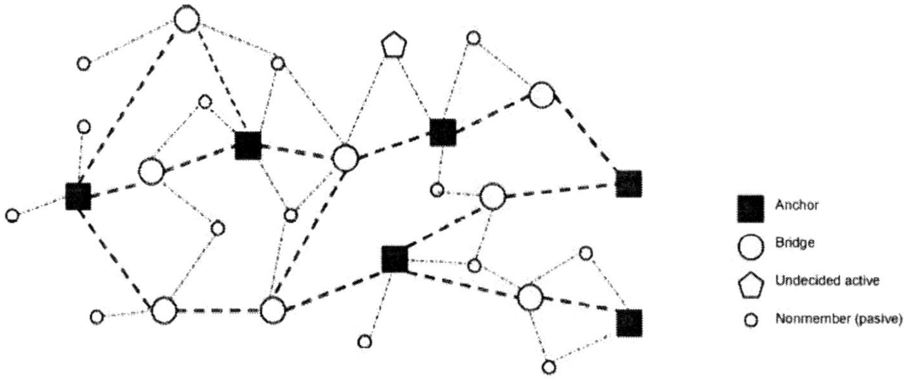

Figure 10. Example of network topology in E-MAC.

allow the spatial reuse of time slots an small table in included in the TC that includes the information of what time slots are been used in the neighborhood. Also, the TC section is used for network omnicast messages. The Data section is used for data transfer of the active node that owns the time slot or the passive node authorized by this active node.

To achieve a self-organizing network, it is necessary a carefully designed algorithm. This algorithm is based in the information contained in the TC section of time slots and it is followed by every node of the network. The network is organized in a mesh-like backbone of active nodes and several passive nodes connected to these active nodes. Nodes can play four roles in the network: anchor, bridge, undecided-active or nonmember (passive) (see figure 10). When a new node is introduced in the network it takes an undecided-active role and tracks neighbor node communication to decide the role that it can take in the network. Since active nodes wastes more energy, a role rotation algorithm is supported in the EMAC schema.

Although E-MAC represents an advance over other MAC protocols and it is very attractive to be used in cases with few device deployments [20], it has several disadvantages:

- Since all nodes must listen during CR and TC periods of time slots, there is a significant energy waste.
- Network setup may take considerable time for large node deployment networks.
- It is difficult to adapt the network to different traffic conditions varying slot assignment.
- There are an overhead of TC section since many network maintenance
- Network organizing and maintenance activities generate an overhead in the TC section and it causes a considerable energy waste.

3.11. B-MAC

B-MAC [21] is an algorithm designed at Berkeley for managing the access to the medium in wireless sensor networks. Although this protocol does not introduce any new and original concept not covered by its predecessors, it is a good example of MAC protocol because it summarizes the innovations studied so far and it eliminates some superfluous mechanisms which had been inherited from the Ad-Hoc network world, thus reducing the amount of resources needed for its implementation. It also removes from the MAC algorithm all knowledge about network organization, that is related to higher communication layers.

This protocol uses a CSMA approach to govern the access the wireless channel. Thus, nodes are not organized in a hierarchy and they transmit when the channel is free and they have information to forward. To do so, it implements a service for deciding if the channel is free: CCA (Clear Channel Assessment). This service samples the value of the RSSI (Received Signal Strength Indicator) indicator given by the radio module. This analog value is directly related to the amount of RF energy available at the radio input and it can be used for detecting if some other node is emitting information: the RSSI samples are stored in a FIFO and a median filter is applied to eliminate the appearance of spurious peaks in incoming RF power. On the other hand, a moving average block with exponential decay is used for determining the position of the noise floor. Thus, comparing the actual value of the incoming power and the noise floor, we can decide if some other node is using the channel.

If another node is using the channel when the CCA procedure is triggered, then B-MAC starts a backoff process, where the node simply puts the MAC interface in a sleep state for a certain (random) lapse of time. Thus, the protocol avoids the situation in which several nodes collide consecutively after they try to access the channel simultaneously.

B-MAC does not follow the typical synchronization scheme present in other CSMA protocols. Instead of waking up all nodes in a certain area at the same time, each node follows its own schedule, without sharing it with its neighbors. Using the LPL (Low Power Listening) technique, already introduced in WiseNET, each node uses the CCA results to detect the presence of other nodes using the channel in a short time. If the channel is being used, the node remains active until activity disappears and then, it takes control of the channel. Otherwise, if no activity is detected, the node will return to the sleep state. Of course, this procedure does not guarantee that all nodes (not even the destination node) are awake when a frame is emitted, but it greatly simplifies the algorithm. Anyway, the B-MAC protocol admits that higher level layers take charge of the synchronization among nodes and implement a schedule exchange protocol. In the design of this kind of unsynchronized protocols, there are some temporal restrictions that have to be met in order not to miss a possible talking neighbor. For instance, the preamble must be long enough to fill the lapse of time between two consecutive LPL procedures, so special care must be taken when deciding the lengths of the packets or the duty cycle.

Another feature present in previous MAC protocols that has been eliminated in B-MAC is the RTS-CTS handshake algorithm. B-MAC, as a MAC protocol, does not provide a mechanism for avoiding the hidden terminal problem and it leaves this functionality to upper communication layers. However, B-MAC does provide a service that previous protocols used to neglect: acknowledging incoming packets. Whenever a packet is successfully delivered to its destination, the latter emits an ACK packet to declare that it has received the information. This prevents higher level procedures from solving the problem of a single packet loss, letting them decide what to do when a major problem (e.g. a fallen link or a time missconfiguration) appears.

After making all these simplifications, the resulting protocol requires less hardware resources than previous protocols, such as S-MAC. However, this difference is only remarkable if the eliminated procedures are not included in the upper layers. The most interesting point in the design is the use of LPL to synchronize the nodes. This technique hides the process of switching the radio interface on and off and the node behaves like in the old Ad-Hoc protocols, supposing that the RF module is idle-listening, but consuming less energy. According to the measurements described in [21], the LPL process takes about 350 μs, consuming about 6 mA. If the time period between LPL checks is about 100 ms where the radio interface is completely off (about 30 μA), the result is a conventional radio interface with a idle listening current consumption of about 50 μA. If we take into account the amount of time needed to power the radio on and off, the final result will be a bit worse, but it will not be much worse. This perspective is interesting because it shows that schedule exchange is not required to successfully enable the wireless link. Thus, the duty cycle of the nodes are adapted to the real throughput needs so that nodes start transmitting and using the channel when a relevant event has appeared, keeping silent the rest of the time. However, in this scheme there are different parameters that must be adjusted to keep the MAC algorithm in its optimum position: the time between LPL procedures, the length of the preamble, the fragment size, etc. On the other hand, more or less unpredictable environmental variables, such as the number of neighbors or the network traffic, do have a remarkable impact on the protocol behavior, so these parameters should be continuously adapted to suit the variable conditions in the network.

3.12. Z-MAC

Z-MAC (Zebra-MAC) algorithm [22] tries to combine the strengths of both flavors for MAC protocol design: TDMA and CSMA. This protocol allows nodes to behave as if they were in a CSMA environment when the contention of the network is low and to follow a TDMA scheme when it is high. Thus, it uses the best part of both worlds, providing a simple, robust and almost collision-free CSMA environment for low traffic situations and an ordered, totally collision-free TDMA environment when the nodes in a certain area have large amounts of information to transmit.

The algorithm is divided into two different phases: the setup phase, where the nodes detect who their neighbors are performs the initialization, and the working phase, where information exchange is performed. In the initialization phase, each node must perform four different tasks: first, all nodes exchange their addresses so that at the end each sensor knows all its neighbors at a two hop distance. Then, using this information, the nodes run a slot assignment algorithm, DRAND [23], which ensures that all the nodes within two hops do not share the same slot and, as a final step, all nodes synchronize, so that they have the same time reference.

Once each node is assigned a unique slot that is not shared with the nodes in the two-hop area, then the working phase begins. In this phase, the nodes behave differently depending on the contention level they suffer and the current slot. If a node is the owner of the current slot it waits a fixed lapse of time T_o, then performs a CCA operation and, if the channel is clear, it begins the transmission. However, if it is not the owner of the current slot and it has a low contention level or if it does have a high contention level and the current slot is not owned by any two-hop neighbor, the node will wait for T_o and then it will perform a second random backoff operation within the $[T_o, T_n]$ window. Then it will check if the channel is clear and if it is, it will transmit the data. This scheme implements a two-level hierarchy while accessing the medium: the node with the highest priority is the owner of the slot and if it does not have information to transmit, the rest of the nodes may gain access to the medium, provided that they do not conflict with other nodes that are within the two hop distance.

For low network traffic load, most nodes will have access to the network at any time, because the owner of the current slot will not usually information to send. Of course, using another node's slot to send one's information will also make that nodes have lesser information to send in their own slots. As a result, the concept of slot gets more diffuse because almost any node is allowed to transmit at any time. However, as traffic load grows, nodes have more information to send and they are more likely to use their own slots to transmit their information. Thus, the scheme turns into a full TDMA protocol, where each node only transmit in certain slots.

The most obvious characteristic of Z-MAC is the fact that time structure depends on network traffic, thus adjusting the behavior of the network to the actual performance needs: when contention is low, nodes behave almost as if they were in a CSMA scheme, reducing latency. Of course, if there are few frames to send, the probability of collision is much lower. On the other hand, with heavier network loads, collisions are much more likely to appear in CSMA schemes. Therefore, Z-MAC's slotted time structure becomes more important until it becomes a traditional TDMA scheme, where collisions are completely avoided. However, there are some drawbacks that must be pointed out when analyzing this protocol:

- The main problem of TDMA schemes, i.e. assigning each node a slot to transmit, is still present in this algorithm. The authors propose DRAND as a distributed method for performing this task, but it still requires some knowledge about the network (e.g. the size of the two-hop neighborhood) that may not be available. On the other hand, a change in network topology will trigger a new slot assignment, at least in certain areas of the network.

- For low contention scenarios, although the medium access procedures are very similar to CSMA schemes, time is still slotted, as new transmissions must wait at least until the first backoff period (T_o) elapses after the slot change. This prevents nodes from sending long frames whose transmission would end in the next slot (which in fact is not a problem for WSNs, as sensor data usually form short frames) and from using time as freely as in traditional CSMA approaches. This effect affects latency, because it forces nodes to wait until the next slot if they have verified that the channel is free too late.

3.13. MERLIN

MERLIN (*Mac Energy efficient, Routing and Location INtegrated*) [24] offers a solution for latency-aware sensor network applications. This approach is not a simple MAC protocol but also integrates data routing and localization capabilities.

MERLIN defines two kinds of elements within the network: nodes and gateways. Sensor nodes are deployed in the working area and are battery powered whereas the gateways are powerful nodes which are the point of entrance for users requests to the network. Due to their high performance requirements, the latter ones need continuous power supply.

The network is divided into as many subnets as gateways are deployed. Each subnet must have only one gateway and each sensor node can only belong to one subnet. When a node joins the network, it will listen to any gateway surrounding and will determine its *time-zone* is 1 (*time-zone* is the same concept that other protocols call *hop*). If at any time, another gateway appears, the node will listen to its announcement. Consequently, the node will compare the distance to the two gateways and change its time-zone and schedule table if the new gateway is closer. Once a node has decided which gateway is its parent, it will store the number of *time-zone* it is in. This *time-zone* indicates the moment when it has to listen and transmit.

Hence, MERLIN performance is based upon these simple statements:

- Nodes knows their own *time-zone* (hop number) and send it to the rest of the nodes in every transmission. Moreover, this value is updated during the network performance.
- Node hop count is calculated as: min(hop count of neighbors) + 1.
- Gateways hop count is zero.
- Information can flow in two directions: from the gateway or toward it.
- "From the gateway": gateway broadcast the data petition from the user to all the nodes. Nodes forward all the packets received from a node with a lower hop number.
- "Toward the gateway": nodes will answer with the requested information to the gateways through several hops. Nodes forward all the packets received from a node with a higher hop number.

MERLIN MAC mechanism is defined as a combination between TDMA and CSMA approaches. Time is divided into time slots in a TDMA fashion and each node has a specific slot in which transmit. Each time slot is composed by a *contention period* (CP), a period for packet transmission and finally a *collision report period* (CR).

However, similar to DMAC, all the nodes belonging to the same *time-zone* shares the same *active/sleep* scheme. Therefore, they will have to contend for the access to the communication medium in a CSMA fashion. This is performed during the *contention period* and the node back-offs randomly, listens to the medium and if nothing is heard, the node starts the transmission. If a node listens to a neighbor that has already started transmitting, it will wait until the next assigned slot to contend again for the medium. Therefore, its functionality is based upon a precise synchronization mechanism.

MERLIN's synchronization mechanism consists on transmitting the time of transmission in each packet. This way, each node can estimate the start of every transmission slot according to the synchronization information received. As gateways are supposed to provide the best time, all the nodes get synchronized with the information received from nodes with lower hop count and therefore, closer to the gateway.

However, MERLIN, in the same way as DMAC, does not implement RTS/CTS mechanism for channel accessing. Carrier sensing in the contend period is supposed to be enough for avoiding collision. It is possible and even probable a collision between two transmission that starts in very same moment, as all the nodes in the same time zone (the same hop number) have the same time slot for transmitting purposes. Moreover, the *hidden terminal* problem can appear as each node that belongs to a hop number may not hear the rest of the nodes of that hop. Therefore it may think that the medium free of transmission when in the same moment other node is starting sending its data packet.

In order to solve the collision problem *a posteriori*, MERLIN offers a collision report system based on the collaboration among nodes. A sender can not know if a collision has occurred due to *hidden terminal* problem or *overemitting*. Every node that listens to a collision sends a CR at the end of the following slots to notify that a collision has taken place. Every node that has sent a packet, has to listen to the CR period of all the following slots in order to know if it packet has suffered a collision.

MERLIN also implements some kind of data gathering as, if a node has more than one packet to forward, it will send only one packet with the information, reducing overhead and power consumption. Nevertheless, the mechanism implemented by MERLIN for forwarding information to the gateway is not optimized as the node are disjointed and the same information can arrive at the gateway through different paths. This provokes a great overhearing and unnecessary power consumption of the nodes involved in those useless forwarding tasks.

As stated before, MERLIN is very similar in concept to DMAC in medium access and data gathering tasks, but also offers downwards communication (from the sink to the nodes in the network) and location facilities.

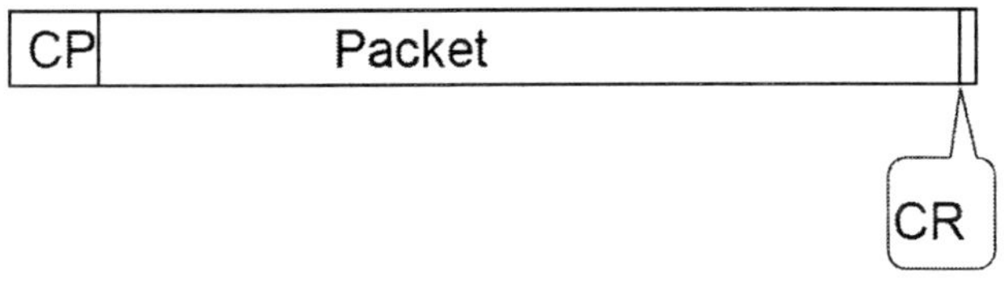

Figure 11. Frame structure for MERLIN protocol.

3.14. SCP-MAC

SCPMAC presents an approach similar to B-MAC and WiseMAC. In the same way as WiseMAC, a transmitter must know the moment when the receiver will be awake, listening to the medium.

SCPMAC belongs to a kind of protocols that are called *Low Power Listening* (LPL) and offer very low power waste due to *idle listening*. Instead of waking up and keeping on listening to the medium during a specified lapse of time, they implement a *medium polling* mechanism in order to detect if the medium is busy. Moreover, this approach is called *scheduled channel polling* (SCP), as it offers a combination of scheduled MAC and LPL.

In a different way to LPL and similar to WiseMAC, SCPMAC synchronizes all the nodes in the network in order to *poll* in the correct moments. This global synchronization allows shorter preambles and therefore reduces overhead. However, SCPMAC needs a very precise synchronization mechanism as its performance is based upon the exactitude of

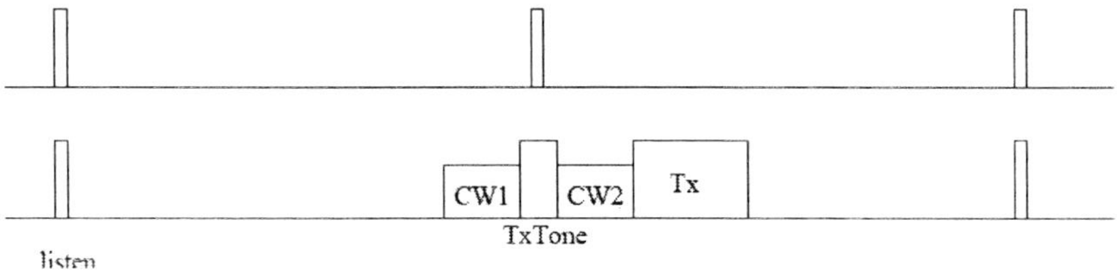

Figure 12. Frame structure for SCP-MAC protocol.

the moment they are awake/transmitting. Nodes have to know in advance the *active/sleep* schedule of each surrounding node in order to send data packets to them.

Figure 12 presents the performance scheme of SCPMAC. When a node has a data packet to send, it waits until the moment the receiver will be awake and listening. Previous to that moment, it starts a carrier sense period within the first contention window (CW1). If it finds the medium free it will send the wake up tone to the receiver. If the medium is busy, it will go to sleep and try it in the next polling moment. Once the wake up tone has been sent, the sender enters in the second contention window (CW2). If the node still detects the medium idle, it will start the data transmission. The receiver will listen to the wake up tone and will keep on waiting for the data transmission.

Instead of a simple contention window, SCPMAC uses two contention window in order to achieve a lower collision probability. Only the nodes that found the medium idle in the first contention window will try their luck in the second. Consequently, the number of nodes a node has to contend with decreases.

The probability of collision is inversely proportional to the contention window size. If *m* is the number of slots in a contention window, the probability of collision is $1/m$. If we divide the contention window in two, the probability of collision during both windows will be proportional to $4/m^2$. If $m > 4$, the performance of two separate contention windows is better than a unique one.

A synchronization mechanism is vital for SCPMAC's correct performance. Therefore the clock drifts have to be closely controlled. SCPMAC defined its synchronization period as T_{sync}, that can be configured and clock drift rate as r_{clk}. The maximum clock difference between the sender and the receiver is presented in equation 2.

$$t_{diff} = 2 \cdot T_{sync} \cdot r_{clk} \tag{2}$$

The relative time difference between two nodes can be in two directions, and therefore, the guard time needs to be twice t_{diff}. Moreover, if a node has N neighbors, each one of them will send a synchronization packet at the period of T_{sync}. All nodes in the network hear the SYNC packet and get re-synchronized. This leads to a clock drift reduction by $(n+1)$ times, due to the $(n+1)$ nodes in the network. Thus, the guard interval becomes

$$t_{guard} = \frac{2 \cdot t_{diff}}{n+1} = \frac{4 \cdot T_{sync} \cdot r_{clk}}{n+1} \tag{3}$$

The duration of the wake-up tone must be equal to the guard time plus a small time necessary to detect the tone (t_{mtone}).

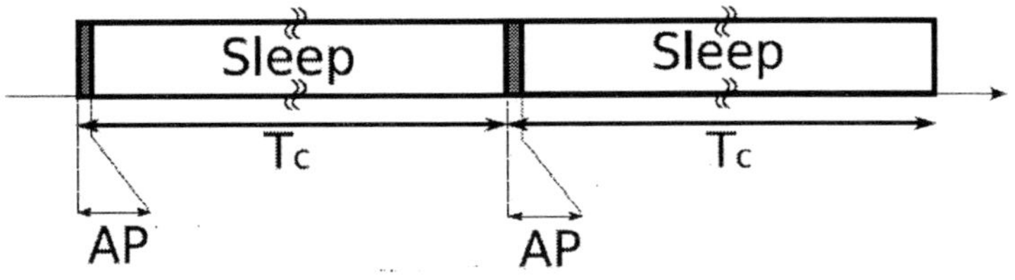

Figure 13. Working cycle for the LL-MAC protocol.

$$t_{tone} = \frac{4 \cdot T_{sync} \cdot r_{clk}}{n+1} + t_{mtone} \quad (4)$$

Taking into account the equation 4, SCPMAC must trade between the synchronization period and the power consumption. If T_{sync} increases, t_{guard} also increases and *idle listening* or *overhearing* may appear. However, a high value of T_{sync} implies a reduction of overhead of control messages as the synchronization is not updated very often.

Synchronization overhead is in part reduced thanks to the mechanism of *piggybacking*, that SCPMAC implements: the SYNC packets can be sent inside the data packets instead of generating additional control traffic.

This approach offers a good solution for low power consumption but trades latency for energy consumption. Moreover, it requires a highly synchronized execution for good performance. Its low duty cycle results are based upon the preciseness of the synchronization mechanism. If the node keeps on listening longer than necessary, its performance will be similar to the rest of the scheduled MAC protocols.

3.15. LL-MAC

Chapter authors would like to present LLMAC, a protocol aimed the specific application in WSN of data collecting from all the nodes to the sink through multi-hop paths (*convergecast*) with very low latency. It offers a brand new performance planning for nodes to transmit information divided into time slots, apart from topology management, energy consumption reduction and global network latency improvements.

Similar to most of the presented protocols, the working cycle T_c is divided into two periods: Active Period (*AP*) and Sleep Period (*SP*). The *AP* is also divided into two intervals: *CONTROL* interval and *DATA* interval. In the *CONTROL* interval topology information is shared, and in the *DATA* interval collected information is routed to the destination. A general overview of a working cycle of LL-MAC protocol is shown in figure 13.

On one hand, *CONTROL* interval in LL-MAC is quite different from most of the presented protocols, as it does not share the medium access, although all the nodes are listening all the time. It is formed by three sub-intervals, named as follows:

1. *Node Advertisements* (NA): Each node publishes its own *advertisement* in this sub-interval.

2. *Child Adoption Request* (CAR): This sub-interval will only be used for nodes that want to join the network because they are new to it or due to a parent loss or change.

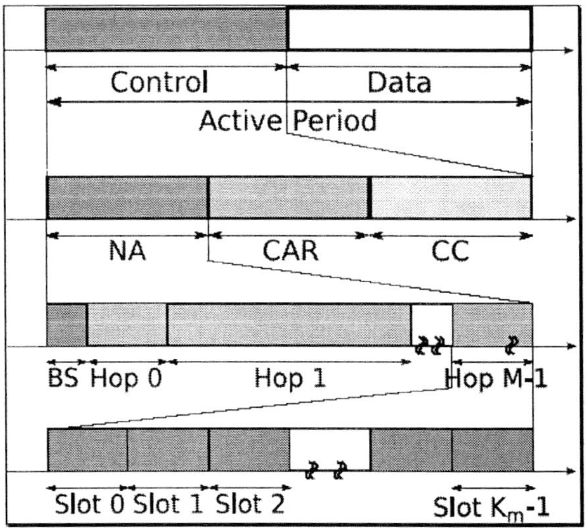

Figure 14. Control procedure in LL-MAC.

3. *Child Confirm* (CC): This sub-interval is used by the parents to confirm to the newly *adopted* children.

Advertisement messages include all the information needed for the nodes to get synchronized and choose a parent. The second and third sub-intervals (*CAR* and *CC*) only have traffic when any node decides to change its parent relationship (due to topology changes or node movement) or new nodes appear in the network. Although each node must wait for its time slot to publish its own information, every node is listening to every transmission in the air. Figure 14 shows the *CONTROL* interval.

Each one of these sub-intervals (*NA*, *CAR* and *CC*) is divided into M non-uniform divisions. As there are different amount of nodes in each hop, there are different amount of time slots reserved in each hop division. K_m represents the number of time slots included in hop m. Equation 5 offers the distribution of the $N+1$ time slots inside each sub-interval.

$$\sum_{m=0}^{M}\prod_{i=0}^{m} P_i = \sum_{m=0}^{M} K_m = N+1 \tag{5}$$

On the other hand, DATA interval is divided into M divisions (one for each hop number), and each of them divided into N time slots subdivisions, each one for each node's transmission.

The basic behavior of every node is quite simple: each node has to publish its *advertisements* to the rest of the nodes, receive information from its children and retransmit it to its parent. When a node wakes up for the first time, it is not synchronized and it has no real parent. As long as a node does not have a functional parent, it cannot send neither *advertisements* nor data because it does not know when nor who send them to. Moreover, *orphan* motes keep awake and listening all the time, waiting for an *advertisement*.

However, Base Station behavior is a bit different: it wakes up with an operative parent (the sink), so it does not have to wait for any *advertisement*. As expected, the Base Station is the node that establishes the working cycle (T_c) and the active/sleep scheme from the beginning.

As indicated before, when a node wakes up for the first time it has no real parent. In order to update its parent-relationships, every time the *CONTROL* interval is executed, all the nodes analyze every *advertisement* received. All packets are analyzed in order to determine if the current parent is the best out of the total of the available parents. Once the node decides which one is the most suitable parent (the best link quality), it will ask for its *child adoption* in the *CAR* sub-interval. Following the process explained before, the parent-to-be decides whether it can *adopt* that child or not, and if so, it will answer a *CC* message. When a parent has its maximum number of children allocated, he will not send an answer.

In a *CC* message, the parent informs to its new child the number of the data slot in which it has to transmit. With this information, the child can calculate the exact moment when it has to transmit the data to its parent and its own hop number. If a child does not receive a *CC* message in the same *CONTROL* interval that it had asked for, that child keeps the current parent and forgets about the parent-to-be.

In every working cycle, each node sends its own data to its parent in the time slot number that had been assigned before. A parent change (and therefore, a topology change) can be completely done within a unique *CONTROL* interval. Thus, in the *DATA* interval of the same working cycle, the node can send its data to its new parent, being the new topology functional in the moment that it has been established.

Every child processes *advertisements* received from its parent in order to extract the synchronization information. Once it has updated its own variables, it will include that information on its own *advertisements* and transmit it. Thanks to this, every information a node sends is updated with the information of its parent. This is possible due to the time slot order: the first node to talk is the base station and its children hear it (*NA* sub-interval, BS division). Then, in the next division (hop 0 division), base station's children talk with its own information, updated with the information received from the base station and so on, as shown in figure 14. The synchronized time stamp allows every node to adjust its own clock to its parents. With this synchronization method, all the nodes in the network wake up simultaneously at the beginning of a working cycle and listen to each other at the right time.

It must be taken into account that the synchronization, as well as topology management, is performed with the information attached in *advertisements* so it can only be done every *CONTROL* interval. C is the ratio that indicates how many times there is a *DATA* interval for one *CONTROL* interval. Figure 15 presents an example of this modified performance for $C = 2$. It can be seen that in the second T_c, when it was supposed to be a *CONTROL* interval, mote keeps on sleeping until the *DATA* interval arrives.

If C is high, there is a *CONTROL* interval for many *DATA* intervals and this means that topology information might not be updated as often as necessary. For this reason, C must be carefully chosen for each network's requirements of topology dynamics.

It must be stressed that the *DATA* interval is not a simple TDMA period of time. Time slotting is been carefully analyzed in order to determine the way to obtain the minimum latency in a working cycle. As previously mentioned, *DATA* interval is divided into M divisions, and each one of them is, in turn, divided into N time slots sub-divisions. Each node

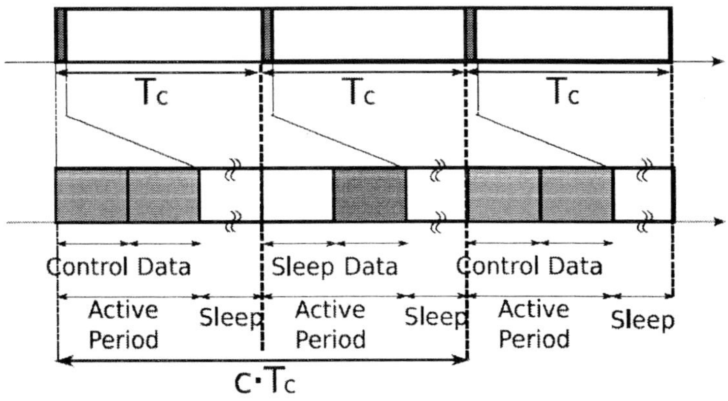

Figure 15. Adjustment example for resynchronization.

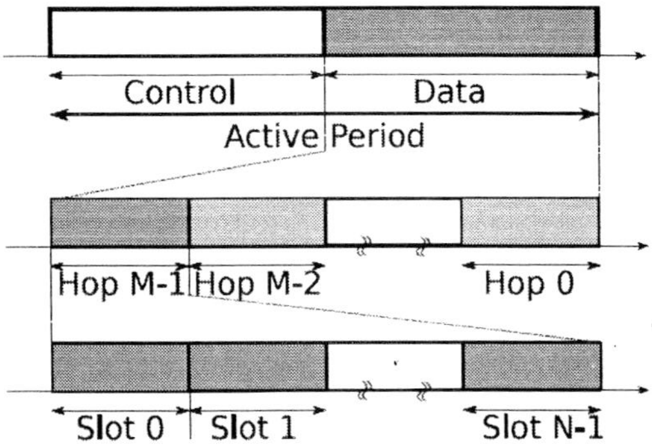

Figure 16. The DATA interval.

will talk to its parent in the time slot sub-division assigned inside division corresponding to the hop number it is in. *DATA* interval is presented in figure 16.

The information from the nodes is transmitted hop-by-hop from the furthest hop to the sink. The first nodes to talk belong to the highest hop number (in the division $M-1$), then the following hop $M-2$ nodes and so on. Every parent listens to all its children, each in a different time slot and them store that information. If a parent is in the hop 1, it will listen to its children in the hop 2 in the time slots preassigned. In the next hop division (hop 1), the parent will send its own data along with the stored data from its children in the previous hop division to its parent, and so on.

LL-MAC is a medium access control protocol specifically designed for wireless sensor networks. Apart from energy efficiency, low end-to-end latency and efficient topology management are the main goals of the protocol design. Together with control messages (RTS/CTS) avoidance, LL-MAC outperforms other sensor networks MAC protocols in latency reduction, topology management and energy consumption, thanks to a *meticulous* time slot structure.

Moreover, LL-MAC evades *hidden terminal* problem and improves channel utilization, becoming a resilient protocol to packet collision and network dynamics, apart from reducing dramatically power consumption and global latency.

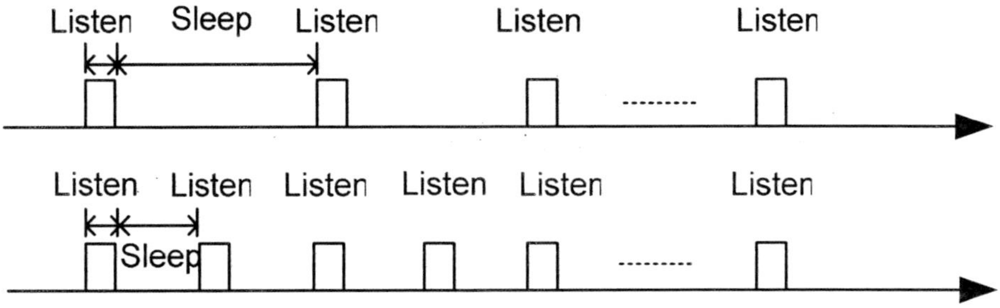

Figure 17. Frame structure for DSMAC protocol.

3.16. DS-MAC

Dynamic Sensor-MAC (DSMAC) [25] offers another modification of S-MAC in order to improve its performance. In this case, DSMAC adapts S-MAC to a traffic variable environment in which dynamic duty cycle is needed in order to deal with latency requirements. DSMAC offers a solution that works as S-MAC when traffic is low and increases the working cycle frequency when traffic becomes higher. Therefore, the power consumption is not increased for all the nodes and all the time, but just for those nodes who need it and for a specific period of time.

As explained above, this approach modifies *active/sleep* scheme frequency regarding traffic flow variations. Authors define *one-hop latency* as the difference between the moment a packet gets into the queue and it is successfully sent out. This value is included by the sender in each packet's header and it will be processed by the receiver. The average latency of receiving will be calculated as the mean value of the latency value received in a SYNC period. This average latency value helps estimating the traffic condition of the network.

On a common basis, nodes work as S-MAC was defined in subsection 3.3.. However, from time to time, higher data traffic could appear and therefore data packet quantity in the network increases. This situation can be checked with the latency value that is received with every packet. When a node, acting as a receiver, determines that the latency of its transmitter is unacceptable, the node assumes that the sender must have its queue full of packet for sending. Having detected this high latency, it decides to divide its sleep period in half, offering double opportunities to the transmitter. Doubling the listen period, the receiver will be listening more times to its sender and more than one packet could arrive in the same period of time that only one packet was sent before. This behavior is shown in figure 17.

In the same way as S-MAC, node exchange their synchronization information in the SYNC packet and every node maintains a table with the schedule of every neighbor. When a node hears a SYNC packet, it will update its table with the new values and will adjust its own clock to get synchronized.

DSMAC's SYNC packets includes a field called *SYNC initiator's duty cycle*, apart from the synchronization information explained above. Once a node has changed its own duty cycle, it publishes the new one in its next SYNC packet. Surrounding nodes that will heard this SYNC advertisement will check its sending queue. If that queue has packets waiting to be sent and the received schedule has a higher duty cycle than its current one, it adopts it and consequently, will double its schedule. Thanks to this change, the transmitter can send data packets quicker and hence empty the queue.

Nevertheless, a node cannot double its working frequency without taking into account other requirements of its behavior such as, the power consumption. Each node has a threshold that indicates the maximum level of energy consumption that it can reach. Only when the power that the node is consuming is below that threshold, a higher operation frequency is allowed. Although it hears a higher duty cycle and has packets to send, if it is consuming too much, it will not be able to adopt the new schedule.

It is important to remark that this change in its duty cycle period does not affect to the rest of the nodes, as the frequency is a multiple of the original duty cycle. This has an obvious advantage: all the nodes that are not involved in the duty cycle change, do not need a new synchronization as the new schedules do include the former behavior.

DSMAC trades power and computational resources consumption for latency. However, as the basic duty cycle can be doubled or even quadrupled, the data latency can be reduced but the power consumption increases dramatically (reaching duty cycles of 20 or even 40%). Moreover, managing these duty cycles modifications needs higher computational resources.

4. Conclusions

In this chapter, we have described the different medium access techniques that have been developed for wireless sensor networks. As emitting information through an RF interface requires a large amount of energy, the access to the channel medium must be regulated so that the life time of the nodes is kept to the maximum. The main idea behind this energy aware medium access control is the fact that wireless sensors do not require large bitrates, so they will be idle most of the time, without making use of the radio channel. In this situation, these radio interfaces may be switched off to avoid unnecessary energy wastage. However, this operation may make it difficult to exchange information between two neighbors because both of them must be active for the information exchange to be successful.

The medium access protocols developed so far try to avoid those situations which require using more energy than strictly necessary, which mainly involve retransmitting a packet or waiting for a packet to arrive. The main solutions proposed so far are the following:

- TDMA schemes: nodes are assigned a slot of time when they can transmit information, so all chances of collision are practically avoided. As a drawback, the algorithm to assign these slots is usually costly and the whole assignment must be recomputed whenever the network topology changes. Latency is also increased because nodes must wait for the next slot before they can transmit their data.

- CSMA schemes: nodes are not assigned a slot when they can transmit, but they do this whenever they need to. To avoid collisions, they check if a neighbor is already using the channel, but this situation does not fully avoid them. The lack of a neighbor hierarchy and time slotting make this algorithms very adaptable to network changes and reduce latency because the channel is used as soon as a node has information to transmit.

All the solutions proposed so far involve a modification of these two approaches, trying to take the maximum profit from the energy efficiency in TDMA schemes and low latency in CSMA approaches. The current trends in MAC algorithms show that hybrid approaches have been used lately to achieve this objective, so that the network reduces its latency at the link layer level by allowing some probability – a small one, anyway – of collisions in order to allow nodes to emit a message at any time when the traffic load is low. A second bottleneck which reduces latency is the fact that a message will not reach its destination in one or two hops, so it will end up stuck at some sleeping node, where it will have to wait until this node becomes active again. The MAC algorithms that will appear in the next years will try to assign slots in such a way that packets always jump from an active node to other, and never get stuck at some point. On the other hand, the nature of events in wireless sensor networks produce a burst-type network load that coincides with the appearance of event detections. These events will trigger relatively large amounts of packets, which contain almost the same information. Identifying these redundant packets and merging them will reduce the traffic load and, therefore, the amount of consumed energy for reporting the event to the sink. In the future, MAC protocols will be designed to offer more information to upper layers, so that higher level procedures, such as routing or compressing information, may be performed at a lower energy cost and higher efficiency.

References

[1] Wei Ye, John Heidemann, and Deborah Estrin. An Energy-Efficient MAC Protocol for Wireless Sensor Networks. In *Proceedings of the International Annual Joint Conference of the IEEE Computer and Communications Societies*, June 2002.

[2] E.P. Harris and K.W. Warren. Low Power Technologies: A System Perspective. In *3rd International Workshop on Mobile Multimedia Communications*, 1996.

[3] F. Douglis, F. Kaashoek, B. Marsh, R. Caceres, K. Lai, and J. Tauber. Storage Alternatives for Mobile Computers. In *Symposium on Operating Systems Design and Implementation*, 1994.

[4] W. Mangione-Smith, P.S. Ghang, S. Nazareth, P. Lettieri, and W. Boring R. Jain. A Low Power Architecture for Wireless Multimedia Systems: Lessons learned from Building a Power Hog. In *International Symposium on Low Power Electronics and Design*, 1996.

[5] Suresh Singh and C.S. Raghavendra. PAMAS - Power Aware Multi-Access Protocol with Signalling for Ad Hoc Networks. *ACM Compputer Communication Review*, 28(3):5–26, 1998.

[6] Wei Ye, John Heidemann, and Deborah Estrin. Medium Access Control With Coordinated Adaptive Sleeping for Wireless Sensor Networks. *IEEE/ACM Transactions on Networking*, 12(3):493–506, June 2004.

[7] Yuan Li, Wei Ye, and John Heidemann. Energy and Latency Control in Low Duty Cycle MAC Protocols. In *Proceedings of the IEEE Wireless Communications and Networking Conference, WCNC*, New Orleans, LA, USA, March 2005.

[8] Rajgopal Kannan, Ram Kalidindi, and S. S. Iyengar. Energy and Rate Based MAC Protocol for Wireless Sensor Networks. In *SIGMOD Rec.*, volume 32, pages 60–65, 2003.

[9] Venkatesh Rajendran, Katia Obraczka, and J.J. Garcia-Luna-Aceves. Energy-Efficient, Collision-Free Medium Access Control for Wireless Sensor Networks. In *Proceedings of the ACM Conference on Embedded Networked Sensor Systems, SenSys*, Los Angeles, California, USA, November 2003.

[10] Tijs van Dam and Koen Langendoen. An Adaptive Energy-Efficient MAC Protocol for Wireless Sensor Networks. In *Proceedings of the ACM Conference on Embedded Networked Sensor Systems, SenSys*, November 2003.

[11] Amre El-Hoiydi and Jean-Dominique Decotignie. WiseMAC: an Ultra Low Power MAC Protocol for Multi-Hop Wireless Sensor Networks. In *Proceedings of the Algorithmic Aspects of Wireless Sensor Networks, ALGOSENSORS*, Turku, Finland, July 2004.

[12] Christian. C. Enz, Amre El-Hoiydi, Jean-Dominique Decotignie, and Vicent Peiris. WiseNET: An Ultralow-Power Wireless Sensor Networks Solution. *IEEE Computer*, 37(8):62–70, August 2004.

[13] Amre El-Hoiydi. Spatial TDMA and CSMA with Preamble Sampling for Low Power Ad Hoc Wireless Sensor Networks. In *Proceedings of Symposium on Computers and Communications, ISCC*, pages 685–692, July 2002.

[14] Huan Pham and Sanjay Jha. An Adaptive Mobility-Aware MAC Protocol for Sensor Networks (MS-MAC). In *Proceedings of the IEEE International Conference on Mobile Ad hoc and Sensor Systems, MASS*, Florida, USA, October 2004.

[15] Gang Lu, Bhaskar Krishnamachari, and Cauligi S. Raghavedra. An Adaptive Energy-Efficient and Low-Latency MAC for Data Gathering in Wireless Sensor Networks. In *Proceedings of the International Parallel and Distributed Processing Symposium, IPDPS*, April 2004.

[16] L. Roberts. Extension of Packet Communication Technology to a Hand Held Personal Terminal. In *Proceedings of the Spring Joint Computer Conference, AFIPS*, pages 293–298, 1972.

[17] Syed Waqar Hussain, Tashfeen Khan, and S.M.H. Zaidi. Latency and Energy Efficient MAC (LEEMAC) Protocol for Event Critical Applications in WSNs. In *International Symposium on Collaborative Technologies and Systems*, pages 370–378, 2006.

[18] Jeremy Elson, Lewis Girod, and Deborah Estrin. Fine-Grained Network Time Synchronization using Reference Broadcast. In *Proceedings of USENIX Symposium on Operating Systems Design and Implementacion, OSDI*, Boston, MA, December 2002.

[19] L.F.W Van Hoesel, T. Nieberg, H.J. Kip, and P.J.M. Havinga. Advantages of a TDMA based, Energy Efficient, Self-Organizing MAC Protocol for WSNs. In *Vehicular Technology Conference*, volume 3, pages 1598– 1602, 2004.

[20] K. Kredo and P. Mohapatra. Medium Access Control in Wireless Sensor Networks. *Computer Networks*, To appear, 2006.

[21] Joseph Polastre, Jason Hill, and David Culler. Versatile Low Power Media Access for Wireless Sensor Networks. In *Proceedings of the Second ACM Conference on Embedded Networked Sensor Systems (SenSys)*, 2004.

[22] Injong Rhee, Ajit Warrier, Mahesh Aia, and Jeongki Min. Z-MAC: a Hybrid MAC for Wireless Sensor Networks. In *Proceedings of the 3rd International Conference on Embedded Networked Sensor Systems, SenSys*, pages 90–101, 2005.

[23] Injong Rhee, Ajit Warrier, Jeongki Min, and Lisong Xu. DRAND: Distributed Randomized TDMA Scheduling for Wireless Ad-hoc Networks. In *Proceedings of the 7th ACM Interational Symposium on Mobile Ad Hoc Networking and Computing, MobiHoc*, pages 190–201, 2006.

[24] Antonio G. Ruzzelli, Richard Tynan, and G.M.P. O'Hare. An Energy-Efficient and Low-Latency Routing Protocol for Wireless Sensor Networks. In *Proceedings of the Systems Communications, ICW*, pages 449–454, August 2005.

[25] Peng Lin, Chunming Qiao, and Xin Wan. Medium Access Control with a Dynamic Duty Cycle for Sensor Networks. In *Proceedings of the IEEE Wireless Communications and Networking Conference, WCNC*, volume 3, pages 1534–1539, March 2004.

In: Wireless Communications Research Trends
Editor: Tong S. Lee, pp. 165-190
ISBN 978-1-60021-674-9

Chapter 5

DISTRIBUTED DATA MANAGEMENT IN SENSOR NETWORKS

Stefano Chessa
Dipartimento di Informatica, Università di Pisa, Italy
Istituto di Scienza e Tecnologie dell'Informazione,
Area della Ricerca CNR di Pisa, Italy
Francesco Nidito
Dipartimento di Informatica, Università di Pisa, Italy
Electrical and Computer Engineering (ECE),
Northeastern University, Boston, USA
Susanna Pelagatti
Dipartimento di Informatica, Università di Pisa, Italy

Abstract

Wireless sensor networks (WSNs) are a recent technology designed for unattended, remote monitoring and control, which have been successfully employed in several applications. WSNs perform environmental data sampling and processing, and guarantee access of the processed data to remote users. In traditional WSN models these tasks consist in transmitting sensed data to a powerful node (the sink) which performs data analysis and storage. However these models resulted unsuitable to keep the pace with technological advances which granted to WSNs significant (although still limited) processing and storage capabilities. For this reason recent paradigms for WSN introduced *data base* approaches to define the tasks of data sampling and processing, and the concept of *data-centric storage* for efficient data access. In this paper, we revise the main research contributions on both sides and discuss their advantages with respect to traditional approaches.

1. Introduction

1.1. Wireless Sensor Networks

A Wireless Sensor Network (WSN) is a computer network formed by a large number of little and inexpensive wireless devices (the *sensors* or *sensor nodes*) that cooperate to monitor the environment using transducers [51] [4] [1] [2] [3]. Recent technology advances have enabled the design and the development of tiny processors and radio systems that can be easily embedded in little multi-purpose and easily programmable sensors. A sensor is a microsystem which also comprise one or more sensing units (transducers), a radio transceiver and an embedded battery. Sensors are spread in an environment (the *sensor field*) without any predetermined infrastructure and cooperate to execute common monitoring tasks which usually consist in sensing environmental data from the surrounding environment. Due to low cost, sensors have poor reliability and are subject to failures and battery exhaustion. Sensors are typically deployed in harsh environments where the nodes' substitution can be impracticable. Due to these reasons protocols and sensors' applications must be highly fault-tolerant and the network must be able to self adjust to fast configuration changes.

WSNs can be employed in a large variety of tasks. They can be used in medicine, agriculture, military, inventory monitoring, intrusion detection and many other fields. In the medical field, they can be used to remotely monitor patients' conditions in a non intrusive way: This enables the patients to move in the medical facility while their life parameters are still monitored [32][21][10]. In agriculture, sensors can be used to enable the so called *precision farming*, in which the fields' conditions are constantly monitored to tune the water or fertilizing quantities to maximize the production. Pollution monitoring can be enhanced by such little sensors. The presence of a large number of non intrusive sensors enables a fine grain monitoring of the environment. Such a fine grain monitoring can be essential in the location of pollution sources. The same sensor systems can be used to monitor the environment to prevent flooding, fire or other natural disasters[47]. A more recent application of the wireless sensor networks is the monitoring of animals in a non intrusive way, due the reduced size of the nodes[48][19][49]. Other applications range from home automation to education[46] to inventory monitoring and machinery status monitoring.

The WSN advantage is in the capability of reporting data every time from everywhere in the sensor field. One of the main issues in sensor networks is how to organize data management and retrieval in a reliable and efficient way.

In the early sensor networks, data collection was performed by human operators who had to physically reach specific positions in the deployment area to acquire data from the environment. This manual operation had three main drawbacks:

1. The data collecting operation was very expensive because human operators had to manually collect data.

2. The operation was error prone due poor automation.

3. The operation could be risky because of environmental hostile conditions (radiations, poison etc.).

The typical sensors' deployment in an environment consists of hundreds of sensors. The deployment can be both random or predetermined by the user of the network. After the deployment, the sensors self-organize to form a multi-hop network to enable communications between nodes that lie out of the communication range of each other. For instance, in structures monitoring, sensors can be deployed on both static structures, as bridges and building, or dynamic structures, as airplanes or cars. Sensors continuously monitor these equipments ensuring their reliability. In particular, they sense the current status of the system and help forecast of the future evolution of that equipment [27].

The user can query a WSN using a special purpose node called *sink*. The network can be queried using different paradigms. In traditional networks, sensors collect data and send to the sink all the data without processing. More recent approaches use the whole network as a database enabling the user to perform complex queries and in-network computation. Data management and the mechanisms due to enable it in sensor networks are the main topics of this chapter.

In the next section, we will focus on the data management in modern systems. We will briefly review the principal approaches to the problem, namely *directed diffusion*, *external storage*, *data centric storage* and *database model*. In the rest of the chapter we will focus on the last two models.

1.2. Network Management Models

All the possible uses of wireless sensor networks deal with the idea of *data acquisition* and *data processing* (by the sensors), and of *data retrieval* (by the user). These concepts are strictly related because data acquisition and data processing are defined by a user query, and data retrieval is the response to a user query. The user should be able to actively program the network, via control programs, and to retrieve data that is considered useful. This can be achieved in various ways, following different approaches.

In all approaches, nodes' cooperation can be essential to provide refined information to the end user. The simplest pattern of node cooperation is *data aggregation*. In data aggregation, the sensor nodes that used to forward the message from the source to the sink(s) (the *relay nodes*) take care of combining incoming data using some pre-defined rules.

For instance, incoming data following different paths, converging on a single relay node, can be combined in a single average value; then this new datum is forwarded towards the sink(s). This reduces the total amount of data transmissions with respect to the plain forwarding. However, these merge rules are very simple and sensors cooperate only for communication and not for computation. This last is left to the sink: Complex operations, such as data correlation, are performed off line.

In more modern approaches, data processing is moved directly into the network [30] [29]: Sensors acquire and process data to extract meaningful information. In this case, sensor nodes actively cooperate to produce this information (such as averages, variances and other statistical measures or spatial and temporal association of data).Stand-alone sensor networks are a natural evolution of these approaches. In these networks, sensors operate independently from the sink(s) to perform complex operations. The sink(s) can arbitrarily connect to the network to inject queries or to read processed data. In this vision, processed data must be stored within the network.

In stand-alone sensor networks should offer three main services:

1. *Network programming*, in which the sensors are programmed to replay to queries (set up of routes, data aggregation strategies etc.).

Table 1. Correspondence between models and services.

	Directed Diffusion	External Storage	Data Centric Storage	Database Model
Network programming	yes	no	no	yes
Data acquisition and in-network processing	partially	partially	yes	yes
Data retrieval	yes	yes	yes	no

2. *Data acquisition and in-network processing*, in which the data is physically acquired by transducers and processed to produce highly meaningful data.

3. *Data retrieval*, that provides the collection of data by the sink(s).

In the literature, we can identify four main models to data management: Namely *directed diffusion*, *external storage*, *data centric storage*, and *database models*. Each of these models implements a part or all the above services. The services provided by these models are summarized in Table 1. External storage is the simplest model and provides node cooperation only by data aggregation and data forwarding. Directed diffusion is able to program the network to optimize the data flow from the sensors to the sink, providing a better usage of the resources with respect to external storage. Data centric storage is able to use the network to easily store and retrieve a datum using its meta-datum (an unique *name* for the datum) as a key. Data centric storage does not need, or provide, network programming facilities for the data management. Such facility can be added on an upper layer laying on this one. Finally, the database model is able to abstract the network as a relational database to perform complex queries that are executed by the sensors themselves.

In the rest of this section, we give more details on these four models. we will focus more on directed diffusion and external storage because the data centric storage and the database models will be the focus of the rest of the chapter.

External Storage model. In the External Storage (ES) model, the nodes send the sensed data, in both raw or more refined forms, to one or more sinks which perform the actual aggregation, processing and storage. In this model, the sensor nodes perform only data acquisition and send sensed data to the sink. To this purpose, they also have some routing capability. Each sink node collects data from the sensors and implements an interface for the user. For instance, a sink might be a PC acting as router between the sensor network and other networks.

The main advantage of this model is that it is simple to implement. Moreover if the network produces few data it can be energy efficient.

The drawbacks are represented by the poor usage of the network's computational resources because the sensors, that can be capable of more complex operation could refine data. Moreover, the data correlation is due to the sink and the off line processing can require multiple sink-to-sensors communications while in an in-network computation the sensors can efficiently self organize to provide correlated data. For instance, an user that wants to correlate various temperature sampling from a region (far from the sink) needs to query all the nodes of that region while an in-network processing strategy is able to program the sensors of the same region to communicate and provide only the correlated data.

Directed Diffusion model. Directed Diffusion (DD)[22] is one of the first approaches proposed for data management in sensor networks. The user requests some specific data with a query including a tuple $\langle key, value \rangle$ and a *data rate*, which specifies the amount of data that must flow in the unit of time. The first step in data retrieval is represented by the *interest dissemination* phase in which the sink broadcasts the query in the network. The broadcast used to disseminate the query sets up a directed acyclic graph (DAG) in the network which is rooted at the sink. A node x is connected to a set of *upstream* nodes (the *gradients*) which are its relay nodes when x sends packets to the sink. Each node receives the interest, records the query data rate and sets up the corresponding *gradient* toward the sink. The nodes that receive or sense some data that match one or more interests forward them up the route created by the gradients according to the data rate. The sink, at the reception of messages with the queried data, can exploit *reinforcement* of the paths that bring data with higher data rates sending a packet down to the data stream. Reinforced paths may augment their data rate, while paths that are not reinforced expire after a given amount of time.

The main advantage of the directed diffusion model is that the interest propagation happens hop by hop and the network does not need to use long communications to set up or modify paths' characteristics. DD is able to provide multi-path routing and delivery using the reinforcement of multiple paths. The nodes are capable to repair or tune the paths in a local way sending local reinforcement messages.

DD presents also two main drawbacks. The first one is related to the poor load balancing of the protocol: the nodes that are closer to the sink are burdened with data and control packets relay. The second drawback is the limited opportunity for in-network data processing because data may pass through different paths and aggregation and processing often happens only at the sink or at its neighbors.

Some recent improvements on DD have been proposed in [33] and [28]. In [33] the authors present Directed Diffusion *Light* (DDL). The improvement acts on the interest propagation: each node can have only a limited set of gradients (while in DD the size of this set was not limited). In this way, the DAG is sparser and it is not formed by all the nodes in the network. When reinforcement is performed the DAG may change to include previously discarded nodes. This offers the opportunity of replacing depleted nodes with fresh spares and thus improves the network lifetime.

In [28], the authors present the Information Directed Routing (IDR). IDR finds paths with maximum information gain at a moderate communication cost. A query message is sent from a source node to a destination node. The network routes the message from the source to the destination through a path that is not minimal in terms of energy but which passes through *high interest areas*. The purpose is to collect as much information as possible to reply to the query.

Data Centric Storage model. In the Data Centric Storage (DCS) [42] model, the sensors network itself provides support to data storage. Data is stored in the network according to keys (meta-data) which are also used for data retrieval. Given a data and its key, DCS stores the data in a subset of sensors which is uniquely determined by the key. DCS addresses all the problems concerning the topology changes due to node failures, energy efficiency and load balancing using Geographic Hash Tables (GHT) [42] based on physical coordinates of the sensors. Other approaches to DCS, such as Graph EMbedding (GEM) [36] uses virtual coordinates instead of real ones.

Cell Hash Routing (CHR) [11] is another DCS system based on hash tables. CHR clusters nodes in cells of predefined and globally known shape using a distributed protocol (e.g. dividing the sensor field in a mesh of squares) and stores each data in a cell.

Database model. The Database model [30] [29] [50] [7] [8] [9] is a more recent model that offers a high abstraction is the one that enable the user to do SQL-like queries on the network. The WSN is abstracted as a distributed database and the user can specify both queries and their duration (to provide temporal aggregation). In these systems the queries are translated in data acquisition, data processing and data transfer operations that are performed in a distributed way by the nodes of the network. This approach provides a network abstraction which is completely independent of the network details, and the sensors do not need to be preprogrammed for specific tasks and the user can change the behavior of the network by injecting new queries.

Specifically programmed applications can be more efficient, but they are very specialized and they are tuned only for a given task. These applications must be programmed from scratch each time a new query is needed. In particular when the query changes, the design, coding and debugging steps must be performed again. On the other hand, in a database abstraction the system is always the same and only the queries change to perform different tasks with only little design, code and debug efforts.

1.3. Organization of this Chapter

This chapter is organized as follows. Section 2. presents protocols and strategies that are used to enable the data centric storage and database models in wireless sensor networks. Section 3. presents the database model, describing some approaches in the implementation of databases on wireless sensor networks. Section 4. presents the techniques used to store data in the network and to provide reliable and fast access to data by the sink(s). In Section 5. we draw conclusions and points out some future research issues.

2. Mechanisms Supporting the Storage and Database Models

Localization and routing are the main support to any activity in sensor networks. Localization is necessary to relate data with the physical location where they have been collected, but it can also provide support to geographic routing (which uses geographical position of the nodes to perform routing). Routing is essential because sensor networks are generally multi-hop, hence data must be routed through multi-hop paths to reach the sink or the storage point. We address localization and routing in section 2.1. and 2.2. respectively.

2.1. Localization Mechanisms

Localization mechanisms are used to find the position of a sensor. The position of a sensor can be useful for two different reasons:

1. A query requests data from specific locations.

2. A datum must be stored in a specific location for later retrieval.

The localization techniques are divided in three main categories: *GPS-based*, *approximate*, and *virtual*.

GPS-based. GPS-based solutions assign real geographical coordinates to sensors. In this model, each sensor is equipped with a GPS antenna and it is able to receive the GPS signal and locate itself.

This solution is not always applicable in nowadays WSNs because GPS receivers are expensive both in terms of money and energy and they may fail if the nodes are located inside buildings or in areas in which obstacles prevent the GPS signal to arrive.

The growing capabilities in hardware could provide, in a near future, cost effective and energy inexpensive GPS receivers. If this happens, solutions that now look poorly applicable could become future standards.

Approximate coordinate systems. The physical coordinates can be approximated using special hardware and/or network topology. In these approaches, the geographical coordinates of some of the nodes must be known (for instance using GPS), and a distributed algorithm exploits this information to infer the approximate position of other nodes. To this purpose the algorithm exploits information about relative distances or angles between nodes which can be obtained equipping the sensors with specialized hardware. More specifically, the nodes can estimate the relative distances between themselves using different techniques: *signal power*, which estimates the distance based on the power of the received signal, and *difference in arrival time* which exploits messages sent on different communication medium (e.g. radio and acoustic) that have different propagation times, and estimates the distances based on the inter-arrival time. A different approach exploits the *angle of arrival* of the messages to estimate the relative position of a sender. This approach however must be supported by directional antennas.

A general classification of these techniques divides them in *range based* methods and *range free* methods.

In range-based methods some nodes (called anchors) know their physical position. Non-anchor nodes compute their position using multilateration techniques which minimize the square of the distance between them and the anchors. This computation returns the position of the nodes.

In [43] a node can use multilateration techniques using both anchors and nodes that previously used anchors to estimate their position. In [35] the anchors are special nodes equipped with powerful radios with narrow and rotating beams. The non-anchor nodes use both the signal strength and the angle of messages sent by the anchors to locate themselves. In [38] and [37] less accurate systems are provided: They use the two hop information to compute their position. The fist technique uses distance and the second the angle.

Range free methods can be used by nodes equipped with cheaper radios that are not able to compute the distance of the received signals. In [16] the nodes compute their position as the centroid of the positions of the anchors that they can hear. This approach requires an high number of anchors. In [34] and [38] the anchors flood their position in the network. The number of hops (that is related to communication range) is used to find a node's position.

Virtual coordinate systems. Virtual coordinate systems define a coordinate space that is unrelated to the physical position of the nodes. In such systems the focus is in providing an efficient support for geographic routing.

Multidimensional scaling (MDS) [45][24] is a technique that maps proximity information of the nodes to the element of a matrix. In this new space the Euclidean distance between nodes is related to the original proximity information. Each node has knowledge of its neighbors at various hops of distance (for energy saving reasons, typically this distance is no more than 2 hops), and it creates a matrix containing the shortest paths between itself and its neighbors. These local matrixes are then merged to produce a global map. The global map can be then used by anchor nodes (which know their position) to give to all the nodes a physical position.

In [41], the authors introduce a family of coordinate assignment protocols with increasing complexity suitable for different scenarios. In the first scenario, the coordinate of the nodes in the network are computed, in an iterative fashion, as the averages of the coordinates of the nodes on the external perimeter of the network because these last nodes are the only ones that know their own position. In the second scenario, the nodes on the border do not know their own position and the first step of the algorithm consists in the construction of a *perimeter vector* with the distance from each node of the network with all the nodes on the borders. In the third scenario, the sensors are not aware to be on the border or in the central part of the network, thus two nodes start a bootstrap phase in which the nodes on the border are identified because they are the most distant from the two bootstrapping nodes. All the three systems in [41], use a large amount of energy and memory to compute the coordinates of the nodes. Energy is used by the iterative systems that are used to compute the average of the coordinates of the nodes. Memory is used to maintain the perimeter vectors used in such computation.

It is possible to provide less expensive solutions that use only the distance in hop from some of the nodes on the border. In [17], a reconfigurable number of nodes are used as anchors. The anchors are chosen randomly, and the coordinates of each node are the hop distances between the node and the anchors. However, since anchors are selected randomly, some of them might be close to each other, and this may result in an unbalanced coordinate system in which too many nodes share the same coordinates. Another result provided for uniformly deployed networks is the one presented in [18], in which the size of the areas of nodes sharing the same virtual coordinate is minimized if the anchors are as far as possible from each other. The authors then propose a distributed algorithm that elect three nodes with these properties and compute in a distributed fashion the virtual coordinates of all the nodes. The authors also show that proactive routing protocols, perform similarly with virtual and physical coordinates.

Virtual Polar Coordinate Space (VPCS) is the virtual coordinate system used by the

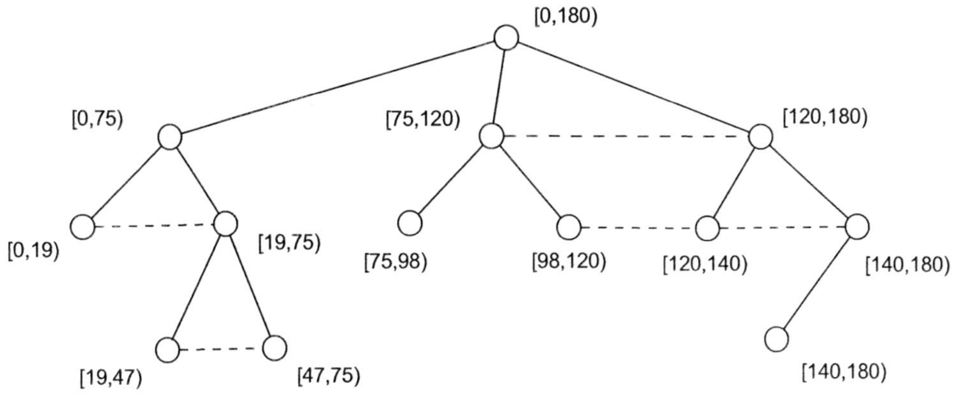

Figure 1. A VPCS angle range assignment. The root has range $[0, 180)$.

GEM storage system [36]. In VPCS a ringed tree graph is embedded in the network topology. Each node of the tree has an identifier that is formed by two values: The first one is the level in the tree and the second one is a virtual angle that identifies the node in the level. The virtual angle is assigned in such a way that it is consistent with the network topology. In this way the coordinates represent a polar system whose origin is a single node (usually the sink). The VPCS algorithm uses several steps to build the polar coordinates. The first step builds the ringed tree. To build the tree, the root node assigns to itself the level 0 of the tree. All the nodes that are at one hop distance from the root have level 1. At this point, the nodes at level 1 broadcast the information to all their neighbors. The procedure continues until all the nodes are assigned with a level. After the tree is built, the size of each sub-tree is sent back to the root of the tree in a recursive way. The nodes at level i collect the information about the size of the sub-trees formed by the nodes of level $i + 1$ or greater. When the root has collected all the sub-tree size for each node of level 1 it begins to assign the virtual angles. The root uses for itself the whole angle range, let us say the range $[0, 2^{32} - 1]$. Then it assigns to each one of its neighbors a sub-range of its angle. The size of the sub-range is proportional to the size of the sub tree. This system gives to larger subtrees a wider angle, balancing the coordinate system. Each level 1 node assigns to its neighbors of level 2 a sub range of its range proportional to the size of the sub-trees. This procedure is repeated until the leaves are reached. An example of angle range assignment is presented in Figure 1. At this point VPCS aligns the computed topology with the geographical information of the network. The alignment provides the routing system to jump from one branch of the tree to another using cross-links in the embedded tree structure. To perform the alignment in a distributed fashion, each parent node must find the order of its children. The order is used to distribute the angle among the children nodes and it is used to locate the nodes in the network. The distance can be found in two ways: The first one uses the distance between nodes which could be computed using signal attenuation of the transmission. This system, however, can introduce errors, and for this reason the authors propose an alternative method based on triangulation. In particular, the global coordinate system is built using the triangulation, which is done starting from the root node. The root node selects two other nodes that are not collinear and using spanning trees from the root and these other two nodes identify the relative position of the other nodes.

2.2. Routing Mechanisms

Traditional reactive approaches [25], [40] to routing in ad hoc networks appear unsuitable to wireless sensor networks due to the limitations of the sensors in terms of energy, memory and processing capabilities. For this reason, recent approaches in WSN focus on geographic, stateless routing which do not burden the sensors with routing tables and caches. Among these protocols, we mention GFG and VPCR.

Greedy Face Greedy. The Greedy Face Greedy (GFG)[15] is the routing protocol used by GHT [42] to route and store messages in the network.* The GFG protocol uses nodes coordinates to route packets in the network. In GFG, the key idea is that a packet is routed in a greedy way, reducing the distance to the receiver at each hop, when possible. When no improvement is possible, the protocol switch to *face* (or *perimeter*) mode that routes the packet around the face until greedy forwarding is possible again.

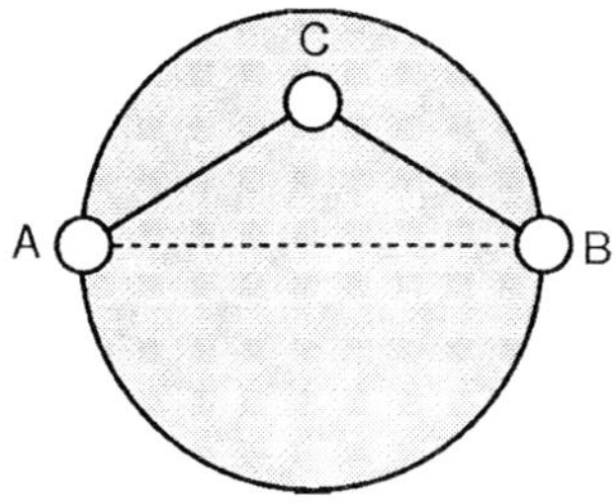

Figure 2. Gabriel Graph (GG) planarization. The arc (A, B) is not inserted in the graph because node C is in the grey area.

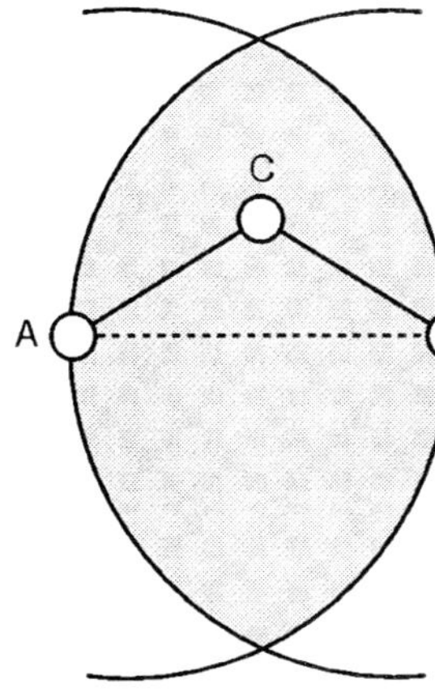

Figure 3. Relative Neighborhood Graph (RNG) planarization. The arc (A, B) is not inserted in the graph because node C is in the grey area.

In the face mode the protocol uses the graph planarization to prevent loops. Graph planarization is a technique that, starting from an input graph and applying rules to prune edges, returns another graph with some wanted properties such as a lower degree per node.

* In [42] the authors call this protocol GPSR [26].

In GFG, the graph planarization is implemented with a distributed algorithm that planarize the graph using the Gabriel Graph (GG) planarization[23] (Figure 2) but it can also use Relative Neighborhood Graph (RNG) planarization (Figure 3). In the GG of a graph G the edge (u, v) is present if and only if the disk with diameter (u, v) does not contain other nodes of G. The GG of a graph G guarantees some property as connectivity: A GG of G is connected if G was connected. The GG is computed by the following distributed algorithm [15].

```
for each u in N(v) do
   if disk(u,v) intersects (N(v)\{u,v}) then
      delete (u, v)
   end if
end for
```

In the algorithm, each vertex v looks in its neighborhood $N(v)$ to find the nodes that are part of the GG. The algorithm complexity is $O(d^2)$, where d is the node degree. Using some additional information, as the Voronoi diagram and Delaunay triangulation the complexity can be reduced to $O(d \log d)$ [39].

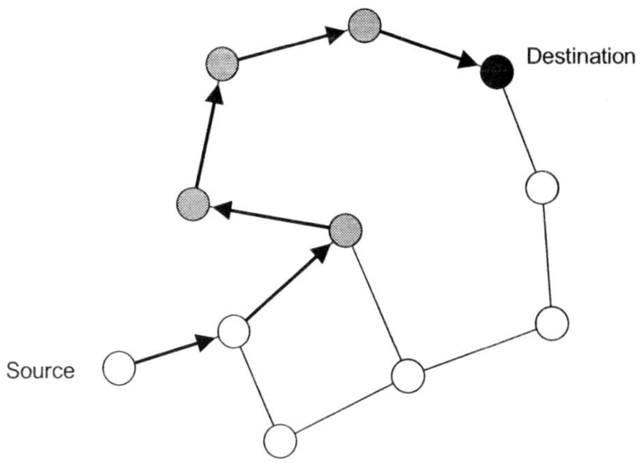

Figure 4. Face mode of GFG protocol. The grey nodes are the ones that belong to the face.

As stated before GFG routing is performed using two modes that are used in two different situations.

The *greedy* routing is used to move quickly the packet from the source to the destination. The greedy routing works as follows. When a node receives a packet (and the node is not the destination) it forwards the packet to one of its neighbors. The selected neighbor is the closest to the destination. If the forwarding node is not the destination and no neighbor is closer than itself to the destination, then it switches to *face* mode (Figure 4).

The *face* routing operates on a planarized graph which should have been computed in a preliminary phase. Planarization partitions the plane into faces that are bounded by the arcs of the planarized graph. Starting from a vertex v on a face f, the boundary of f can be traversed in counterclockwise (or clockwise) direction. The face traversal routing routes the packet along the nodes on a face. For instance, let us consider a packet with destination d which is routed with greedy routing until it reaches a local minimum on a node v, placed on a face f. Node v begins the face traversal routing selecting a node which is the next hop in a counterclockwise visit of the nodes in f. The packet is routed along the face boundary until it intersects the line (v, d) in a point different from v. At this point the packet is routed again using greedy forwarding.

Virtual Polar Coordinate Routing. GEM [36] system uses the Virtual Polar Coordinate Routing (VPCR) to route messages on its VPCS located network. VPCR uses three different routing techniques: the *naive-tree* routing which does not use cross-links, the *smart-tree* routing that introduces optimizations on the naive routing and the *greedy* routing which uses greedy forwarding on cross-links. We explore the three modes in the following.

In the naive-tree routing, a packet is forwarded by a node upward to its parent in the tree in a recursive way until the packet reaches the root. However, if along this path the packet reaches an ancestor of the destination node, then the ancestor forwards the packet downward in the tree, directly to the destination. During the journey from the ancestor to the destination two factors are taken in account: The first is that the level must increase at each forwarding and the other is that each node sends the packet in the sub-tree whose angle range contains the virtual angle of the receiver.

In the smart-tree routing, each node checks if an ancestor of the destination is in its neighborhood. In this case, the packet is forwarded directly to that ancestor. In some cases, this may save in some cases the cost of reaching a common ancestor in the tree. This method can be improved using 2-hop neighborhoods. In this case, if the ancestor is as far as two hops the packet is directly forwarded to the ancestor using the connecting neighbor. The method can be generalized using a proactive area of n-hop neighborhoods. However, as n grows, the cost in terms of space (required to store the proactive area) and messages (to keep updated this information) also grows, and this approach results impractical even for limited values of n. In general, the smart-tree optimization is a good optimization if the source and the destination are close to each other, but it does not improve significantly the efficiency for arbitrary pairs of nodes.

VPCR uses the *greedy* routing which can be effectively when the source and the destination are far away from each other. Because in the ringed tree's rings, the virtual angles are assigned in strictly increasing, or decreasing, order, when the smart-tree routing is forced to route the message upward in the tree the node checks if some of its neighbors hare a virtual angle range that is closer to the destination than its own angle range. If such node exists the packet is greedily forwarded to the that neighbor which, in the tree, is closer to the destination. If the packet reaches a local minimum, the routing algorithm switches to the smart-tree routing, and if this routing also fails then it switches to naive tree routing. However, it can always switch back to greedy or smart tree whenever possible.

3. Database Models

The database model enables the user of the network to abstract the WSN as a database. Using this abstraction, the user can perform queries that program the sensors to retrieve and refine data.

3.1. TinyDB

TinyDB [31] is a WSN database implementation developed at UC Berkeley. TinyDB provides a SQL-like language extended to use both duration and sample ratings, to provide averages on time spans. The database is able to run queries over a single table that contains all the data collected by the network. Each sensor is represented by a row of the table and continuously updates its own data. TinyDB supports a large set of operators [29]: Spatial aggregation, filtering based on patterns and union between row sub-sets.

TinyDB provides power aware optimizations of the queries and the execution plan of the query is performed on the basis of the type of datum (its meta-data) and of the parameters needed by the operators. Then, the operators are scheduled to provide a suitable environmental sampling. The query is distributed using Semantic Routing Trees (SRTs). SRTs are routing trees rooted at the sink. An SRT reaches all nodes needed to cover the attributes of interests for a given query. In general, it may cover the entire network.

Although SRT provides very efficient routing, their use forces a limitation of TinyDB because the queries can only go from the sink to the leaves. This is an obstacle to more complex queries that involve more complex communication patterns. For instance, this prevents comparison of data produced in different subtrees of the SRT. Also, for this reason, TinyDB is a parallel query processor that does not perform in-network joins. All the information flow to the sink and the main operators that are supported by the in-network query processor are selection and the union of the streams. However, joins can be performed at the sink.

An example of TinyDB query is the following one (presented in [31]):

```
SELECT nodeid, light, temp
FROM sensors
SAMPLE INTERVAL 1s FOR 10s
```

This query specifies that each sensor of the WSN, that is abstracted as the virtual table *sensors*, must report its id, light and temperature readings (*nodeid*, *light* and *temp*) once per second for 10 seconds. The results of the query are sent to the requester as a stream of records.

3.2. Cougar

Cougar [13][14][50] is a WSN database system developed at Cornell University and exhibits many similarity with TinyDB. The query language is an SQL dialect in which the nodes of the network are represented by Abstract Data Types (ADTs) with interface methods (as in object oriented programming) to retrieve data stored in sensors.

In Cougar's queries the FROM clause can refer to a WSN's relation. The relation includes nodes attributes and nodes ADTs. The SELECT and WHERE clauses can refer to specific node's data and can use methods defined in the node's ADT. For instance the temperature of a sensor `s` in a relation `R` is denoted by `R.s.getTemp()`.

The query optimizer is run on a workstation and generates the query execution plan that specifies communication patterns, computational activities and tree operations as joins. For each method invoked on ADTs in the query a *virtual relation* is created. A virtual relation is a tabular representation of a method that contains the input values and the output argument of the method it is associated with it.

An example of Cougar query is the following one (similar to the one presented for TinyDB):

```
SELECT R.s.getNodeID(), R.s.getLight(), R.s.getTemp()
FROM R
WHERE $every(1);
```

This query specifies that each sensor of the WSN must report its id, light and temperature readings (in ADT notation) once per second (`$every(1)`).

3.3. MaD-WiSe

MaD-WiSe [7] [8] is a recent database model for WSN. Differently from TinyDB and Cougar, its query processor is fully distributed, so that any query can be completely processed in-network. MaD-WiSe defines an SQL-like query language, a query algebra, and a data model based on streams. It also defines heuristics for query optimization which take into account the transducer sampling costs, predicate selectivity and transmission costs. In the optimized query plan each sensor involved in the query is assigned with a subset of operators. For this reason, the communication pattern between sensors involved in a query can be arbitrary, and it requires general routing strategies.

The MaD-WiSe architecture comprises a user-side module and a network-side module. The user-side module implements the query parser, the query optimizer, and the tools to inject the queries and collect the results. The network-side module implements a layered architecture with a *network* layer, a *stream system* layer and a *query processor* layer.

The *network layer* implements both connection oriented and connectionless services and it implements a general point-to-point routing. It also embeds an energy efficiency module which governs the duty cycle of the sensor. The *stream system* layer [9] provides the stream abstraction to the query processor. In particular the streams abstract the transducers, the remote communication between nodes and the local data exchange between operators allocated on the same node. The *query processor* layer implements all the operators of the algebra, which include selections, projections, spatial and temporal aggregates as well as unions and joins. Note that the ability to perform joins within the network is unique to MaD-WiSe and allows comparison of data from different sources.

An example of MaD-WiSe query is the following:

```
SELECT roomB.Temperature
FROM roomA, roomB
WHERE roomA.Temperature > roomB.Temperature and
      roomA.Temperature > 50
EVERY 100 seconds
```

This query specifies that the temperature of the sensors placed in room A and room B should be compared and only when the condition in the where clause is true then it outputs the temperature in room B. The sampling of the temperatures is performed every 10 seconds.

4. Storage Models

Storage systems can be used either with simple data acquisition mechanisms, or in combination with more sophisticate models such as databases. They provide support for efficient storage of sensed/processed data and for data retrieval, and they exploit data redundancy to ensure data reliability.

4.1. Data Centric Storage Models

Geographic Hash Tables. A Geographic Hash Table (GHT) provides a $\langle key, value\rangle$ associative memory abstraction of the sensor network. Data are represented by $\langle key, value\rangle$ pairs and GHT offers two operations: *put* and *get*. $Put(k, v)$ stores the value v according to a geographic position (x, y) that is computed using the key k. $Get(k)$ retrieves the value v that was stored using the the key k looking in the position (x, y) computed using k.

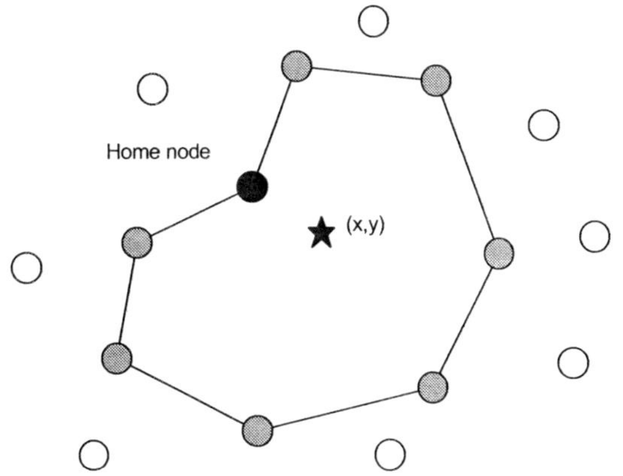

Figure 5. The home node (in black) and the home perimeter (in grey).

The central part of the GHT is a deterministic and uniform hash function that starting from the key k, enables both put and get to find the same geographical location starting from the same key. The hash of the key k ($h(k)$) is a point (x, y) of the WSN deployment area: A $\langle key, value\rangle$ pair is stored on the nodes that are closer to the point (x, y). To this purpose GHT exploits GFG to route a pair $\langle key, value\rangle$ towards the point $(x, y) = h(key)$. In general, (x, y) is a generic coordinate in the deployment area and, with high probability, does not match with any real node. To deal with this problem GHT uses the concept of *home node*. The home node of (x, y) is defined as the node that is closest to the point (x, y), and it is the rendez-vous point for both put and get operations. GHT finds the home node of (x, y) using the face mode of the GFG protocol. Once a packet arrives in the proximity of the point (x, y), it is routed around that point. At the end of the face mode the packet returns at the starting node. This event starts the *home perimeter* building procedure that replicates the pair $\langle key, value\rangle$ on all the nodes that surround the point (x, y). The sensor that is closest to (x, y) becomes the home node. Figure 5 shows, in gray, the nodes belonging to the home perimeter of point (x, y) (represented by a star) and in black the home node.

GHT provides a DCS system that is scalable and robust because in a WSN, which nature is unreliable, these two characteristics are fundamental. In particular GHT features *data persistence*, *data consistency*, *load balancing*, *database increase* and *topological generality*. Data persistence means that a pair $\langle key, value\rangle$ that is stored in the system is available to queries, despite of sensors failures. Data consistency means that a query for a key k, is correctly routed to a node that hosts the right pair $\langle key, value\rangle$. Load balancing means that the pairs $\langle key, value\rangle$ are stored in fair way such that data is not concentrate in any particular node. Database increase means that when new nodes are added to the system data is spread also on the new nodes. Topological generality means that the system does not need a particular topology to work properly.

To the purpose of data availability, all the nodes in the home perimeter store the pair $\langle key, value\rangle$. However, when one or more topology changes of the network occur, a home node can disappear or home perimeters can be broken. In this case, data consistency can fail because a request for a datum can be taken to the new closest node to the point (x, y), that is not the home node nor part of the home perimeter. For this reason, GHT implements the *Perimeter Refresh Protocol* (PRP). PRP periodically generates refresh packets. Each t seconds, the home node performs a put operation to the (x, y) point. The packet enters in face mode and refresh the data on the home perimeter. If the perimeter broke, a new perimeter is built. If new nodes where deployed the PRP is able to find a new home node and the new home perimeter. Each time that a node receives a refresh packet it caches the data and sets a timer to the value t seconds. If no refresh messages arrives after t seconds the node starts its own refresh procedure. This ensures that a home node failure is recovered as soon as t seconds expire.

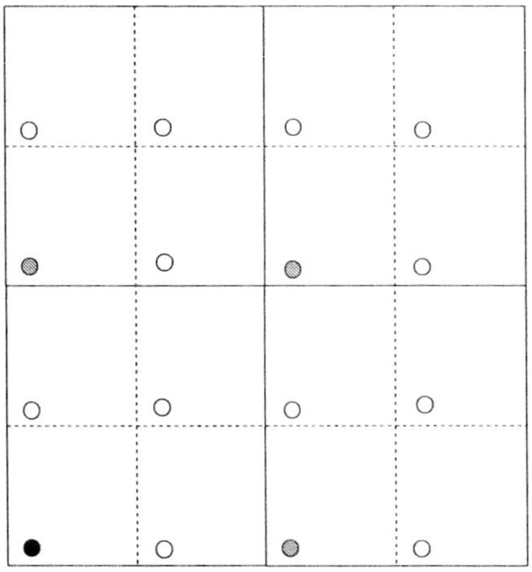

Figure 6. Structured Replication (SR). The black node is the original hashed coordinate, the grey nodes are the level one copies and the white nodes are the level two copies.

In GHT, if a pair $\langle key, value\rangle$ is very popular the nodes that store that value, and the routes to them are heavily stressed. To resolve this scaling problem, GHT uses the *structured replication* (SR) strategy (Figure 6). SR uses space decomposition. The plane is divided into a square grid and groups of contiguous areas are grouped in larger squares in a hierarchical fashion. The hierarchy is represented as a tree. The root of this tree is the whole area and the leaves are the smaller squares. For instance, the hierarchical decomposition depicted in Figure 6 shows a sensing area divided in 16 squares. The squares are aggregated in 4 macro-squares (solid lines). With SR a pair $\langle key, value\rangle$ is not stored only in the position $(x, y) = h(key)$ but also in the $4^d - 1$ *mirror points* of the location (x, y). The point (x, y) is located in a squared sub-area of the grid and occupies a relative position in it, let us call it (x', y'). A mirror point is a point, belonging to a different sub-area, that has the same relative coordinate (x', y') of the point (x, y). In Figure 6 the black node represents the hashed position, the gray nodes are the mirrors of the second level and the white nodes are the mirrors of the third level. A node that performs a *put* operation thus store the data on a mirror, and informs all the ancestor mirrors about the actual position of the data. The *get* operation starts querying the root mirror and propagates to its child mirrors until it reaches the mirror actually storing the data.

Cell Hash Routing. The Cell Hash Routing (CHR) is an evolution of GHT. CHR divides the space in *cells* of equal size. Cells are squared and divide the space in a grid. Each cell defines a *cluster* of sensor nodes. CHR uses cells to solve the problem of node mobility. A cell is considered a super-node that stores some data. If a node moves in a cell, it is informed of the data that it will store entering that cell. CHR uses the physical location of the cells to perform geographic routing of the messages to the cluster. Once a message arrives in a cell, it is replicated in all the nodes of the cell.

The size of the cells is based on the communication range of the nodes. Each node in a cell must be able to listen to all the nodes in its cell and in all the adjacent cells. With this restriction CHR guarantees that, if the initial network was connected, the graph interconnecting the cells is connected too. Let r be the communication range of a node of the network. To enable the node to communicate with all the nodes in its cell and with the eight adjacent cells, CHR requires that the the side of a cell is at most $r/\sqrt{8}$.

The nodes need to communicate to the nodes in the adjacent cell for three main reasons:

1. The routing protocol needs to know if the cells around the actual node are empty.
2. There is no need to elect a leader of the cell because all the nodes are reachable.
3. The next hop in a cell is found using a simple randomization scheme between all the nodes in the cell.

CHR uses the GFG-like routing protocol to route the messages from one node to another. GFG is modified to work on cells and not on the single nodes. This also provides two benefits: (i) geographic routing performs better in low density networks because each hop covers a larger space: The clustered network is a less dense version of the underlying network because each cell may contain more nodes; (ii) geographic routing performs better in networks which have a smaller number of edges: Also in this case the clustered network is a good choice because each cell is connected with all the neighbor cells. The GFG-like approach is able to route message outside local minimums: In CHR the face mode of the routing protocol is applied to route messages around the empty faces.

As in GHT, also CHR uses hash functions to select the cell in which store the pair $\langle key, value\rangle$. In CHR, however the cells have a globally known global logical address, and the hash function returns the index of the cell. At this point, the centroid of the cell is found simply knowing the side of the cell and the total number of the cells. The message is then routed using the GFG-like protocol to the centroid of the target cell.

When a *put* message arrives to a cell the data is replicated in all the nodes in the cell. If the *put* message arrives to an empty cell then CHR uses a strategy that is very similar to the one used in GHT. The GPG-like protocol is able, using face mode, to route messages around empty cells. CHR uses two concepts that derive directly from GHT: the *home cell* and the *home perimeter*. The home cell is the cell that is the closer to the hashed cell and the home perimeter is the ring of cells that surround the empty one. Using the GFG face mode, the packets always reach the home cell. At this point, the packets are routed around the perimeter of the empty cell and the information are stored in all the cells that belong to the home perimeter.

Graph EMbedding. Graph EMbedding (GEM) is an infrastructure for data centric storage that does not need the use of geographical coordinates. GEM is based on VPCS and the routing protocol used on this coordinate system is the VPCR.

In GEM, all the node have a label that enable them to (i) route the messages from one node the other and (ii) to map data names to the existing labels. The nodes are labeled using the VPCS algorithm, presented in Section 2.2..

In GEM, data centric storage is enabled using an hash function that maps the pair $\langle key, value\rangle$ to the embedded graph's labels. The system uses a function $f(key)$ to achieve this result. The function $f(key)$ depends only on the parameter passed enabling a consistent output for all the possible senders.

The node-to-node communication is enabled using a lookup mechanism based on data centric storage itself. If a node x wants to communicate with a node y, it needs to know the label of that node. Let us call that label $L(y)$. At network setup, a node y computes both $L(y)$ and $f(y)$. Then it routes the message $\langle y, L(y)\rangle$ to the node with label $f(y)$. When node x wants to communicate with y it sends a message to the node $f(y)$ to query the value $L(y)$. After this step the node x can send the message to y routing it to $L(y)$.

With GEM, the user can easily add new nodes and the network is able do reconfigure to use the newcomers. When a new node enters in the network, GEM uses VPCS to assign the new node a level and an angle. In particular, a new node selects a parent node in its neighborhood and the assigned level is the level of the parent plus one. At this point, the parent removes part of the angle range from one of its children and assigns it to the new node. This change affects all the child of the node whose range was decreased to keep the whole tree consistent. The node adding procedure can be used to recover the network from previous faults. The new node can become the new parent of previously disconnected nodes. At this point, the previously disconnected nodes start again the procedure to acquire a level and an angle.

K-D tree based Data-Centric Storage. The K-D tree based Data-Centric Storage (KD-DCS) [6] is an in-network data storage that addresses the problem of load balancing. The authors claim that the unbalancing state can be due to non-uniform sensors location or non-uniform sensor readings. The general idea of KDDCS is that data are stored using a distributed K-D tree [12]. A K-D tree is a data structure which allows logarithmic storage and retrieval of data that are identified by geographical coordinates. The K-D tree uses the coordinates to build a balanced tree and to guarantee poly-log access. In KDDCS, the K-D tree is used to provide virtual coordinates to the nodes. A virtual coordinate is a bit string of the form 11001 in which each one of the bits represents the branch of the K-D tree in which the nodes is located.

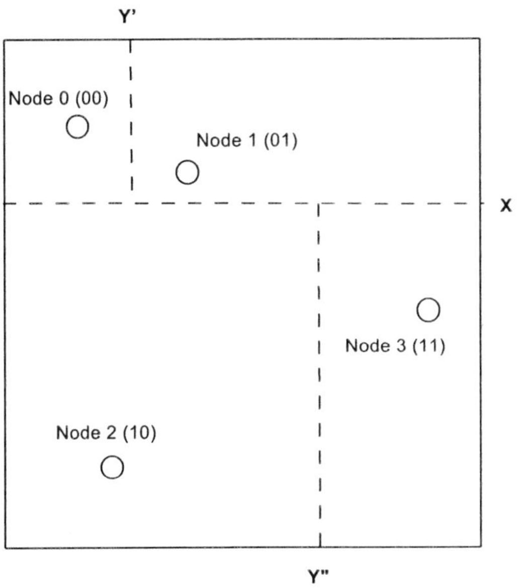

Figure 7. K-D tree built on top of a four nodes network.

KDDCS builds the K-D tree in a distributed fashion. In particular it builds the levels of the tree using partition lines which are constructed by a *weighted split median* algorithm. This algorithm divides the area in regions that contain the same number of sensors. The weighted split median algorithm is built on top of a distributed *Breadth First Search* (BFS) algorithm [20]. A node is first elected root of the BFS algorithm in a distributed way. Then, the root proceeds in the retrieval of the coordinates of the nodes in the network, computes the median value, and communicates it to the other sensors. The algorithm starts partitioning whole area horizontally. Once the median value is computed and spread on the network, the sensors with y coordinate lesser than the median value assign to themselves the coordinate 0, and the sensors with y coordinate larger than the median value assign to themselves the coordinate 1. At the second step the weighted split median algorithm is applied locally in each sub-region. The sensors with x coordinate lesser than the median value left-shift to their coordinate the value 0, the sensors with x coordinate greater than the median value left-shift to their coordinate the value 1. The algorithm stops when each node has an unique id. The tree that is built using this procedure is balanced because at each step the sensors set is split in two sub-sets having half of the nodes each. Figure 7 shows an example of network that is divided in a K-D tree structure: The algorithm first step divides the plane in two halves following line X, thus both $Node$ 0 and $Node$ 1 have the first bit of their address set to 0 and both $Node$ 2 and $Node$ 3 have the first bit of their address set to 1. The second step divides each one of the previous halves in two, following line Y' and Y'' respectively. This last step sets the second bit of the address.

The logical coordinates built using the K-D tree are used to balance the load of the network. To provide load balancing the system uses a mapping between the events that must be notified and this coordinate system. The mapping requires the events' probability distribution. This distribution is computed during the network lifetime and the tree is adjusted to keep the load balanced. At the beginning the network assumes that the events' generation probability is uniform. The mapping assigns the coordinates to specific event ranges. Consider, for instance, a network that must control temperature. KDCCS initially guesses a possible range of values for the temperature and this range is mapped on the the virtual coordinates: When the situation evolves, and the actual range is discovered and the mapping is corrected to balance the load again.

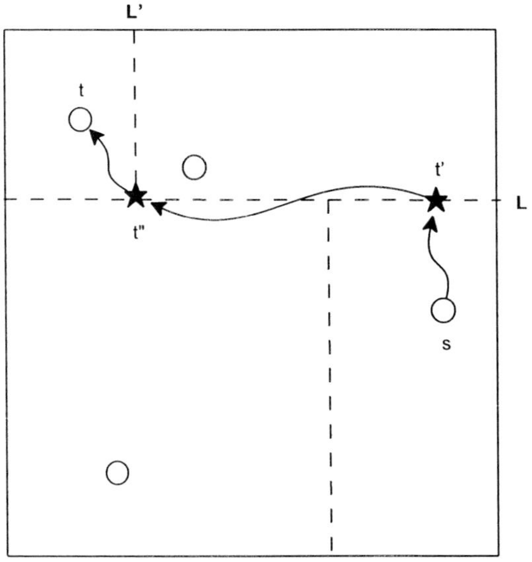

Figure 8. LSR routing from node s to node t using points t' and t'' belonging to L and L' respectively.

KDDCS uses the Logical Stateless Routing (LSR) to route the messages on the network. LSR uses the tree structure of the network to perform GFG routing in multiple *rounds*. LSR uses $\log n$ rounds to route a message from one sensor to another. A sensor s that wants to send a message to a node t must identify the least common ancestor (LCA) of both itself and the receiver. Let R be the region that contains the LCA and let L be the line that splits R in the two sub-regions containing s and t. s knows the position (only one component) of the line L because part of its address belong to this line. The message is routed from s to the point t' of line L that is perpendicular to that line using GFG. Once arrived at the node closer to the point t', the procedure is repeated, and the message is forwarded to a point t'' belonging to L' that is the LCA of t' and t. Figure 8 shows an example of routing performed by LSR. Node s, to send a packet to node t, must follow the tree structure and must use the points t' and t'' to route the message. Both t' and t'' are point belonging to the division structure of the tree.

4.2. Redundancy Models

Redundancy models are used to assure data availability, i.e. the system ensures that the stored data remains available despite sensors faults. The fundamental redundancy technique used in sensor networks is the *pure replication*.

Pure replication is the simplest way to achieve data availability. It ensures that a datum is available until all the sensors storing a copy of that datum fail. In a storage system, pure replication can be used in an *uncontrolled* or *controlled* way.

Uncontrolled replication. In uncontrolled replication the number of replicas of a datum is not known a priori. This is the modality used for instance in GHT. In fact GHT copies a datum on its home node and on all the nodes belonging to its home perimeter. Since the perimeters can be arbitrary long, it is not possible to determine the number of replicas which will be stored. In some cases, this leads to excessive replication while while in other cases it results in a very poor replication of a datum. In particular, a coordinate hashed on the border of the network can have as perimeter the whole extern perimeter of the network. On the other hand, in very dense networks, perimeters can be very small (down to three nodes). This high variability can not grant that the datum will be available for a long time, or for an expected time. This also produces states of load unbalance in the network.

Controlled replication. Controlled replication systems are based on the concept of Quality of Service (QoS). Each datum has an associated QoS value. This value can also be the same for all the nodes. The number of copies of each datum are a function of its QoS level. Let us suppose a datum d needs $n = QoS(d)$ copies. A system using pure replication will find n nodes according to the data distribution rules that will store a copy of the datum. QoS enabled non-uniform GHT (Q-NiGHT) [5] uses controlled pure replication using geocasting protocols [44]. The system is an improvement of the classical GHT approach. Once the home node for a datum is found, the protocol, using the knowledge of the nodes' distribution and position is able to find around the home node the required number of nodes that will host the replica of the datum. In this way, each datum can be stored in a predictable number of nodes and both the reliability and load balancing are acquired.

5. Conclusions

In this chapter, we have presented a survey of the principal approaches of data management in WSN. Early WSN were able only to perform data acquisition and to send raw data, without processing, to a special node (the sink) which task was to collect these streams and provide them to the user. More recent data management strategies use the network itself to provide data computation. In this way, the user is able to program the network as a database. The network self organizes to manage sensor-to-sensor communications to compute more complex functions and to enable in-network data storage.

We focused on two main models of the network data storage: The database and the data centric storage.

The database model enables the user to abstract the way in which the network is structured to provide an uniform model for data collection, access and computation. The use of the database model allows the user to change the behavior of the network from the outside. The database abstraction provides in this way a meta-application layer to specialize the database application on some more specific tasks. We presented different approaches that provide different aspects of the database logic on a WSN, namely *TinyDB* [31], *Cougar* [50], and *MaD-WiSe* [7] [8] [9]. All these systems provide the relational abstraction over a WSN, but they provide different sub-sets of databases' primitives.

The data centric storage enable the network to work as a load balanced storage for the data that are collected by the sensors. This use of the network is suitable for applications in which the sink node can be unavailable for long periods of time and between such accesses the network must store the data in a reliable way. We presented different protocols that implement such model, namely *Geographic Hash Tables* [42], *Cell Hash Routing* [11], *Graph EMbedding* [36], and *KDDCS* [6]. All these systems are able to use a meta-datum associated to an acquired datum and to find a position in the network to store the data.

Both these models enable a more accurate use of the network's resources with respect to traditional ones. The capability to store and compute data in a network is essential to provide only meaningful data to the user. The database model provides also an easy way to program the network for different tasks simply changing queries. The data centric storage model enables the network to store data for later retrieval and usage by the user or by the network itself.

These models are far to be complete and we can easily identify research issues that must be investigated to provide more solid tools. Two important issue are the *redundancy* and the *load balancing*. The redundancy techniques need to be improved to provide QoS strategies based on more sophisticated systems than pure replication. The load balancing techniques need to improved to guarantee an almost perfect balancing in both storage and computational load among all the sensor in order to optimize the resources usage.

References

[1] K. Akkaya and M. Younis, *A survey on routing protocols for wireless sensor networks*, *Ad Hoc Networks* 3 (2005), no. 3, 325–349.

[2] I. F. Akyildiz, W.lJ. Su, Y. Sankarasubramaniam, and E. Cayirci, A survey on sensor networks, *IEEE Communications Magazine* 40 (2002), no. 8, 102–114.

[3] ______, Wireless sensor networks: a survey, *Computer Networks* 38 (2002), no. 4, 393–422.

[4] K. Al-Karaki, Routing techniques in wireless sensor networks: a survey, *IEEE Wireless Communications* 11 (2004), no. 6, 6–28.

[5] M. Albano, S. Chessa, F. Nidito, and S. Pelagatti, *Q-NiGHT: Adding QoS to data centric storage in non-uniform sensor networks*, Tech. Report TR-06-16, Dipartimento di Informatica, Università di Pisa, 2006.

[6] M. Aly, K. Pruhs, and P. K. Chrysanthis, *KDDCS: a load-balanced in-network data-centric storage scheme for sensor networks*, *Proc. of the 15th ACM international con-*

ference on Information and knowledge management (New York, NY, USA), ACM Press, 2006, pp. 317–326.

[7] G. Amato, P. Baronti, and S. Chessa, *MaD-WiSe: Programming and accessing data in a wireless sensor networks*, *Proc. of IEEE Eurocon* (Belgrado, Serbia and Montenegro), November 2005, pp. 300–303.

[8] ______, *MaD-WiSe: a distributed query processor for wireless sensor networks*, Tech. Report 2006-TR-39, Istituto di Scienza e Tecnologie dell'Informazione del CNR, Pisa, Italy, November 2006.

[9] G. Amato, P. Baronti, S. Chessa, and V. Masi, *The stream system: a data collection and communication abstraction for sensor networks*, *Proc. of IEEE International Conference on Systems, Man, and Cybernetics,* (Taipei, Taiwan), October 2006.

[10] G. Amato, S. Chessa, F. Conforti, A. Macerata, and C. Marchesi, Health care monitoring of mobile patients, *Ercim news* 60 (2005), 6.

[11] F. Araujo, L. Rodrigues, J. Kaiser, C. Liu, and C. Mitidieri, CHR: a Distributed Hash Table for Wireless Ad Hoc Networks, *Proc. of the 25th IEEE International Conference on Distributed Computing Systems Workshops* (ICDCSW'05), June 2005.

[12] Jon Louis Bentley, Multidimensional binary search trees used for associative searching, *Commun. ACM* 18 (1975), no. 9, 509–517.

[13] P. Bonnet, J. Gehrke, and P. Seshadri, Querying the physical world, *IEEE Personal Communications* 7 (2000), no. 5, 10–15.

[14] ______, Towards sensor database systems, *Proc. of 2nd International Conference on Mobile Data Management (MDM 2001)* (Hong Kong, China), January 2001, pp. 3–14.

[15] P. Bose, P. Morin, I. Stoimenoviċ, and J. Urrutia, Routing with Guaranteed Delivery in Ad Hoc Wireless Networks, *Wireless Networks,* 7 (2001), no. 6, 609–616, Also in *Proc. of Workshop on Discrete Algorithms and Methods for Mobile Computing and Communications (DialM'99)*, Seattle, Washington, August 1999, 48–55.

[16] N. Bulusu, J. Heidemann, and D. Estrin, "gps-less low cost outdoor localization for very small devices, *IEEE Personal Communications Magazine* 7 (200), no. 5, 28–34.

[17] Q. Cao and T. Abdelzaher, Scalable logical coordinates framework for routing in wireless sensor networks, *Proc. of 25th IEEE International Real-Time Systems Symposium (RTSS 2004)* (Lisbon, Portugal), December 2004, pp. 349–358.

[18] A. Caruso, S. Chessa, S. De, and A. Urpi, GPS free coordinate assignment and routing in wireless sensor networks, *Proc. of 24th Joint Conference of the IEEE Computer and Communications Societies (Infocom 2005)* (Miami, FL, USA), March 2005, pp. 150–160.

[19] A. Cerpa, J. Elson, D. Estrin, L. Girod, M. Hamilton, and J. Zhao, Habitat monitoring: Application driver for wireless communications technology, *Proc. of 1st ACM SIGCOMM Workshop on Data Communications in Latin America and the Carribean* (San Jose, Costa Rica), 2001, pp. 20–41.

[20] T. H. Cormen, C. E. Leiserson, and R. L. Rivest, *Introduction to algorithms*, 1990.

[21] T. Gao, D. Greenspan, and M. Welsh, Improving patient monitoring and tracking in emergency response, *Proc. of International Conference on Information Communication Technologies in Health*, 2005.

[22] C. Intanagonwiwat, R. Govindan, and D. Estrin, Directed Diffusion: A Scalable and Robust Communication Paradigm for Sensor Networks, *Proc. of the 6th International Conference on Mobile Computing and Networking (MobiCom 2000)* (Boston, MA, USA), August 2000, pp. 56–67.

[23] J. Jaromczyk and G. Toussaint, Relative neighborhood graphs and their relatives, *Proceedings of IEEE* 80 (1992), no. 9, 1502–1517.

[24] X. Ji and H. Zha, "sensor positioning in wireless ad-hoc sensor networks using multidimensional scaling, *Proc. of 23th Annual Joint Conference of the IEEE Computer and Communications Societies (Infocom 2004)* (Hong Kong), March 2004, pp. 2652–2661.

[25] D. Johnson, D. Maltz, and J. Broch, DSR: The dynamic source routing protocol for multihop wireless ad hoc networks, *Ad Hoc Networking* (C. E. Perkins, ed.), Addison-Wesley, 2001, pp. 139–172.

[26] B. Karp and H. T. Kung, GPSR: Greedy Perimeter Stateless Routing for Wireless Networks, *Proc. of the 6th International Conference on Mobile Computing and Networking (MobiCom 2000)* (Boston, MA, USA), August 2000, pp. 243–254.

[27] M. Lin, A. Kumar, X. Qing, S. J. Beard, S. S. Russell, J. L. Walker, and T. K. Delay, Monitoring the integrity of filament wound structures using built-in sensor networks, *Proc. of SPIE – Smart Structures and Materials 2003: Industrial and Commercial Applications of Smart Structures Technologies* (San Diego, CA, USA), vol. 5054, 2003, pp. 222–229.

[28] J. Liu, F. Zhao, and D. Petrovic, Information-directed routing in ad hoc sensor networks, WSNA '03: *Proceedings of the 2nd ACM international conference on Wireless sensor networks and applications* (New York, NY, USA), ACM Press, 2003, pp. 88–97.

[29] S. Madden, M. J. Franklin, J. M. Hellerstein, and W. Hong, TAG: a Tiny AGgregation service for ad-hoc sensor networks, *Proc. of 5th Symposium on Operating Systems Design and Implementation (OSDI 2002)* (Boston, MA, USA), 2002.

[30] S. Madden, R. Szewczyk, M. J. Franklin, and D. Culler, Supporting aggregate queries over ad-hoc wireless sensor networks, *Proc. of 4th IEEE Workshop on Mobile Computing Systems and Applications (WMCSA 2002)* (Callicoon, NY, USA,), June 2002, pp. 49–58.

[31] S. Maddenand, M. J. Franklin, J. M. Hellerstein, and W. Hong, The design of an acquisitional query processor for sensor networks, *Proc. of the 2003 SIGMOD Conference (San Diego, CA, USA)*, June 2003, pp. 491–502.

[32] D. J. Malan, M. Welsh, and M. D. Smith, A public key infrastructure for key distribution in tinyos based on elliptic curve cryptography, *Proc. of 1st IEEE communications Society Conference on Sensor and Ad-Hoc Communications and Networks*, 2004, pp. 71–80.

[33] A. Marcucci, M. Nati, C. Petrioli, and A. Vitaletti, Directed diffusion light: low overhead data dissemination in wireless sensor networks, *Proc. of Vehicular Technology Conference,* VTC 2005-Spring. 2005 IEEE 61st, vol. 4, 2005, pp. 2538–2545.

[34] R. Nagpal, H. Shrobe, and J. Bachrach, Organizing a global coordinate system from local information on an ad hoc sensor network, *Proc. of 2nd International Symposium on Information Processing in Sensor Networks* (IPSN 2003) (Paolo Alto, CA, USA), April 2003, pp. 333–348.

[35] A. Nasipuri and K. Li, A directionality based location discovery scheme for wireless sensor networks, *Proc. of 1st ACM International Workshop on Wireless Sensor Networks and Applications* (WSNA 2002) (Atlanta, GA, USA), September 2002, pp. 105–111.

[36] J. Newsome and D. Song, GEM: Graph EMbedding for Routing and Data-Centric Storage in Sensor Networks Without Geographic Information, *Proc. of the First International Conference on Embedded Networked Sensor Systems* (Los Angeles, California, USA), Nov. 2003, pp. 76–88.

[37] D. S. Niculescu and B. Nath, Ad hoc positioning system (APS) using AOA, *Proc. of 22nd Joint Conference of the IEEE Computer and Communications Societies* (Infocom 2003) (San Francisco, CA, USA), March–April 2003, pp. 1734–1743.

[38] ______, Dv based positioning in ad hoc networks, *Telecommunication Systems* 22 (2003), no. 1–4, 267–280.

[39] A. Okabe, B. Boots, and K. Sugihara, *Spatial tesselations: Concepts and applications of voronoi diagrams*, Wiley, 1992.

[40] C. E. Perkins and E. M. Royer, Ad-hoc on-demand distance vector routing, *Proc. of the 2nd IEEE Workshop on Mobile Computer Systems and Applications* (New Orleans, LA, USA), February 1999.

[41] A. Rao, S. Ratnasamy, C. Papadimitriou, S. Shenker, and I. Stoica, Geographic routing without location information, *Proc. of 9th International Conference on Mobile Computing and Networking* (MobiCom 2003) (San Diego, CA, USA), 2003, pp. 96–108.

[42] S. Ratnasamy, B. Karp, S. Shenker, D. Estrin, R. Govindan, L. Yin, and F. Yu, Data-centric storage in sensornets with GHT, a geographic hash table, *Mob. Netw. Appl. (MONET)*8 (2003), no. 4, 427–442.

[43] A. Savvides, C. Han, and M. B. Strivastava, Dynamic fine-grained localization in ad-hoc networks of sensors, *Proc. of 7th International Conference on Mobile Computing and Networking (MobiCom 2001)* (Rome, Italy), July 2001, pp. 166–179.

[44] K. Seada and A. Helmy, Efficient and robust geocasting protocols for sensor networks, *Computer Communications* 29 (2006), no. 2, 151–161.

[45] Y. Shang and W. Ruml, Improved MDS-based localization, *Proc. of 23th Joint Conference of the IEEE Computer and Communications Societies* (Infocom 2004) (Hong Kong), March 2004, pp. 2640–2651.

[46] M. Srivastava, R. Muntz, and M. Potkonjak, Smart kindergarten: Sensor-based wireless networks for smart developmental problem-solving environments, *Proc. of 7th International Conference on Mobile Computing and Networking* (MobiCom 2001) (Rome, Italy), 2001, pp. 132–138.

[47] D. C. Steere, A. Baptista, D. McNamee, C. Pu, and J. Walpole, Research challenges in environmental observation and forecasting systems, *Proc. of 6th International Conference on Mobile Computing and Networking (MobiCom 2000)* (Boston, MA, USA), 2000, pp. 292–299.

[48] R. Szewczyk, A. Mainwaring, J. Polastre, J. Anderson, and D. Culler, An analysis of a large scale habitat monitoring application, *Proc. of 2nd International Conference on Embedded Networked Sensor Systems (SenSys 2004)* (Baltimore, MD, USA), 2004, pp. 214–226.

[49] H. Wang, J. Elson, L. Girod, D. Estrin, and K. Yao, Target classification and localization in habitat monitoring, *Proc. of IEEE International Conference on Acoustics, Speech, and Signal Processing (ICASSP 2003)* (Hong Kong), 2003, pp. 844–847.

[50] Y. Yao and J. Gehrke, The Cougar Approach to In-Network Query Processing in Sensor Networks, *SIGMOD Record* 31 (2002), no. 3, 9–18.

[51] F. Zhao and L. Guibas, *Wireless sensor networks an informationprocessing approach*, Morgan Kaufman Publisher, S. Francisco, 2004.

In: Wireless Communications Research Trends
Editor: Tong S. Lee, pp. 191-227
ISBN: 978-1-60021-674-9

Chapter 6

ON RELIABILITY OF MOBILE AD-HOC WIRELESS NETWORKS

Jason L. Cook
US Army Armament Research, Development, and Engineering Center (ARDEC),
Quality Engineering & System Assurance Directorate
Jose Emmanuel Ramirez-Marquez
Stevens Institute of Technology,
Systems Engineering and Engineering Management Department

Abstract

The advancement of the self-forming, multi-hop Mobile Ad-hoc Wireless Networks (MAWN) created the need for new analysis methods which enable the accurate determination of the reliability and availability of these networked systems. Accordingly, a set of new and innovative methods has been developed and further research is on-going. The need for these methods is because contrary to hardwired networks, the MAWN is a scalable network without infrastructure. Along this line, the MAWN's configuration forms dynamically and probabilistically and as such no singular graphical depiction or mathematical function is able to describe its reliability. It is this feature that precludes the use of traditional methods. These new analytical methods are progressing in parallel with the proliferation of this technology so that reliable performance may be realized. Due to its flexibility, this new network scheme is often deployed in critical applications such as military and first responder applications. In such cases, reliability becomes a paramount system attribute. The primary contribution of this method is to describe and summarize the published methods in a single location. Taken together the tools are now available to a practitioner to analyze and optimize MAWN reliability. Mostly, this work is motivated by the application of MAWN technology in the DoD tactical networks and the import of their reliable operation when employed for this use. Throughout the chapter, DoD network examples will be used to demonstrate the reliability methods developed for the MAWN. The methods include both closed form analysis and Monte Carlo simulation techniques to establish terminal-pair reliability of the MAWN under a random waypoint mobility model; the metrics include two-terminal, k-terminal, and all-terminal reliability.

Introduction

The reliability of a Mobile Ad-hoc Wireless Network (MAWN) MAWN is paramount in its prevailing applications; such as for DoD and First Responder networks. For these applications, ad-hoc networking allows for the deployment of tactical networks in areas that are either too harsh or unstable to emplace permanent network infrastructure. It provides the life-line for the users of this technology which employ it in safety critical applications and life threatening situations. MAWN is rapidly becoming the preferred solution for flexible and low cost networking. In addition, due to its flexibility, this new network scheme is often deployed in critical applications such as military and first responder applications. In such applications, the network enables data and voice communications. The loss of the network to one or more members drastically impacts the capability of that member and thus the group as a whole. Thus, reliability becomes a significant system attribute which is critical to mission success and also safety. However, to achieve system reliability the system engineer must first be able to define and measure this metric. This is a prerequisite to improve the reliability of any system. The challenge is that the MAWN does not conform to the basic assumption on which existing network reliability methods are founded. So, these same methods are inappropriate and incapable of measuring the reliability of the MAWN.

The specific feature that precludes the use of traditional methods is the MAWN's absence of infrastructure items (e.g. routers and base stations). For hardwired and infrastructure based wireless networks (cellular networks as an example) the configuration of a network is known and mostly constant. In other words, the structure of the network in terms of component connectivity is known *a priori*. Accordingly, the component-wise relationship to reliability can be depicted graphically with methods such as: reliability block diagrams (RBD) and fault tree analysis (FTA). Similarly, it is possible to rigorously develop a closed form expression to express this relationship mathematically or for more complex systems it is possible to develop a mathematical approximation based on cut-sets and other techniques [1]. On the other hand, a MAWN consists of only mobile nodes that dynamically form and reform the configuration of the network as they move. The result is that any permutation of node pair links may result when the network forms and reforms. In effect, the reliability block diagram or reliability expression that represents the system changes with time due to the mobility of the nodes.

This chapter will review the existing methods for analysis of network reliability. The research will include two, *k*, and all-terminal reliability analysis techniques. In addition, a review of the literature in fields related to this topic will be presented, including: mobility modeling, wireless communications capacity, reliability optimization, and ad-hoc networking. This extensive literature review is performed to identify the current gaps and drawbacks in analyzing MAWN reliability with current methods that exist in the related fields of research. Moreover, it will stress the importance of the new methods as a set of tools that may be applied directly for MAWN reliability analyses and design. In addition, these methods will be applicable, with minor modifications, to other types of distributed and scalable systems of systems. Each method will be compared with existing methods, e.g. RBD and FTA to show its originality and value in terms of expanding the state-of-the-art. Each analysis method or model will be described algorithmically and demonstrated via illustrative examples that closely represent applications expected to employ a MAWN and require its reliable operation.

The set of methods to perform such analyses will be described within this chapter and will contribute to the field by way of their utility to designers of MAWN. The first of three focus areas is a method for MAWN reliability which develops the MAWN configurations probabilistically. This method is developed by fully enumerating the configurations and also implemented via a Monte Carlo (MC) simulation for a computationally efficient approximation. The second focus area combines mobility and reliability techniques to determine the impact of varied mobility characteristics (e.g. speed) on reliability. The third focus area develops a method to determine the reliability of a MAWN that is constrained by capacity requirements. Finally, the chapter concludes with a summary and description of the future direction of this on-going research.

General Problem Formulation

The following mathematical description will represent a general problem formulation that will be adopted and applied for all technical sections of this research. Additional, definitions and corresponding notation will be developed as each method or area may require unique considerations.

Let **G** = (**N**,**L**) represent a MAWN where **N** the set of nodes and **L** is a matrix that represents the links between the nodes. The elements of **N** shall be n_i for i=1,2..n where n is the number of nodes in **N**. Then, the nodes' operational status at time, t, shall be $n_i(t)$ where $n_i(t) = 1$ if node i is operational, else $n_i(t) = 0$. The elements of **L** shall represent the wireless links between nodes i and j as $l_{ij}(t)$ for every combination of arcs $i,j = 1,2,..,n$. Let $l_{ij}(t) = 1$ if the link between the nodes exists else let $l_{ij}(t) = 0$. The reliability associated with each node of network is represented by $r_i(t)$ and let $v_{ij}(t)$ represent the probability associated with $l_{ij}(t)$ existing (i.e. $v_{ij}(t)=P(l_{ij}(t)=1)$). The model chosen to imitate node mobility is RWMM and thus, the resulting distribution is uniform [23] resulting in $v_{ij}(t)=\lambda\ \forall\ i,j$ where $i \neq j$ and at all points in time.

Critique of Existing Methods

Much research has been done in areas related to MAWN reliability but very little focus has been paid to combining these adjacent fields for the purpose of reliability evaluation. The literature review to follow clearly shows this void and reviewed works are classified into four main areas: network reliability analysis, node mobility, ad-hoc networks, and wireless link capacity. These areas are all pertinent to this research and this fact will be demonstrated as the chapter evolves.

Network Reliability

Research in the field of network reliability is prevalent. Traditional fixed infrastructure networks where the interest lies in communication between a pair of specified nodes are commonly known as two-terminal networks. These networks are considered operational if there is a path between a pair of nodes usually labeled: source and destination. As an example, in a communications network the source node is the terminal that sends a message while the destination node, is the terminal intended to receive that message. Thus, the probability of a message successfully reaching the destination node from the source node is termed 2-terminal reliability (2TR). Two other popular metrics for network reliability are k-terminal reliability (kTR) and all-terminal reliability (ATR). All-terminal is defined as the probability that all terminals (nodes) within the network can communicate with each other node through some existing path, where as k-terminal the probability that at least k terminals are connected. If the variable k is set as 2 or as n, it then is equivalent to 2TR and ATR; respectively.

A method to calculate terminal reliability for a fixed network was presented by Fratta & Montanari [2] using Boolean algebra and the assumption that nodes are completely reliable and links are only available for a percentage of time. This percentage of time is assigned as the "arc reliability". Subsequently, these researchers provided a recursive method [3] that removed their initial assumption of completely reliable nodes. Hansler [4] claimed that algorithms assuming perfect node reliability are generally unrealistic and developed an algorithm, again leveraging recursive methods to split the problem into smaller problems and thereby developing a more efficient method. The author does this more efficiently by identifying any identical sub-problems and solving them only once. Netes & Filin [5] also proposed a method to include imperfect nodes in existing terminal reliability algorithms; specifically in the modified Dotson [6]. However, these authors also consider imperfect edges. These authors define their metric as terminal-pair reliability but defined it consistent with this paper's use of 2TR. Torrierri [7] proposed a method to include node reliability in the Dotson algorithm. Kuo *et al* [8] provided a 2TR analysis method that significantly decreases computation time by using edge expansion diagrams. Marseguerra *et al* [9] developed an approach to incorporate uncertainty into reliability calculations by using Monte Carlo simulation and Genetic Algorithms.

Network reliability methods have been expanded to account for 2TR when node and link capacity are incorporated into the model. Rocco & Muselli [10] compared machine-learning techniques to develop approximate reliability expressions for the capacitated case. Ramirez-Marquez & Coit proposed a heuristic method to address both multi-state and capacitated network reliability [11].

Due to the analytical complexity and computational cost of developing a closed form solution, Monte Carlo simulation is often used to analyze network reliability [12-18]. Fishman [12] provides a method of Monte Carlo Sampling for a general class to determine the all-terminal reliability of networks. This method uses upper and lower confidence bounds to provide a smaller variance than otherwise realized without estimating such bounds. Elperin et al [13] published another Monte Carlo method used to solve network reliability problems. Here the authors leverage graph evolution models to increase the accuracy of the resultant approximation. Rocco & Moreno [14] propose two methods based upon Monte Carlo methods that were developed to analyze two-terminal reliability of networks. These methods leverage cellular automata to refine the approach and avoid unnecessary recalculations. Next, Rocco & Zio [15] expanded these methods to address k-terminal and all-terminal reliability. The term k-terminal reliability is defined as a subset of all-terminal where k is a number less than the total node count. Cancela & Khadiri [16] also propose a method to solve the k-terminal problem, specifically they propose a formulation using recursive variance-reduction Monte Carlo estimator. Again, the benefit of their method is reduction of simulation time while providing and accurate approximation. Konak *et al* [17] propose additional methods that leverage Monte Carlo simulation techniques to solve network reliability problems. In this paper, the authors make use of geometric sampling and block sampling to efficiently analyze the network states while reducing the variance of the results. Finally, Ramirez-Marquez & Coit [18] expand upon traditional network reliability analysis by using Monte Carlo methods to explore the multi-state problem.

There has been a relatively few attempts in analyzing reliability of cellular and other infrastructure based wireless networks. Chen & Lyu [19] illustrated the process of handoff in a mobile cellular network; the transition of a mobile cellular phone's linkage from one cell tower to another. These transitions happen as a cellular user moves from the coverage area of one tower to the area covered by the other. Markov models were used to represent this configuration change and expressed network reliability as a function of the reliability of each node active in the configuration and the percentage of time that each configuration exists. However, this method is not directly applicable to a MAWN the major assumption is that the failure of any active node in the message's route results in failure, and as such, the reliability model is always represented by a configuration in series. This is not the case in a MAWN, because redundant paths may exist between source and destination. AboElFotoh *et al* [20] address two-terminal reliability calculations for Radio-Broadcast Networks where only nodes can fail and demonstrated that the method could be expanded to include the case of imperfect links. Unfortunately, there is no connection made between the probability of link failure and the relative movements of the nodes.

So, despite the published methods that apply to wireless networks, they still do not address the need for methods for MAWN. Specifically, they do not address the unique characteristics that make these methods inappropriate. These wireless methods still focus on a network configuration that is relatively constant and stable.

Node Mobility

The modeling of node mobility for wireless networking has also been a topic of extensive research [21,22,23]. Zonoozi & Dassanayake [21] reviewed mobility models to characterize the motion of mobile nodes in a communications network. Turgut & Chatterjee [22] reviewed the effect of mobility models on the longevity of communication path existence in mobile ad-hoc networks. Camp *et al* [23] surveyed current mobility models; describing the mobility patterns of each model and comparing several metrics relevant to MAWN performance. One of the mobility models is the Random Waypoint Mobility Model (RWMM).

The RWMM has been widely applied for modeling the mobility of a MAWN. The model describes the motion of a Mobile Node (MN) that travels in a randomly selected direction, at a randomly selected speed, for a certain amount of time. The MN then selects a new direction and speed. This model is run within a simulation boundary representing the expected coverage of the MAWN. Once a MN reaches this specified boundary, it changes direction and moves back into the area.

These models have been used to evaluate different network protocols for implementation in MAWN. Bhatt *et al* [24] describe the mobility effects on the network performance metrics Bit Error Rate (BER) and the minimum required node spatial density of an ad-hoc wireless network for full connectivity. However, Bhatt *et al* [24] do not incorporate the potential for node failure when accounting for total network connectivity. The mobility model described closely resembles the Reference Point and Pursue Group Mobility models detailed by Camp [23]. Chung [25] reviewed the finite life of a route (set of links making a path) wihin the MAWN and found that the life is exponentially distributed.

. The existence of these mobility models will aid in the development of reliability techniques however the quantitative application of these models to measure reliability has not yet been done.

MAWN Networks and Protocols

To understand the reliability of a system is to understand the ways in which it can fail and succeed. To gain this insight, it is necessary to examine the features and attributes of the MAWN. Several authors have explored these features. Toh [26] describes the MAWN by comparing it to a traditional network. Rather than enumerating the unique features of a MAWN, the author describes the features that are absent in a MAWN, namely, the infrastructure. Pahlavan & Krishnamurthy [27]go further and describe that in the MAWN the formation of links and paths is accomplished by the nodes only and without infrastructure items such as routers, switches, wiring, and base-stations. This freedom from infrastructure is what provides the MAWN its advantage; a self-forming and dynamic topology. In a MAWN, the nodes and the links they form comprise the network. The creation and termination of links is dependent on the position of a node relative to the other nodes in the network. The result is a dynamic network configuration that is subject to variation with every node movement and free from the constraints associated with infrastructure based networks. For these reasons, the Department of Defense (DoD) employs the MAWN to enable tactical communications. Freebersyser & Leiner [28] report on several MAWN developed for military use – the DARPA Packet Radio Network (developed in 1972) and the 1997 Task Force XXI Advanced Warfighting Experiment are two of these.

Currently, MAWN are mostly employed in safety critical operations that depend on the reliable operation of the network. To date, research on ad-hoc networks has been focused on the modification of network protocols, such as Transmission Control Protocol/Internet Protocol (TCP/IP), to accommodate for the mobility of the nodes and make network performance more robust in these applications [29,30]. The proliferation of MAWN began early in the last decade, most notably with the development of IEE802.11 "Bluetooth" technologies [31]. Abolhasan *et al* [31] compared and contrasted several protocols and their effectiveness when employed in a MAWN. Ye *et al* [29] proposed a deployment strategy to increase the probability of a 'reliable path'. The increase in path reliability was accomplished through strategic node placement, limiting its application to the few instances where node mobility can practically be controlled. Luo *et al* [30] presented a protocol to accommodate the probabilistic reliability of an ad-hoc network but do not explicitly measure network reliability.

After reviewing the existing literature, it becomes apparent that there is still much research focus on developing MAWN that are useful and robust for intended applications. The technology is still progressing from a breakthrough to a commodity and as this maturation continues a distinguishing factor for a MAWN may be its reliability. As seen in this section, most researchers disregard this feature in favor of expanding or extending the performance of MAWN.

Wireless Link Capacity

Reliability is a statistic regarding performance, specifically the probability of the system performing its intended function under specified conditions over a specified period of time. Since, a key performance parameter of a network is its transmission and the quantity it can transmit, the capacity of a MAWN is a key measure in this research. Some of the fundamental principles in this area are reviewed and will be utilized and applied in the methods that comprise this chapter. Specifically, several mathematical relationships will be reviewed to outline the basis on which the new methods will be developed.

Pahlavan & Krishnamurthy [27] describe the phenomenon of wireless transmission and its degradation by path loss. As described by these authors [27], Equation 1 illustrates the relationship between distance and received power; where d is the transmission distance; p_0 is the received power at a 1 meter distance; p_r is the received power at distance d; and α is the constant determined by the physical properties of the wireless medium; free space propagation yields $\alpha = 2$.

$$p_r = \frac{p_0}{d^{\alpha}} \tag{1}$$

The determination and optimization of capacity over a wireless link (channel) is a specialized field and a topic of much research [32, 36]. Shannon's Equation provides the theoretical maximum rate at which error-free digits can be transmitted and is therefore widely accepted as an appropriate model of the capacity of a wireless link, see Equation 2.

$$c = b * \log_2(1+s) \tag{2}$$

Within Shannon's Law, b is the bandwidth in Hz, and s is the signal to noise ratio (SNR). Equation 3 may used to calculate SNR at the receiver, with N_0 representing the noise at the receiver [27].

$$s = p_r/N_0 \tag{3}$$

Many researchers expand or adapt Shannon's law [32]. Yao & Sheikh [33] proposed a method to include Nakagami fading models into the signal to noise ratio parameter of Shannon's law. Lee [34] fills a gap left by Shannon. Pinkser *et al* [35] approach the problem of capacity approximation from a different angle. These authors develop closed form expressions to describe the sensitivity of wireless channels to non-Gaussian contaminating noise. Wyner [36] provides a clear and exhaustive survey of published works in the area of Shannon Theory (65 journal articles are referenced). Wyner concludes this survey with a summary of the areas requiring additional research. This brief synopsis is not intended to be exhaustive – certainly since the time of Wyner's review (1974) much more work has been done.

The purpose of the preceding section is to summarize some of the existing work in the area of capacity approximation and, in general, to draw out the breadth and depth of research in this area. Mainly, the preceding discussion is provided to demonstrate the wide use of Shannon's Equation and validate its use herein. However, further discussions of wireless channel capacity research are outside the scope of this chapter.

Wireless link capacity is the final area of technical literature that will be reviewed. Like the previous technical areas, capacity is one of the fundamental understandings required to develop appropriate measures of reliability and methods to determine them. For this reason, a brief review of research in this area is presented in order to draw out related concepts and equations that will be applied.

Probabilistic MAWN Formation Analysis Method

This section will describe a new method for the analysis of MAWN reliability. Specifically, it includes a description of the problem, the formulation of the analysis method, and examples of its application. This method will serve as the fundamental basis for this chapter. Methods that follow will build upon the basic principles and techniques described herein. The goal is to provide a method that, unlike those that presently exist, acknowledges a probabilistic formation of the MAWN topology and its susceptibility to change. This method is the first of its kind and does not require the systems engineer to know the configuration of the network *a priori*.

The method includes: a closed form analysis and two simulation based methods to accomplish this goal. The methods are useful to the system engineer because they enable the analysis of a MAWN's reliability when certain system and node attributes are known. These include: node reliability, node count, and the probability of link existence between node pairs. Similarly, with this method the designer is now capable of setting and manually optimizing these input parameters such that reliability objectives are met. This method enumerates all the potential configurations, the probability each would exist, their reliability, and the resultant MAWN two-terminal reliability ($2TR_m$). This method has published by Cook & Ramirez-Marquez [37,38].

Problem Definition

In addition to the general problem described in the introduction of this chapter, also let $\mathbf{C}$ define the set of possible network configurations. Finally, $2TR\alpha_k$ defines the two-terminal reliability of configuration α_k, $k = 1, 2 \ldots |\mathbf{C}|$.

Method Description

The first step towards developing this method is to define the metric of interest, $2TR_m$, as the probability that a communication path exists between the source and destination nodes. For MAWN, a path between two nodes will exist only if:

1. Links between the nodes form a path.
2. Every node along the path is operational.

The existence of a link in **L** is probabilistic and the number of potential network configurations is given by $|\mathbf{C}| = 2^{n*(n-1)/2}$. That is, the permutations of existing and non-existing links generate a set of all possible network configurations. The probability of each existing, in turn, is a function of link probability of existence, λ, the number of linked node pairs, η_l, and the number of unlinked pairs, η_u, in the configuration. The probability associated with each possible configuration is given by:

$$P(\alpha_k = 1) = \lambda^{\eta_l}(1-\lambda)^{\eta_u} \tag{4}$$

Finally, the $2TR_m$ can be obtained as a weighted average of the probability of existence for each configuration and the associated reliability. The result is the two-terminal reliability for the MAWN. Mathematically, this can be expressed as shown in Equation 5:

$$2TR_m = \sum_{k=1}^{|\mathbf{C}|} 2TR\alpha_k * P(\alpha_k = 1) = E[2TR\alpha_k] \tag{5}$$

The first step is to consider complete enumeration of the possible states of the network. The approach that follows, enumerates all possible configurations that a MAWN can take then, each configuration is assigned a probability of existence based on the RWMM. The method follows:

Initialization: Define n, r_i, and λ

Step 1: Enumerate all possible configurations of **G(N,L)** and stack them in set **C**.

Step 2: Determine $P(\alpha_k=1)$ based on Equation 2.

Step 3: For $k=1,..|\mathbf{C}|$, obtain $2TR\alpha_k$ based on r_i considering links in the configuration to have perfect reliability.

Step 4: Apply Equation 5 calculate $2TR_m$.

Illustrative Example

To maintain clarity of illustration, a three node MAWN has been analyzed in this section. This network is considered working if there exists a communication path between source node 1 and destination node 3. Moreover, the nodes in **G(N,L)** are assumed to be identical having a known r_i=0.9. The application of a RWMM provides the constant probability of link existence, $\lambda = 0.7$.

Step 1: For this network the total number of possible configurations |**C**| = 8. The set of all possible configurations is shown in Figure 1.

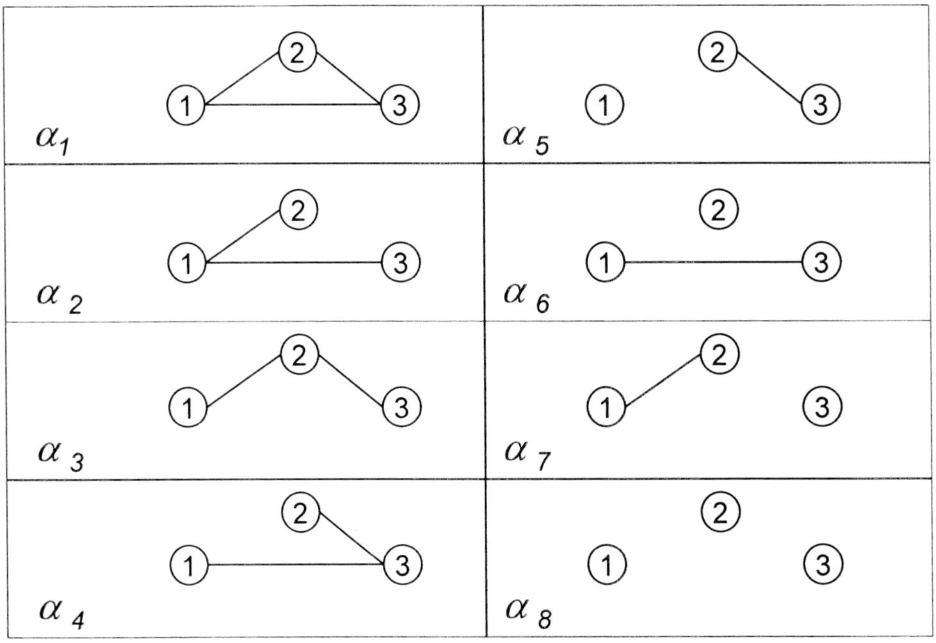

Figure 1. Set *C* for 3 Node MAWN

Step 2:Once the configurations have been obtained, the probability of existence for each configuration is calculated using Equation 2. Table 1 contains these results.

Table 1. Probability of α_i for 3 Node MAWN

α_i	l_{12}	l_{13}	l_{23}	**P(α_i = 1)**
1	1	1	1	0.34
2	1	1	0	0.15
3	1	0	1	0.15
4	0	1	1	0.15
5	0	0	1	0.06
6	0	1	0	0.06
7	1	0	0	0.06
8	0	0	0	0.03

Step 3: Once each configuration (and corresponding probability of existence) is enumerated, the 2TR of each configuration, $2TR\alpha_k$, is calculated. The reliability expression of each configuration is derived and the result is shown in Table 2.

Table 2. 2TRα_i for 3 Node MAWN

α_k	Path	$2TR\alpha_k$
1	$r_1 * r_3$	0.81
2	$r_1 * r_3$	0.81
3	$r_1 * r_2 * r_3$	0.73
4	$r_1 * r_3$	0.81
5	None	0.00
6	$r_1 * r_3$	0.81
7	None	0.00
8	None	0.00

Step 4: After the reliability and probability of existence are determined for each configuration, $2TR_m$ can be calculated using Equation 6 as illustrated in Table 3.

Table 3. 2TRm for 3 Node MAWN

α_i	l_{12}	l_{13}	l_{23}	$P(\alpha_i = 1)$	$2TR\alpha_i$
1	1	1	1	0.34	0.81
2	1	1	0	0.15	0.81
3	1	0	1	0.15	0.73
4	0	1	1	0.15	0.81
5	0	0	1	0.06	0.00
6	0	1	0	0.06	0.81
7	1	0	0	0.06	0.00
8	0	0	0	0.03	0.00
					$2TR_m$=0.6742

The complete enumeration method was implemented on a Windows based laptop running on a 1.6 GHz processor. The program was run for networks of four, five, and six nodes with the same input values for r_i and λ. The results obtained are shown in Table 4.

Table 4. Network Analysis Results

# Nodes	# Links	# Configurations	$2TR_m$	CPU Time
3	3	8	0.6742	10 sec
4	6	64	0.7461	90 sec
5	10	1024	0.7837	600 sec
6	15	32768	0.8001	7200 sec

Computational Efficient Alternatives

The enumeration method provides an exact solution of $2TR_m$, yet the method becomes computationally expensive for the analysis of large networks. Thus, a simulation technique comparable to that of Ramirez-Marquez & Coit [18], only modified for a MAWN and the non-capacitated case, has been developed as a means to estimate $2TR_m$. This MC simulation approach includes simulating the operational state of the nodes and links as described in pseudo-code below. The nodes are simulated first and then the links, acknowledging that no failed node can be linked.

Procedure to Simulate Node Status

for i=1,2…n

  $test \leftarrow$ select random number from uniform distribution between (0,1)

  if $test \le r_i$ then $n_i = 1$

  else n_i=0

  $\mathbf{N} \leftarrow n_i$

Procedure to Simulate Link Status

for i=1,2…n

  for j= i+1,2…n

  $test \leftarrow$ select random number from uniform distribution between (0,1)

  if ($test \le \lambda \cap n_i = 1 \cap n_j = 1$) then l_{ij}=1

  else; l_{ij}=0

  $l_{ji} = l_{ij}$

  $\mathbf{L} \leftarrow l_{ij}$ and l_{ji}

After simulating node and link status, the resulting link and node states are compared against the success criteria; a path between source and destination exists. This is done by analyzing the link configuration matrix, **L**, to determine all the nodes that have a path to and from the source node. A connectivity vector is defined with a length n as $\boldsymbol{\Lambda}$. Let $\Lambda_i = 1$ if node i is connected to the source node; therefore if a path exists from source to destination node which is numbered n then $\Lambda_n = 1$. $\boldsymbol{\Lambda}$ is then populated by performing a breadth first search on **L** to see if any combination of l_{ij} creates a path from between source and destination node.

These procedures are used to generate the following MAWN simulation approach, where Q is the number of runs in the simulation.

Calculation of $2\hat{T}R_m$

for q=1,2…Q

Simulate Network $\rightarrow$ $\mathbf{L}(q)$

Find Connectivity $\rightarrow$ $\Lambda_n(q)$

$$2\hat{T}R_m = \frac{\sum_{q=1}^{Q} A_n(q)}{Q} \tag{6}$$

After an estimate of the actual $2TR_m$ is generated, it is necessary to obtain the uncertainty of the approximation. When the sample is large enough, the variance associated with $2\hat{T}R_m$ can be approximated by:

$$Var\left(2\hat{T}R_m\right) = \frac{2\hat{T}R_m\left(1 - 2\hat{T}R_m\right)}{Q} \tag{7}$$

Simulation Approach 1 Tests

To test the accuracy of this simulation and the resultant approximation, the results of simulations of 10,000 runs were compared to the complete enumeration results. Table 5 shows the error and efficiency of the simulation method relative to the complete enumeration technique. This comparison is limited to networks of 6 nodes due to the computational time required to perform the complete enumeration technique on larger networks. However, the simulation approach allows for the analysis of larger MAWN and these are included in Section 4.

Table 5. Network Simulation Results

# Nodes	$2TR_m$ Results			
	Complete Enumeration	**Simulation**	**Simulation CPU Time**	**Relative Simulation Error**
3	0.6742	0.6770	10 sec	0.42%
4	0.7461	0.7498	23 sec	0.50%
5	0.7837	0.7882	29 sec	0.57%
6	0.8001	0.8083	32 sec	1.02%

It is important to note that the number of simulation runs Q, only captures a small percentage of the network configurations for large networks. However, since the contribution to $2TR_m$ of any given configuration is weighted based upon the P(α_k=1), the configurations with the largest contribution are most likely to be captured by the simulation. Table 6 shows that for larger values of Q the variation in the results is small.

Table 6. Simulation Results with varied Q

# Runs (Q)	$2TR_m$	CPU Time
10000	0.8049	5 sec
50000	0.8032	10 sec
100000	0.8023	19 sec
250000	0.8024	63 sec
500000	0.8025	240 sec
1000000	0.8028	600 sec

Monte Carlo Simulation of $2TR_m$ (Path Method)

This method provides two benefits. First, the method will compare the simulated network state to the possible successful paths. Enumeration of all paths saves computational time because rather than analyzing the link matrix via the breadth first search for each run, the enumeration process occurs only once; before the execution of the simulation's runs. This simplifies the steps that get repeated in each run to simply comparing the link matrix to the group of possible successful paths. Second, this method provides another capability that the above method does not, the ability to exclude successful paths if their length exceeds a pre-defined limit. For performance reasons such as latency (message transit time), it is often necessary to limit the path length between source and destination. Toh [26] describes that path (also termed route) hop limits are usually employed to decrease latency by limiting the number of intermediate nodes between source and destination. For the remainder of this paper path length will be referred to by the number of intermediate nodes between the source and destination for the path in question.

The same method to simulate the node and link status previously presented is employed in for this approach. However, rather than determining all nodes linked to the source, only paths that meet a maximum path length constraint are considered. Step 1 in this method includes the enumeration of the possible node permutations that can form a path of a length that does not violate the length constraint. For example, for a network of 10 nodes where n_1 is the source and n_{10} is the destination, all permutations of m or fewer elements of 2,3...9 form potential paths when m represents the limit on intermediate nodes. The number of paths with m intermediate nodes is then given by Equation 8 while the number of paths of length less than or equal to m is given by Equation 9.

$$\#path_m = \frac{(n-2)!}{(n-2-m)!} \tag{8}$$

$$\#path = \sum_{i=1}^{m} \frac{(n-2)!}{(n-2-i)!} \tag{9}$$

Each defined path is enumerated and then stored in the form of an n-by-n matrix, **P**. **P** is comparable to the link matrix **L**, where $p_{ij}=1$ if that link is contained in the path, else $p_{ij}=0$. Finally, for each run (q) the simulation of **L** is then compared to each **P** matrix. To make this comparison, define **L>P** if $l_{ij} \geq p_{ij}\ \forall\ i,j$. Then, if **L>P** for any **P**, then $\Lambda_n = 1$, else $\Lambda_n = 0$. That is, if **L** includes any path identified by a matrix **P**, then a successful path meeting the length constraint exists. So an approximation of $2TR_m$ is given by:

$$2\hat{TR}_m = \frac{\sum_{q=1}^{Q} \Lambda_n(q)}{Q} \tag{10}$$

Table 7. *2TR$_m$* Results

	***2TR$_m$* Results**		
# Nodes	**Complete Enumeration**	**Simulation**	**Simulation (Path Method)**
3	0.6742	0.6770	0.6816
4	0.7461	0.7498	0.7492
5	0.7837	0.7882	0.7906
6	0.8001	0.8083	0.7990

The results from this method have been appended to the results obtained in Table 5 for comparative purposes. Table 7 depicts the results for the same network parameters ($r_i = 0.9$ and $\lambda = 0.7$) and sizes previously analyzed. The number of runs used for each simulation method to gather the results in Table 7 was $Q = 10{,}000$. For these comparisons, no limit was placed upon maximum path length. However, analyses that include limits on maximum path length are presented in Section 3.4.5.

Parametric Sensitivities

To evaluate a MAWN and its reliability from a systems engineering perspective, it is important to understand the sensitivity of $2TR_m$ to the various input parameters. Figure 2 and Figure 3 show two simulations that both approach a limit of r_i^2 as the network size increases.

This natural limit is the result of the existence of redundant paths between source and destination, therefore the only single point failures become the terminal nodes, and the $2TR_m$ of any MAWN is limited by r_i^2 and will approach this value as λ and n are increased. However, the lower the probability of link existence, the greater the number of nodes required to reach this limit. Note that this limit is reached at a network of 8 nodes for $\lambda = 0.7$ but, at a network size of 18 nodes for $\lambda = 0.25$.

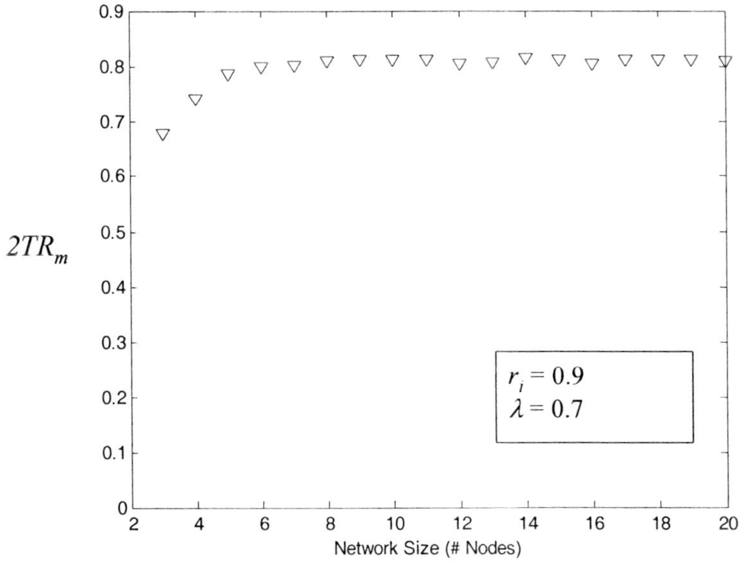

Figure 2: Simulation results of *2TRm* for Various Network Sizes

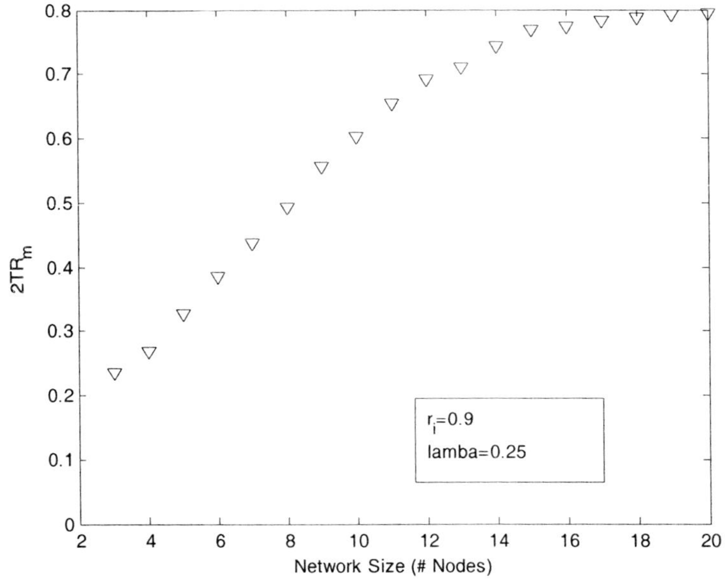

Figure 3: Simulation results of *2TRm* for Various Network Sizes

The primary reason for developing the alternate method of Monte Carlo (MC) simulation for the MAWN is to explore the effect of limited path lengths on $2TR_m$. Reviewing the results in Table 4, two relationships appear. First, as the probability of link existence increases so too does the reliability. Similarly, the reliability of the network increases as the allowable path length increases. Both of these relationships are the result of increasing the total number of paths between source and destination nodes. The result is important as it pertains to throughput or delay constrained message traffic that will have limitations on path length (i.e. number of hops). Figure 4 graphically depicts these relationships.

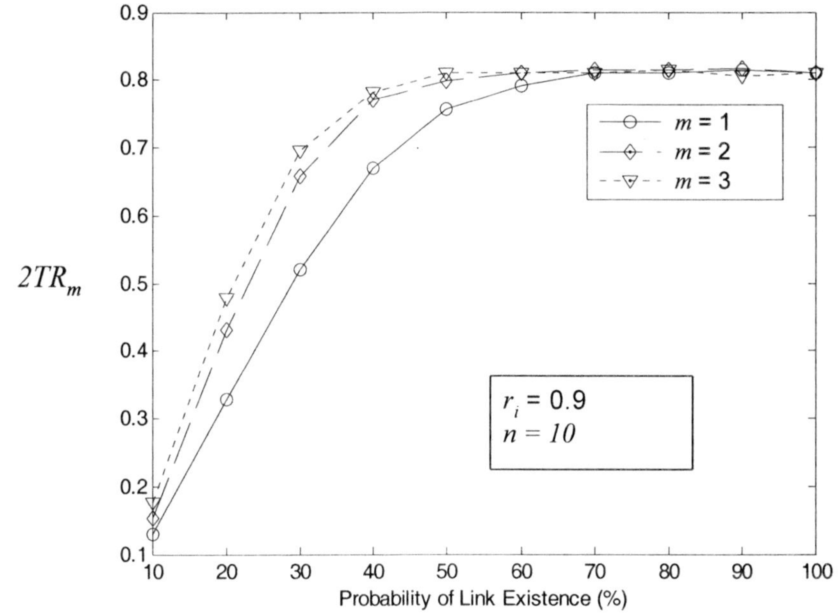

Figure 4: Effect of Path Length with Varying λ

Summary

These sensitivity analyses provide valuable knowledge to the systems engineer faced with balancing performance, reliability, operational and cost constraints. These methods provide the system engineer the ability to identify optimal combinations of path length, n, λ, and r_i. They may also be used qualitatively while comparing design and deployment decisions. Prior to the development of this method to quantify MAWN reliability, design decisions could only be made qualitatively. This method allows improved design by providing the systems engineer with the ability to quantify reliability.

MAWN Mobility & Reliability Modeling Method

In this and the following section, the goal of each subsequent method is to remove at least one assumption from the primary probabilistic method. The removal of assumptions and replacement by more exact numerical values derived from system attributes is key to enhancing the accuracy and utility of the methods described in the chapter. In this method, the assumption or input that the probability of link existence is known is addressed. In fact, this is removed and replaced by a mobility model. The mobility model is then used to determine node location in time and therefore determine whether a given link exists.

The added value with this method is the understanding gained when mobility parameters (e.g. speed or coverage area) and transmission range can also be analyzed and their impact on the reliability of an ad-hoc network may be determined. In addition, this method can quantify the impact of radio transmission distance on reliability. This method is described in [39]; which is currently under peer review.

Problem Definition

The notation from Section 1.2 is also retained for this method. Also, define node separation distance, d_{ij}, as the distance between node i and j for all nodes in G. This value, d_{ij} is calculated by using Equation 11.

$$d_{ij} = ((x_i - x_j)^2 + (y_i - y_j)^2)^{1/2} \tag{11}$$

Then, let a link exist between nodes i and j if their separation distance, d_{ij}, is not greater than their transmit/receive range, t_{ij}.

$$l_{ij} = 1 \text{ if } d_{ij} \le r_{ij}\text{; else let } l_{ij} = 0 \tag{12}$$

Because the nodes' position changes due to their mobility, then the connectivity matrix **L** representing the network configuration also changes. Similarly, the nodes operational status is not constant; rather the nodes are subject to failure over time. For the problem definition the RWMM is implemented as follows: The mobility of a given node is defined by the linear velocity, v_i, and heading, φ_i. Considering the nodes' position in time increments, Δt, the position of all nodes at each increment of time must be determined. The next position will be calculated by obtaining the linear distance traveled along the x (horizontal) and y (vertical) axes during the previous time increment, that is, the change in x and y position. Equations 13 and 14 are used.

$$x_i(t+\Delta t) = x_i(t) + \Delta t * v_i(t) * \cos\varphi_i(t) \tag{13}$$

$$y_i(t+\Delta t) = y_i(t) + \Delta t * v_i(t) * \sin\varphi_i(t) \tag{14}$$

Define the connectivity of the network, Λ, where $\Lambda_i(t)$ is the connectivity state of the ith node at time t with respect to the source node. That is, $\Lambda_i(t) = 1$ if the ith node has a path to the source node at time t, else $\Lambda_i(t) = 0$. Then the two-terminal reliability (2TR) of the MAWN, notated at $2TR_m$, is given by Equation 15.

$$2TR_m(t) = P(\Lambda_i(t) = 1) \tag{15}$$

Another metric to quantify the reliability of a network is the proportion of nodes for which a path to and from the source exists; this shall be termed coverage. This metric is important because it provides a quantitative measure of the overall quality of the network on a aggregate scale. Define χ as this proportion and refer to this proportion as network coverage henceforth. Then, $\chi(t)$ is the coverage of the network at time t given the connectivity due to mobility and operational status of the nodes. Equation 16 can be used to quantify network coverage as:

$$\chi(t) = \frac{\sum_{i=1}^{|n|} \Lambda_i(t)}{n} \quad (16)$$

Finally, define the metric ϖ as network volatility, as the percentage of all the possible links for which a status change occurs in each time increment and it is calculated as in Equation 17.

$$\varpi(t) = \frac{|L(t) - L(t+\Delta t)|}{n(n-1)} \quad (17)$$

Method Description

A Monte Carlo simulation approach has been developed to calculate the $2TR_m(t)$, $\chi(t)$, and $\varpi(t)$. The following steps make up the procedure.

Initialization: Define network input parameters: number of nodes, |N|; transmission range, $r_{i,j}$; node reliability, θ and β; maximum and minimum velocity, v_{max} and v_{min}; network coverage area; mission duration, t_{max} and time increments, Δt.

Step 1: Simulate the operational status of the nodes. The Weibul distribution is selected for this model due to its flexibility, however any distribution may be used by modifying Equation 18 accordingly.

$$\mathrm{P}(n_i(t) = 1) = e^{(-t/\theta)^{\beta}} \quad (18)$$

Therefore, step 1 selects from this distribution to determine operational status for each node. Note that once a node fails it remains failed for the remainder of that simulation run. That is, the simulation considers the node failures non-repairable.

$n_i(t) \leftarrow$ Weibul distribution defined by θ and β

Step 2: Given the nodes' position, determine connectivity of the network.

$l_{i,j}(t) \leftarrow$ output of Equation 12 for all node pairs i and j

$\mathbf{L}(t) \leftarrow l_{i,j} \; \forall \; i \;\&\; j = 1,2...n$

Step 3:Analyze **L** to determine connectivity to source, Λ. The matrix **L** is analyzed via a breadth-first search. Pseudo-code to describe this BFS of **L** and determine Λ follows with the source as the first node.

if $n_1 = 1$ then $\Lambda_1 = 1$

for *hop*=1,2...*n*-1
for i=1...*n*
if $\Lambda_i = 1$
for j=1,2...*n*
if $l_{ij} = 1$; then $\Lambda_j = 1$, else $\Lambda_j = 0$
$\Lambda(t) \leftarrow \Lambda_i \ \& \ \Lambda_j$

Step 4: Calculate $\chi(t)$ from $\Lambda(t)$ via Equation 17.

Step 5: Simulate the mobility of the network's nodes according to a mobility model that accurately represents the expected behavior of the nodes intended application and host platform.

$v_i(t) \leftarrow$ uniformly distributed velocity between v_{max} and v_{min}

$\varphi_i(t) \leftarrow$ uniformly distributed direction between 0 and 2π radians ($0 \ldots 360^{\circ}$)

Step 6: Next, calculate the position of each node in the next time increment.

$x_i\ (t+\Delta t) \leftarrow$ determine next position using previous position and Equation 3.

$y_i\ (t+\Delta t) \leftarrow$ determine next position using previous position and Equation 4.

Step 7: Perform steps 1 through 6 for time increments of Δt from $t = 0 \ldots t_{max}$.

$$t \leftarrow t + \Delta t$$

Step 8:Repeat simulation steps 1 through 7 for Q runs. Calculate an estimate of two-terminal reliability from the source to a defined destination node or nodes, n_i, and the network coverage metrics as functions of time by averaging the results for each simulation run, q, at each time increment of the total mission duration.

$$2\hat{TR}_m(t) = \frac{\sum_{run=1}^{Q} \Lambda_i(q,t)}{Q} \tag{19}$$

$$\hat{\chi}(t) = \frac{\sum_{run=1}^{Q} \chi(q,t)}{Q} \tag{20}$$

$$\hat{\varpi}(t) = \frac{\sum_{run=1}^{Q} \varpi(q,t)}{Q} \tag{21}$$

Illustrative Example

The flexibility provided by a MAWN is employed to meet the unique requirements for the tactical networks of the DoD. For the purposes of illustration, such an implementation will be detailed and analyzed using these methods. The application is defined as follows:

The network is composed of eighteen dismounted infantry (soldiers on foot) outfitted with identical man-portable radios capable of ad-hoc networking connectivity. Each of the soldiers represent an individual network node. Each radio has a transmission range of 3 miles with its reliability dictated by a Weibull distribution with $\theta = 1000$ $\beta = 1.5$. Coverage area of 64 mi^2 squared with a maximum and minimum velocity of 6 and 3 mph. The mission time is 72 hours.

Initialization

$$n = 18,\ r_{i,j} = 3 \text{ miles } \forall\ i,j\ \theta = 1000\ \beta = 1.5,\ v_{max} = 6 \text{ mph and } v_{min} = 3 \text{ mph}$$

network coverage area = 64 mi^2, t_{max} =72 hours , $\Delta t = 1$ hour

Results

The results of the implementation of the simulation approach for these settings are presented graphically and quantitatively. First, the results of a single run are presented graphically. depicts the results for $\Lambda_i(t)$ (i=18 the destination node status is then given by n_{18}).

As defined previously, $\Lambda_i(t)$ exhibits a binary nature; either a path from source to destination exists or it does not. For example, the network exhibits a binary nature because as the nodes move paths between the source and sink node are created or terminated. Figure 5 illustrates the results of this process. As the mission time continues more node failures are suffered and the frequency of occurrence for path failure increases. That is, $\Lambda_i(t) = 0$ more frequently as t increases. At $t = 64$ the source node failed in simulation run $q = 15$, resulting in $\Lambda_i(t) = 0$ for all $t > 64$.

Figure 6 depicts the results of Equations 9 (reliability) and 10 (coverage) applied against the results of all Q simulation runs.

These two metrics, $2TR_m$ and χ, are closely related because the probability of a path from source to destination is comparable to the probability of a path to any other non-source node. This is only the case for mobility patterns that result in a uniform spatial distribution of nodes, such as RWMM. The systems engineer developing a MAWN or a system of systems that uses a MAWN as its network may apply these metrics and these methods to understand how mobility and other factors will influence the reliability and availability of the functions provided by the network.

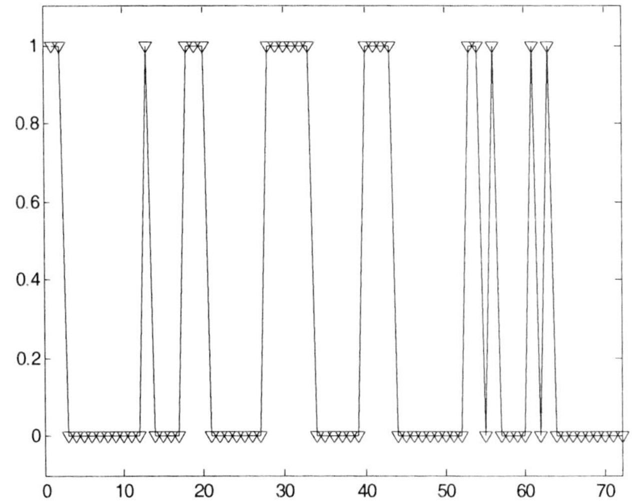

Figure 5: Simulation Run Results

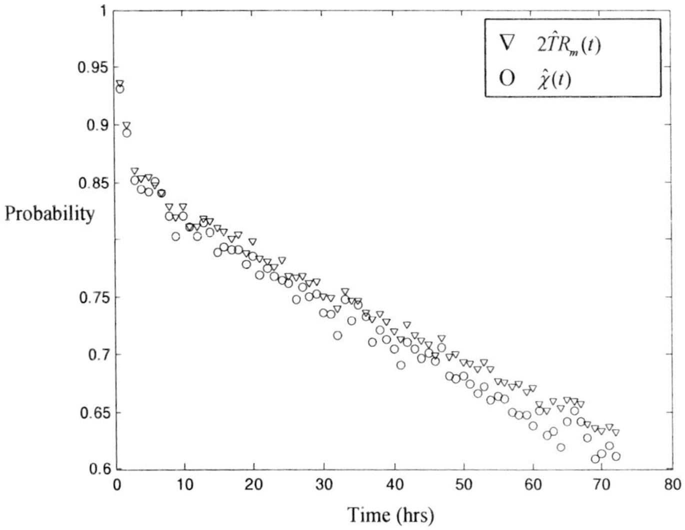

Figure 6: Reliability and Coverage

Results for Equation 11 (volatility) are displayed in Figure 7. The simulation included 2500 runs; Q = 2500. For this case, the volatility is fairly constant indicating that the frequency of state change for the links is not impacted by time but will remain approximately the same as nodes move and the mission time elapses.

Result Accuracy

The results described above and captured within the figures describe the statistics captured when Q = 2500. Increasing the number of simulations will increase the accuracy of the approximations for $2TR_m$ and χ, however this also increases the computational time required to employ this Monte Carlo method. The variance associated with $2\hat{TR}_m$ can be approximated by Equation 22:

$$Var\left(2\hat{TR}_m\right) = \frac{2\hat{TR}_m\left(1 - 2\hat{TR}_m\right)}{Q} \tag{22}$$

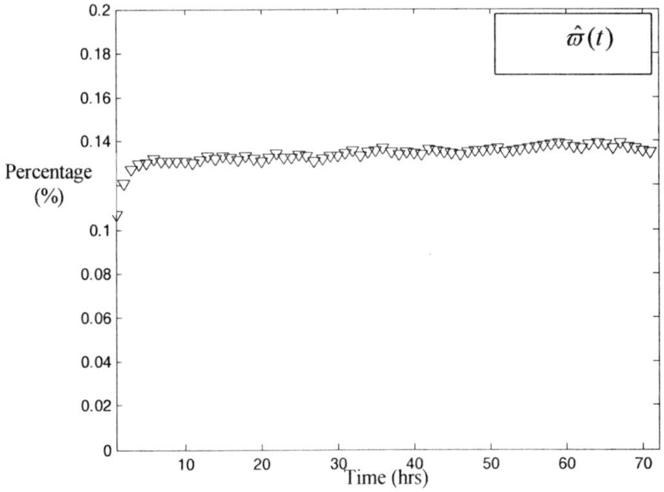

Figure 7: Volatility

Q is varied to display the effect on accuracy and computational time. The results and variance of associated with varied run quantities are presented in Table 8. The method allows the systems engineer to select number of runs so that the accuracy (variance) of the approximation is sufficient.

Table 8. Effect of Varied Q

Q	$2\hat{TR}_m(72hrs)$	**Variance**	**Computation Time**
100	0.6600	0.00224	30 sec
1000	0.6010	0.00024	60 sec
2500	0.6152	0.00009	120 sec
5000	0.6152	0.00005	300 sec
10,000	0.6152	0.00002	750 sec

Parametric Sensitivities

The primary purpose of the method and metrics developed is to allow for the accurate approximation of two-terminal reliability for a MAWN. In so doing, the mobility, reliability, and performance (e.g. range) of the individual nodes are considered. These individual mobile communication systems (the nodes) may also be considered components of the networked system enabled by MAWN technology. As such, the development and leverage of the

MAWN schema must also consider a required level of reliability. In this light, the sensitivities and relationships between reliability and the individual component parameters must be understood. This method is used to expand upon these interactions and describe how it may be utilized to aid in the system engineering processes.

The first relationship explored is how sensitive is $2\hat{TR}_m$ to the number of nodes within the network. Table 9 includes the results for different values of n, when all input parameters remain constant. For this case, the relationship is clear: the number of nodes increases the network reliability by providing more potential paths from source to destination. Thus, the methods developed can be used to optimize the networks scale (node count) to meet reliability targets.

Table 9. Impact of Network Size on MAWN Reliability

n	$2\hat{TR}_m(72hrs)$
9	0.4136
15	0.5804
18	0.6152
22	0.6520
27	0.7120

The coverage area of a mission can change unexpectedly as such; it is intuitive that the reliability of the network will decrease as a function of the area because as the nodes disperse over a greater distance, more links break, and the number of connecting paths from source to sink will decrease. The results depicted in Table 10 confirm this hypothesis.

Table 10. Impact of Network Coverage Area on MAWN Reliability

Area (mi^2)	$2\hat{TR}_m(72hrs)$
64	0.6152
100	0.6024
144	0.4976
225	0.2909
400	0.1272

Similarly, Table 11 shows the relationship between range and MAWN reliability. Based on these results, it seems intuitive that the network reliability will be proportional to the ratio of transmission range to coverage area.

Table 11. Impact of Transmission Range on MAWN Reliability

$t_{i,j}$ **(miles)**	$2\hat{TR}_m(72hrs)$
1	0.0444
3	0.6152
5	0.6846
7	0.7134
8	0.7652

Finally, the impact to network volatility, due to changes in node mobility is explored. The results presented in Table 12 show that the volatility of the network links increases as the maximum velocity of the nodes increases.

Table 12. Impact of Node Speed on Volatility

v_{max} (mph)	ϖ
1	0.0361
2	0.0619
4	0.0937
6	0.1336
8	0.1425

*Coverage range of 400 mi^2; $v_{min} = 0.5v_{max}$

Summary

These results illustrate how several input variables impact the reliability of the network. This is important because it provides multiple possible solutions to meet an overall MAWN reliability requirement. For example, when developing a design one may choose to deploy more nodes, slower nodes, nodes with a greater reliability, greater transmission range, or most likely some optimized combination of all these attributes. This method allows the analyst to quantify the impact of the four parameters explored; none of which are traditionally related to reliability.

These options provide the context for further analyses and these new methods provides the foundations to perform it. Along with these analyses, this method can be used in the development of any MAWN with reliability requirements or goals. It can be used to determine minimum node reliability characteristics. It may also be used to determine the network size with respect to both coverage area and node count. Finally, once the network components and host platforms are developed, this method may provide requirement verification where a reliability test of an entire network may prove too costly.

Similarly, the ability to understand these complex interactions provides for value for operational use. Retaining the DoD example, the commander can make decisions about deployment by considering the impact of coverage area, mobility patterns, and the number of deployed nodes.

Mobility & Reliability of Capacitated MAWN

With this method, the goal is to provide even greater resolution about the state of the links within the MAWN. In the previous section, the knowledge of the link state was abstracted from a mobility model by determining separation distance and comparing that to a fixed transmission distance. Here, the assumption of a fixed transmission distance is removed from the inputs and replace by parameters that describe the power and capacity of the nodes along with the demands of capacity on the MAWN.

In addition to this new feature, this method will measure $2TR_m$ along with k-terminal and all-terminal reliability.

Problem Definition

Let the capacity demanded be $c_d(t)$ and the capacity available between a pair of nodes, i and j, be $c_{ij}(t)$. Therefore a link exists between nodes i and j if the available capacity c_{ij}, is greater or equal to the demand, mathematically:

$$l_{ij}(t) = 1 \text{ if } c_{ij}(t) \geq c_d(t); \text{ else } l_{ij}(t) = 0 \tag{23}$$

The nodes' mobility is simulated via RWMM and by combining Equations 1 through 3, the separation distance of each node pair is translated to available capacity via Equation 24.

$$\begin{aligned}
&c = b * \log_2(1+s) \\
&s = \frac{p_r}{N_0} \\
&\Rightarrow c = b * \log_2(1+\frac{p_r}{N_0}) \\
&p_r = \frac{p_0}{d^2} \\
&\Rightarrow c_{ij}(d_{ij}(t)) = b * \log_2[1+\{(\frac{p_0}{d_{ij}(t)^2})'/N_0\}]
\end{aligned} \tag{24}$$

The link matrix, $\mathbf{L}(t)$, is then analyzed to determine connectivity to the source node. This connectivity to the source node, is represented by $\Lambda(t)$, as follows: $\Lambda_i(t) = 1$ if the ith node has a path to the source node at time t, otherwise $\Lambda_i(t) = 0$. Then, if node n is the destination node, the probability of a successful path from source to destination, $2TR_m$, is given by Equation 25. Similarly, the k-terminal reliability (kTR_m) is given by Equation 26 (for all-terminal $k = n$). Equation 26 should be interpreted as follows, for each node I there is a corresponding Λ_i representing the connectivity of that node to the source through one or more paths. Taking the sum of $\Lambda_I \,\forall\, i=1,2\ldots n$ provides the number of nodes linked. The probability that this value is greater than k is then the k-terminal reliability.

$$2TR_m(t) = \mathrm{P}(\Lambda_n(t) = 1) \tag{25}$$

$$kTR_m(t) = \mathrm{P}\{\sum_{i=1}^{n} \Lambda_i(t) \geq k\} \tag{26}$$

Monte Carlo Method

In this method, with each run of the simulation, the operational status of the nodes is simulated. Then, matrix $\mathbf{L}(t)$ is populated by calculating the capacity between each node pair from the distance between them and comparing the result with the demanded capacity. Once

L(t) has been obtained, a breadth-search-first (BFS) approach is used to determine the connectivity between the source and destination. These steps are implemented iteratively for each run and time increment. Each run is indexed by q and the total number of runs is Q.

Initialization: Define network input parameters: number of nodes, $|\mathbf{N}|$; bandwidth, b, transmission power, p_0, and spectral noise, N_0, node reliability, $r_i(t)$, maximum and minimum velocity, v_{max} and v_{min}; coverage area; mission duration, t_{max} and time increments, Δt. Also, to define the required capacity demand upon each link this metrics will be selected at random from a normal distribution defined by a mean (μ_d) and standard deviation (σ_d) in each time increment. Finally, set k as the number of terminals which is desired to be connected to the source.

Step 1: Simulate the operational status of the nodes. The Weibul distribution is selected for this model because it provides a good model of reliability for most systems, however any distribution deemed appropriate may beused in its place. The scale (θ) and shape (β) parameters which define the Weibul distribution may be varied to represent the mobile nodes expected reliability, reference Equation 27.

$$P(n_i(t) = 1) = e^{(-t/\theta)^{\beta}} \tag{27}$$

Therefore, step 1 selects from this distribution to determine operational status for each node. Note, once a node fails it remains failed for the remainder of that simulation run.

n_i (t)$\leftarrow$ Weibul distribution defined by θ and β

Step 2: Simulate the capacity required of the links, $c_d(t)$.

$c_d(t) \leftarrow$ Random number from normal distribution with μ_d and σ_d

Step 3:Given the position of the nodes, determine capacity offered and compare with the demanded capacity.

l_{ij} (t) $\leftarrow$ Equation 5 applied

L(t) $\leftarrow l_{ij}\ \forall$ combinations $i, j = 1,2... n$ except $i=j$.

Step 4: Analyze **L**(t) to determine $\Lambda(t)$ via a BFS. Determine if k-terminals are connected by analyzing $\Lambda(t)$:

if $|\Lambda(t)| \geq k$

then, $test(t) = 1$; else $test(t) = 0$.

Step5: Simulate the mobility of the nodes of the network according to the RW defined by the following:

$v_i(t) \leftarrow$ uniformly distributed velocity between v_{max} and v_{min}

$\varphi_i(t) \leftarrow$ uniformly distributed direction between 0 and 2π radians (0, 360°)

Step 6: Calculate the position of each node in the next time increment.

$x_i\ (t+\Delta t) \leftarrow$determine next position using previous position and Equation 3.

$y_i\ (t+\Delta t) \leftarrow$determine next position using previous position and Equation 4.

Step 7: Repeat steps 1 through 6 for time increments of Δt from $t = 0 \ldots t_{max}$.

$$t \leftarrow t + \Delta t$$

Step 8:Repeat simulation steps 1 through 7 for Q runs. Calculate an estimate of two-terminal reliability from the source to a defined destination node or nodes, n_i, and the network coverage metrics as functions of time by averaging the results for each simulation run (indexed by q) at each time increment (indexed by t)of the total mission duration.

$$2\hat{TR}_m(t) = \frac{\sum_{q=1}^{Q} \Lambda_i(q,t)}{Q} \tag{28}$$

$$k\hat{TR}_m(t) = \frac{\sum_{q=1}^{Q} test(q,t)}{Q} \tag{29}$$

Illustrative Example

For the purposes of illustration and comparison the same network used to demonstrate the previous method is used here. The input parameters are as follows:

$n = 18$
$t_{max} = 72$ hours
$r_i(t) = e^{(-t/\theta)^\wedge\beta}$ $\theta = 1000$ and $\beta = 1.5$
coverage area of 64 mi^2
$v_{min} = 3$ miles per hour
$v_{max} = 6$ miles per hour

In this context, fully connected means that all terminals can communicate with each other as opposed to full connectivity where all nodes are linked directly.

So the results of this model may be compared to the results of the previous method, the radio performance characteristics (b, p_0, N_0) and capacity demands are set so that they equate to a transmission range of 3 miles or 4.8 km. To do so, the capacity demand is defined within the model with $\mu_d = 320$ bits per second (bps) and $\sigma_d = 0$ bps.

$b = 50$ MHz
$p_0 = 100$ dB
$N_0 = 1$ dB
$\rightarrow c_{ij}(3 \text{ mi}) \cong 320$ bps

The results of this model are shown in Figure 8 and they compare closely reported, see Table 13.

Table 13. Model Results

Method	$2\hat{TR}_m(72hrs)$
Binary Links	0.6152
Capacitated Links	0.6050

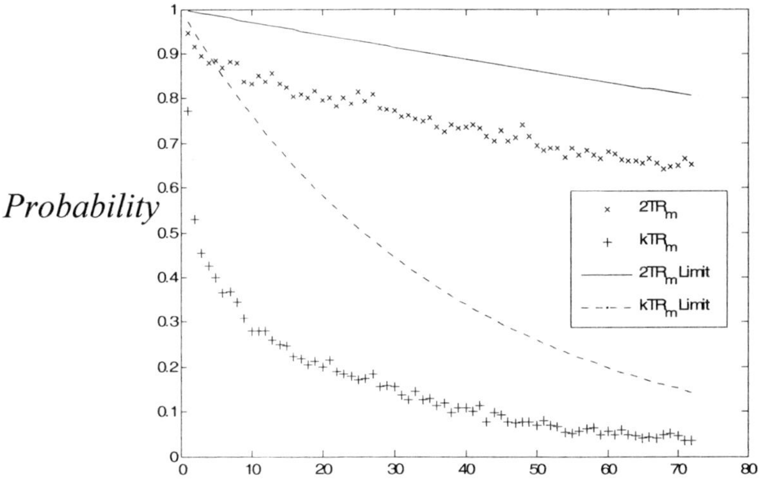

Figure 8: Capacitated MAWN Results

When comparing the results, the reliability of the binary link network and the equivalent capacitated network correlate closely. These implementations of the two methods were performed to validate the model. However, the model is further tested by running it with zero demanded capacity, $\mu_d = 0$ bps and $\sigma_d = 0$ bps. The impact of removing this constraint is all operational nodes are linked and therefore a path from source to destination exists unless one or both nodes have failed. Similarly, all terminals are connected if all terminals are operational. In this case the two and all terminal reliability are expected to approach the limits as shown in Equations 30 and 31. When k-terminal is the special case, all-terminal, Equation 31 is reduced to Equation 32. Figure 9 shows the close correlation of the simulation results with these limits.

$$2TR_m(t) = P\{n_1(t) = 1 \cup n_n(t) = 1\} = r_i(t)^2 \tag{30}$$

$$kTR_m(t) = \sum_{i=k}^{n} \frac{n!}{i!(n-i)!} r_i(t)^i * (1 - r_i(t))^{n-i} \tag{31}$$

$$ATR_m(t) = r_i(t)^n \tag{32}$$

As intended, this method provides some additional utility. To demonstrate this, networks with different capacity demands are analyzed. Table 14 shows the inverse relationship between mean capacity and reliability (both two and all terminal).

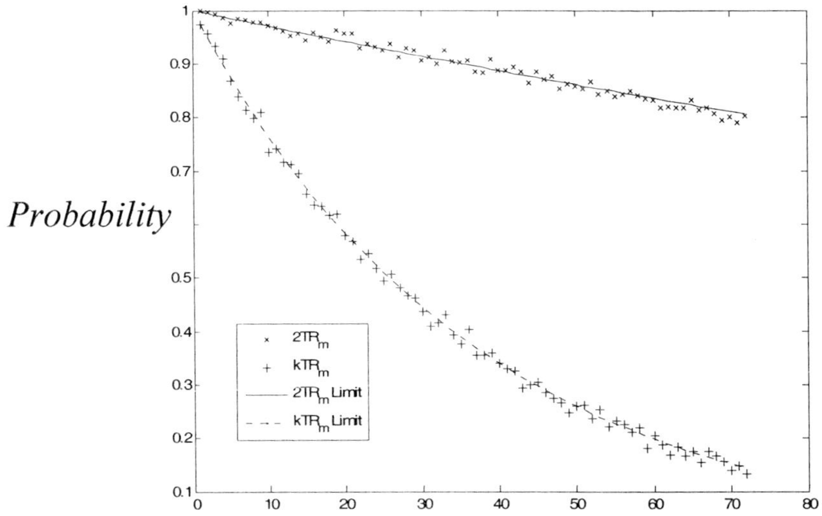

Figure 9: Zero Capacity Demand MAWN Results

The volatility of demanded capacity is simulated by changing the standard deviation of the capacity, σ_d. The effect of this is demonstrated graphically. It is seen that the decreasing trends on $2TR_m$ and ATR_m are similar. However, the variation about that trend line is increased as the σ_d is increased; the variation is further compounded when the standard deviation goes from 25% to 100% of the mean, see Figures 10 and 11. Note that the normal distribution results in some negative capacity demands; within the simulation model these are considered zero demand because negative demand has no practical meaning.

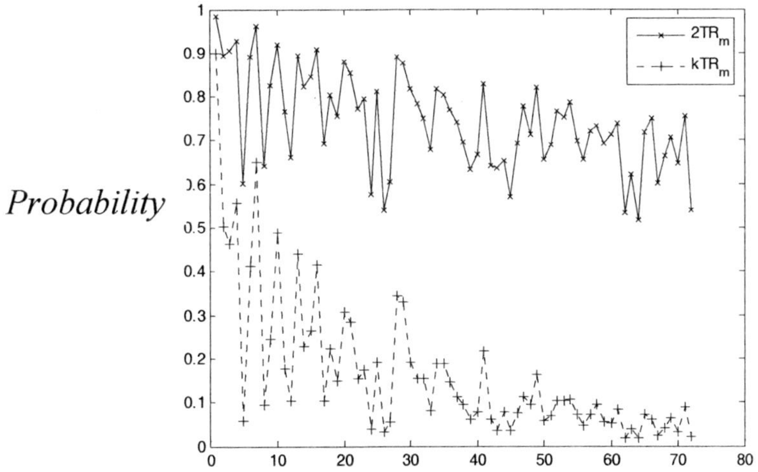

Figure 10: Varied Capacity Demand MAWN Results

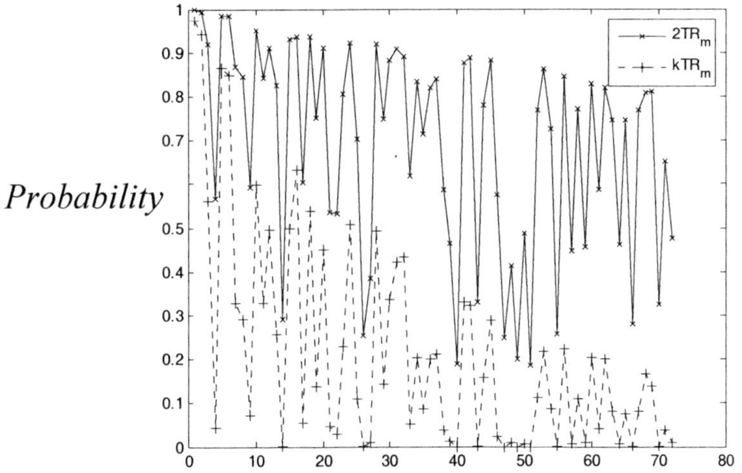

Figure 11: Varied Capacity Demand MAWN Results

Table 14. Capacity Demand Impact on Reliability

Mean Capacity (μ_d) Bps	$2\hat{TR}_m(72hrs)$	$A\hat{TR}_m(72hrs)$
0	0.8140	0.1280
100	0.7530	0.0680
200	0.7060	0.0630
320	0.6050	0.0280
500	0.4220	0.0070

The next examination is to determine the manner in which *k*-terminal reliability varies as the value of *k* increases. The results for kTR_m are presented in Table 15. As expected, the kTR decreases as the success criteria is made more demanding by increasing *k*.

Table 15. *k*-Terminal Reliability

Number of Destination Nodes (*k*)	$k\hat{TR}_m(72hrs)$
2	0.8140
5	0.7750
10	0.7080
15	0.4460
16	0.2730
17	0.1350
18	0.0280

μ_d = 320 bps; σ_d = 0 bps

Future Work

A MAWN of practical scale often requires some segmentation, typically termed clustering, for successful implementation. The reasons are primarily related to network bandwidth, frequency reuse, and interference [27]. The cluster-based network is best described as many smaller networks joined together by a back-bone network to form a single and cohesive network of networks. The network that joins the clusters together is usually known as a back-bone. This chapter and area of research will allow the system engineer to analyze the reliability of cluster based ad-hoc networks with a Monte Carlo based approach. It will also provide insight into the affects to reliability due to design decisions on cluster size and gateway assignment.

The work in this area is still in the preliminary stages but the research and development envisioned will build upon the methods described within this chapter.

Conclusion

The contribution of this body of work is the capability these methods provide the system engineer. That is, the ability to define, understand and assess the reliability of this emerging network scheme. The methods developed allow for a complete systems view of MAWN reliability by providing quantitative methods to assess the impact of the system attributes that impact reliability. The system engineer will be given the tools to analyze and optimize MAWN reliability and therefore the ability to influence and improve it. The MAWN is still a break through technology; however, as it matures reliability will be expected and demanded by its users. Mostly, this work is motivated by the application of MAWN technology in the DoD tactical networks and the import of their reliable operation when employed for this use.

The results included herein and that will emerge from the continuation of the research agenda described will define the key contributors to MAWN reliability. The research performed to date has already identified the following key system parameters when considering reliability.

Key Reliability Drivers
Number of nodes in the network, n
Coverage area of network
Node reliability, r_i
Probability of link existence, λ
Capacity demands on the network, σ_d and μ_d
Performance of the nodes in terms of
Transmission range, t_{ij}
Performance of nodes in terms of transmitted power, p_0

Notation

N	Set of nodes
n	The number of nodes in the network
n_i	Binary variable representing the state of the ith node
$r_i(t)$	Reliability of node i
G(N,L)	Network G, composed of nodes and their links
C	Set of possible configurations from network $G(N)$
α_k	$G(N,L)$ configuration α_k, $k=1,\ldots\|C\|$
l_{ij}	Virtual link between nodes i and j
L(t)	Link configuration matrix
n_l	Number of linked node pairs
n_u	Number of unlinked node pairs
$2TR\alpha_k$	Two-terminal network reliability of configuration α_k
$2TR_m(t)$	Two-terminal MAWN reliability
λ	Constant probability of link existance
e	Average neighboring nodes
$\Lambda(t)$	Connectivity vector
$\Lambda_i(t)$	Connectivity of ith node to the source
Q	Number of runs in MC simulation
m	Limit on number of intermediate nodes in a path
v_{ij}	Probability of link between nodes i and j
$d_{ij}(t)$	Distance between nodes i and j
t_{ij}	Radial transmission range
$x_i(t)$	Postion of the ith node along the x-axis
$y_i(t)$	Postion of the ith node along the y-axis
$\chi(t)$	Connectivity of the network
θ	Scale parameter of Weibul failure distribution
β	Shape parameter of Weibul failure distribution
$v_i(t)$	Velocity of ith node
$\varphi_i(t)$	Direction of ith node

$\varpi(t)$	Volatility of network link status
$c_d(t)$	Demanded capacity
μ_d	Mean demanded capacity
σ_d	Standard deviation of demanded capacity
$c_{ij}(t)$	Available capacity between nodes i and j
b	Bandwidth
$N0$	Spectral Noise
$p0$	Transmitted power
p_r	Received power
λ_l	Probability of local link existence for subnet
λ_b	Probability of backbone link existence
ζ	Number of subnets in the network
s_l	Local subnet l; where $l=1,2\ldots\zeta$

Acronyms

2TR	Two-Terminal Network Reliability
RWMM	Random Waypoint Mobility Model
BER	Bit Error Rate
DoD	Department of Defense
MAWN	Mobile Ad-hoc Wireless Network
MC	Monte Carlo
MN	Mobile Node
TCP/IP	Transmission Control Protocol/Internet Protocol
WLAN	Wireless Local Area network
TCP/IP	Transmission Control Protocol/Internet Protocol
RW	Random Waypoint

*The singular and plural of the acronym are used indistinctively.

References

[1] Ebeling, Charles E. (1997) An Introduction to Reliability and Maintainability Engineering. Waveland Press, Inc., Copyright 1997. p. 5

[2] Fratta, Luigi and Montanari, Ugo G. A Boolean Algerbra Method for Computing Terminal Reliability in a Communication Network. *IEEE Transactions on Circuit Theory*. Vol. 20, No. 3, May 1973, pp. 203-211.

[3] Fratta, Luigi and Montanari, Ugo G. A Recursive Method Based on Case Analysis for Computing Network Terminal Reliability. *IEEE Transactions On Communications*. Vol. Com-26, No. 8, August 1978

[4] Hansler, Eberhard. (1975) Comments on 'A Fast Recursive Algorithm to Calculate the Reliability of a Communication Network'. *IEEE Transactions on Communications. May* 1975 pp. 563-566.

[5] Netes, Victor A.; Filin, Boris P. (1996) Consideration of Node Failures in Network-Reliability Calculation. *IEEE Transactions on Reliability*. Vol 45, No. 1, 1996 March, pp 127-128.

[6] Y.B. Yoo, N. Deo, "A comparison of algorithms for terminal-pair reliability", *IEEE Transactions on Reliability*, vol 37, 1988 Jun, pp 210-215.

[7] Torrierri, D. (1994). Calculation of Node-Pair Reliability in Large Networks with Unreliable Nodes. *IEEE Transactions on Reliability*, Vol 43, n 3, pp. 375-379.

[8] Kuo, S., Lu, S. and Yeh, F. Determining Terminal Pair Reliability Based on Edge Expansion Diagrams Using OBDD. *IEEE Transactions on Reliability*, Vol 48, n 3, pp. 234-246

[9] Marseguerra, Marzio; Zio,Enrico; Podofillini,Luca; and Coit,David W. Optimal Design of Reliable Network Systems in Presence of Uncertainty. *IEEE Transactions on Reliability*, VOL. 54, NO. 2, JUNE 2005

[10]Rocco, Claudio M. and Muselli, Marco. Empirical Models Based On Machine Learning Techniques For Determining Approximate Reliability Expressions. *Reliability Engineering and System Safety*. Volume 83, Issue 3, 1 March 2004, Pages 301-309.

[11]Ramirez-Marquez, J.E. and Coit, D. A Heuristic for Solving the Redundancy Allocation Problem for Multistate Series-Parallel Systems. *Reliability Engineering & System Safety*, Vol. 83, no.3, pp 341-349.

[12]Fishman, George S. (1986) A Monte Carlo Sampling Plan for Estimating Network Reliability. *Operations Research*. Vol 34. No. 4. pp 581-594

[13]Elperin, T., Gertsbakh, I. and Lomonosov, M. (1991) Estimation of Network Reliability Using Graph Evolution Models. *IEEE Transactions on Reliability*. Vol. 40. No. 5. pp 572-581.

[14]Rocco, Claudio M. and Moreno, Jose Ali. (2002) Network Reliability Assessment Using a Cellular Automata Approach. *Reliability Engineering and System Safety*. Vol. 78. pp 289-295.

[15]Rocco, Claudio M. and Zio, Enrico. (2005) Solving Advanced Network Reliability Problems by Means of Cellular Automate and Monte Carlo Sampling. *Reliability Engineering and System Safety*. Vol. 89. pp 219-226

[16]Cancela, Hector and El Khadiri, Mohamed. (2003) The Recursive Variance-Reduction Simulation Algorithm for Network Reliability Evaluation. *IEEE Transactions on Reliability*. Vol. 52, No 2. pp 207-212

[17]Konak, Abdullah; Smith, Alice E.; and Kulturel-Konak, Sadan. (2004) New Event-Driven Sampling Techniques For Network Reliability Estimation. Proceedings of the 2004 Winter Simulation Conference.

[18]Coit, David and Ramirez-Marquez, Jose Emmanuel. (2005) A Monte Carlo Simulation Approach to Approximating Multi-State Two-Terminal Reliability. *Reliability Engineering and System Safety*. Vol. 87. pp 253-264.

[19]Chen, Zinyu and Lyu, Michael R. Reliability Analysis for Various Communication Schemes in Wireless CORBA. *IEEE Transactions on Reliability*. Vol. 54, No 2. June 2005, pp. 232-242

[20]AboElFotoh, Hosam M. and Colbourn, Charles J. Computing 2-Terminal Reliability for Radio-Broadcast Networks. *IEEE Transactions On Reliability*, Vol. 38, No. 5, December 1989, pp. 538-555

[21]Zonoozi, Mahmood M. and Dassanayake, Prem. User Mobility Modeling and Characterization of Mobility Patterns. *IEEE Journal On Selected Areas In Communications*, Vol. 15, No. 7, September 1997, pp.1239-1252.

[22]D. Turgut, S. K. Das, and M. Chatterjee. Longevity of routes in mobile ad hoc networks. *Proc. of IEEE Vehicular Technology Conference(VTC)*, vol. 4, pp. 2833-2837, Spring 2001.

[23]T. Camp, J. Boleng, and V. Davies. A Survey of Mobility Models for Ad Hoc Network Research. Wireless Communication & Mobile Computing (WCMC): Special issue on Mobile Ad Hoc Networking: *Research, Trends and Applications*, vol. 2, no. 5, pp. 483-502, 2002.

[24]Bhatt, M.; Chokshi, R.; Desai, S.; Panichpapiboon, S.; Wisitpongphan, N.; Tonguz, O.K. Impact Of Mobility On The Performance Of Ad Hoc Wireless Networks. Vehicular Technology Conference 2003. IEEE 58th Volume 5, 6-9 Oct. 2003 Page(s): 3025 – 3029.

[25]Chung,Wei-Ho. Probabilistic Analysis of Routes on Mobile Ad Hoc Networks. *IEEE Communications Letters*, Vol. 8, No. 8, August 2004.

[26]Toh, C-K (2002). Ad Hoc Mobile Wireless Networks. ISBN: 0-130-07817-4. Prentice Hall PTR © 2002.

[27]Pahlavan, Kaveh and Krishnamurthy, Prashant. (2002) *Principles of Wireless Networks. Prentice Hall PTR.* ISBN: 0-13-093003-2. Copyright 2002. pp. 224-229.

[28]Freebersyser, James A. and Leiner, Barry (2001) A DoD Perspective on Mobile Ad Hoc Networks. *Ad Hoc Networking*. ISBN: 0-201-30976-9 Chapter 2, © 2001. pp. 29-51.

[29]Ye, Zhenqiang; Krishnamurthy, Srikanth V.; and Tripathi, Satish K. A Routing Framework For Providing Robustness To Node Failures In Mobile Ad Hoc Networks. *Ad Hoc Networks*. Vol. 2. Issue 1. January 2004 pp. 87-107.

[30]Luo, Jun; Eugster, Patrick Th.; and Hubaux, Jean-Pierre. Probabilistic Reliable Multicast In Ad Hoc Networks. *Ad Hoc Networks*. Vol. 2. Issue 4. October 2004. pp. 369-386.

[31]Abolhasan, Mehran; Wysocki, Tadeusz; and Dutkiewicz, Eryk. A Review Of Routing Protocols For Mobile Ad Hoc Networks. *Ad Hoc Networks*. Vol. 2. Issue 1. January 2004 pp. 1-22.

[32]Shannon, C.E. and Weaver, W. (1949) The Mathematical Theory of Communication. Urbana, IL: *University of Illinois Press*, 1949.

[33]Yao, Y.-D.; Sheikh, A. (1993) Evaluation of channel capacity in a generalized fading channel; *Vehicular Technology Conference*, 1993 IEEE 43rd 18-20 May 1993 Page(s):134 – 137

[34]Lee, William C.Y. (1990) Estimate of Channel Capacity in Rayleigh Fading Environment; *IEEE Transactions on Vehicular Technology*, Vol 39, No. 3, August 1990.

[35]Pinsker, Mark S.; Prelov, Vyacheslav V.; and Verdu, Sergio (1995) Sensitivity of Channel Capacity; *IEEE Transactions of Information Theory*, Vol. 41, No. 6, November 1995.

[36]Wyner, Aaron D. (1974) Recent Results in the Shannon Theory. *IEEE Transactions on Information Theory*, Vol IT-20, No. 1, January 1974.

[37]Cook, Jason L. and Ramirez-Marquez, Jose E. (2006) Reliability Method for Ad-hoc Networks. Proceedings of Institute of Industrial Engineers Annual Conference; 20 May 2006.

[38]Cook, Jason L. and Ramirez-Marquez, Jose E. (2006) Two-terminal Reliability Analyses for a Mobile Ad-hoc Wireless Network. Accepted for publication by Reliability Engineering and System Safety

[39]Cook, Jason L. and Ramirez-Marquez, Jose E. (2006) Mobility and Reliability Modeling for a Mobile Ad-hoc Network. Stevens Institute of Technology, Systems Engineering and Engineering Management Department, Working Paper # EO-06-006. (under review IIE Transactions).

In: Wireless Communications Research Trends
Editor: Tong S. Lee, pp. 229-266
ISBN: 978-1-60021-674-9

Chapter 7

MEMS MICRO-ANTENNAS FOR WIRELESS BIOMEDICAL SYSTEMS

P. M. Mendes and J. H. Correia
Department of Industrial Electronics, University of Minho
Campus de Azurém, 4800-058 Guimarães, Portugal

Abstract

Invasive and implantable biomedical devices used for diagnostic and therapy, ranging from neural prosthesis to video-capsule endoscopy (VCE) systems, are emerging innovative technologies and they are expected to originate significant business activity in the near future. The success of such systems is in part due to the advent of microtechnologies, which made possible the miniaturization of several sensors and actuators, as well their integration with readout and communication electronics.

The new biomedical devices offer the possibility of improved quality of life, as well cost savings associated with health care services. However, one open challenging is to communicate to and from a biomedical device placed inside the human body with devices outside the human body. The lack of antennas, small enough to be integrated with the sensing microsystem, is a difficult task to overcome because such communications must be made at relatively low frequencies, due to live tissue signal attenuation. The straightforward solution is to increase the devices size to dimensions where it becomes possible to integrate an antenna. Up to now solutions, use conventional antennas together with miniaturization techniques to achieve the smallest antennas possible. However, the size of such devices is usually limited by the antenna and, is some cases, also by the batteries size.

Micro-Electro-Mechanical Systems (MEMS) are becoming an available option for RF communication systems since they can offer, simultaneously, devices with improved performance and they use IC-compatible materials, allowing their integration in a silicon chip, side by side with semiconductor circuits. Up to now, MEMS have been used for antenna applications to obtain non-conventional front-ends with improved, or new characteristics. However, some preliminary tests have shown that some MEMS structures could have the ability to operate as an antenna itself and this solution would have the potential to be smaller than the conventional antennas.

In this chapter, it is first discussed the need for small wireless biomedical devices. This requires the use of a microsystem completely integrated, from sensors to communications, thus requiring the use of integrated antennas. The electrical properties of substrates available in integrated circuit technology are very important for antenna design and one method used to characterize wafer materials is presented. Moreover, the antenna integration requires the availability of an electrically small antenna fabricated on materials compatible with the fabrication of integrated circuits. This integration requires the use MEMS techniques, like micromachining and wafer level packaging.

Finally, MEMS structures previously used for non-conventional front-ends will be introduced and investigated, having in mind a new application, the MEMS structure itself will be operating as an antenna. The development of new integrated antennas using MEMS solutions has the potential to make the devices smaller and more reliable, which will make them cheaper and adequate for mass production, resulting in a key advantage for competitors in the RF market. Also, the availability of smaller biomedical wireless devices can lead to new applications not yet fully envisioned. The new solutions envisions power saving, smaller volume, lower cost, and increased system lifetime, which are very important features in biomedical microsystems for diagnosis and therapy.

Wireless Biomedical Devices

Introduction

Sensor networks are expected to be the 21st century holly Graal in sensing. Wireless sensor networks are an emerging technology that brings not only numerous opportunities but also many technological challenges. Ranging from automotive to home applications, sensor are expected to become part of our daily live. There are various physical quantities in different environments that, when measured and recorded, could bring new quality to our lives. If, e.g., soil properties like humidity, temperature, chemical composition, could be measured in a distributed way, providing detailed local data for an entire farm field, the watering and nutrition supply could be adapted locally resulting in optimum growing conditions and thus significant environmental savings.

To be effective and to have the ability to adapt itself to complex environments, like a farm field, an office building, or a human body, each sensor node has to be autonomous. This means that, desirably, it should be self-powered and provided with wireless communications. This type of systems usually requires low power consumption (battery life-time) and uses low data-rate communications (small bandwidth), requiring special design. The wireless sensor networks technology is also moving to biomedical applications. Due to the high number of microsensors and microactuators, the monitorization of several physiological parameters is now available. Moreover, application of distributed sensing systems will highly be facilitated if cheap and easy-to-use 'on-chip' or 'in-package' solutions, equipped with short-range wireless communication capabilities, would be available.

With a widespread and increased sophistication of medical implants, new solutions will be required for flexible and small modules to communicate with the implant. Today's most common solution is to use an inductive link between the implant and an external coil. The main drawback of this solution is the small range achieved (not more than a few centimetres). However, this link can be used also to power the implanted device.

Application of wafer-level chip-scale packaging (WLCSP) techniques like adhesive wafer bonding and through-wafer electrical via formation, combined with the selected radio frequency (RF) structures allows a new level of antenna integration. However, these new techniques require the combination of new materials with the standard materials used for integrated circuits fabrication. In this way, together with substrate processability, the knowledge of accurate electrical parameters is of extreme importance when designing for RF or microwave applications.

Applications

The traditional endoscopic techniques shows limitations in their range for the different segments of the digestive tract, namely in the small intestine. Through the gastroscopy, it is possible to access the gastrointestinal proximal tract (gullet, stomach, and duodenum). In the other hand, through colonoscopy the access is almost limited to the colon, leaving inaccessible some parts of the small intestine. When a patient suffers from bleeding in the gastrointestinal tract and the endoscopy doesn't answer the diagnostic needs, it is necessary to use the traditional radiographic or cintilographic techniques. However, with these techniques is very hard to detect bleeding sources in the small intestine [1]. Moreover, the endoscopic diagnostic is a very uncomfortable procedure for the patients and requires highly skilled medical doctors.

One possible solution to overcome those limitations is to use an innovative technique known by video capsule endoscopy, also known by smart pill, capsule camera, or wireless capsule camera. The patient swallows the endoscopic capsule and it takes photos from the different places during its travel through all the digestive tube. In this way, it is possible to obtain access, even in a limited way, to areas in the small intestine previously accessible only by surgical and invasive procedures [2]. The camera-in-a-pill is a very promising technique since after its availability the approx. 5 meters of the small intestine become accessible [3]. Since its development in the eighties, the endoscopic capsule is changing the way we deal with the diseases in the small intestine, and its use is now being extended to the gullet [4].

The endoscopic capsule is neither provided with locomotion nor with stop mechanism, making its way due to peristaltic movements. The control of the capsule locomotion would allow new procedures like biopsies (detection of tumours) or drug delivery, and is an ongoing research topic. Together with locomotion, before we can fully benefit from this technology, miniaturized modules are required to implement the wireless communications to and from the capsule with devices outside the human body.

This is an open challenge, not only for VCE, but also for all the biomedical devices that are implanted inside the human body, like FES systems [5]. As an application example, several people from all ages suffer from incontinence or other urinary pathologies. The bladder and the intestines perform their function in an autonomous way, independently from the voluntary control. However, any disorder in the healthy behaviour leads to the problem of urinary incontinence, bladder infections, low bladder capability and faecal incontinency. The International Continence Society (ICS) defines incontinence as the involuntary loss of bladder or bowel control. Urinary incontinence (UI) is a stigmatised, underreported, under-diagnosed, under-treated condition that is erroneously thought to be a normal part of aging. One-third of men and women ages 30-70 believe that incontinence is a part of aging to accept [6]. The social costs of UI are high and even mild symptoms affect social, sexual, interpersonal, and professional function [7].

The healthy working of the urinary tract is essential for health and well-being in general, and even more critical for patients with lesions in the spinal cord. In this situation, catheters are commonly used to control the daily volume of urine inside the bladder. However, the complications related to the use of catheters, together with the fact that, most of the times, the spinal segments which controls the bladder are intact, are driving the development of several devices to improve the control the inferior urinary system [8].

From the anatomy of the spinal cord, the microsystem could be designed to operate, or in the epidural space, or in the subarachnoid space, allowing the duramater to be completely closed after surgical intervention. The available subarachnoid space varies between 3 mm and 9 mm [9]. This is room enough to accommodate a small microdevice, but the antenna may become a problem.

Other field becoming very popular, where the nanotechnologies are giving a contribution, is related with neural signal recording and stimulation. The accurate record of neural signal requires the use of very small wireless bio-devices for invasive biopotential monitoring [10].

The full system integration is difficult to achieve because the communications must, preferably, be made at low frequencies, due to live tissue signal attenuation, and there is a lack of antennas small enough to be integrated with the sensing microsystem. The adopted solution is to increase the devices size to dimensions where it becomes possible to integrate an antenna. Up to now solutions, use conventional antennas together with miniaturization techniques to achieve the smallest antennas possible [11, 12]. However, the size of such devices is usually limited by the antenna and, is some cases, also by the batteries size.

System Requirements

Frequency and Bandwidth

A wireless microsystem uses the surrounding environment as the communicating channel. This channel is available in different spectral regions, which corresponds to a different channel frequency, bandwidth, and attenuation. For biomedical applications, the surrounding environment is the human body, a highly heterogeneous environment.

In this environment, the attenuation may become highly severe when the antenna miniaturization is required because a smaller antenna is commonly obtained with an increase in the operating frequency. Fig. 1 shows the signal attenuation for a tissue with a high content of water [13]. The figure shows the penetration depth, i.e., the depth where the power density is 13.5 % below the incident power density.

This figure must be used carefully and only as an indication for attenuation since it can be very different for different water concentration in the tissue. However, it can be observed an increase in the attenuation for higher frequencies. This means that the biomedical devices should use low frequencies if low power is required for communications. Despite all freedom of choice about the frequencies that should be used, there is an adoption of ISM (*Industrial, Scientific and Medical*) bands. These bands plays a very important role in communications since they allow the development of wireless devices, where the constraints in terms of power level, frequency, and bandwidth are very well defined. The most common frequency bands we can find are in the 433 MHz range and in the 13 MHz range. These frequencies are very popular since they allow a good signal range in free-air and they are able to reach a few centimeters inside the human body. Moreover, the design of microdevices operating at such frequencies is a well-known process that requires low cost technology.

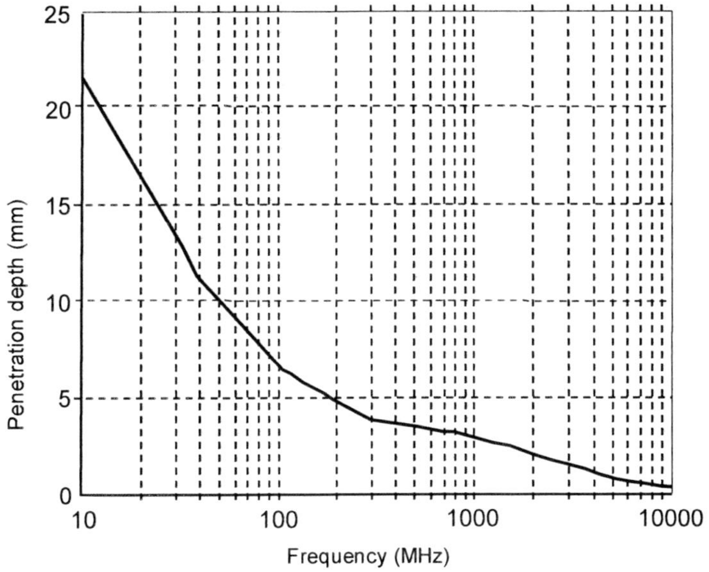

Figure 1. Penetration depth for a radio-frequency signal in a human body tissue.

Besides the radio-frequency communications, optical or ultrasonic communications are also very attractive options due to their possibilities for miniaturization. In optical communications, the main drawback are the high losses inside the human body and for ultrasonic the main drawback are the interfaces of the different tissues, which results in several reflections of the ultrasonic wave. Besides that, there is the interface air-human body that may kill the chance of a reliable communication. Notwithstanding all the drawbacks associated with signal attenuation and antenna integration, the radio frequency is one of the favourites for use in biomedical devices.

Despite all the deregulation that dominates the wireless communications for biomedical devices, except for the definition of ISM bands, the European Telecommunications Standards Institute (ETSI) has made an effort for standardization and specified the Medical Implant Communication System (MICS) [14]. The ETSI document applies to devices wiling to communicate between a base station and an implanted device and for devices wiling to communicate between medical implants within the same body.

The MICS uses the frequency band from 402 MHz to 405 MHz, with a maximum emission bandwidth of 300 kHz. Moreover, the maximum power limit is set to 25 μW Equivalent Radiated Power (ERP), i.e., the maximum field-strength in any direction should be equal to, or lower than, what a resonant dipole would give in its maximum direction at the same distance, with the dipole being fed with a signal of 25 μW.

Power

Like for frequency requirements, it's very difficult to define what are power requirements for a biomedical device. Depending on the device power requirements, the power is provided either remotely or locally. Despite the MICS power level constrain, the power requirements for wireless biomedical devices are highly dependent on the target device. Some systems may have local power, using batteries or energy harvesting, and other systems require the use of remote power.

For devices that are power hungry, like functional electrical stimulating devices, the power, together with the required stimulus information, is provided by the wireless link. Depending on the use, a FES device for bladder control may require five 9 Volt batteries for one day of operation. On the other extreme are, e.g., hearing prosthesis, which use local power where power consumption must be below 1 mW. Also power hungry are the emerging VCE technology. These devices use local batteries since they are moving inside the human body and it is difficult to deliver power using the common solution with coils. In this system, the wireless link is only used for data transmission.

Antenna Design Options

From the previous sections it can be concluded that there is an emergent need for small biomedical devices fully integrated, antenna included. Moreover, the traditional coil does not provide a solution for systems requiring short-range communications [12]. This means that one must look for new solutions to integrate the antenna inside the biomedical microdevices.

Merging of antenna and circuitry leads to innovative RF front-end designs possessing several desirable features such as compactness, lower power consumption, and added design flexibility. Several approaches have been used to achieve antenna integration as summarized in Fig. 2. Antenna integration is a hard task to accomplish since it requires joining the knowledge from antennas, microwaves, circuit design, and materials. Moreover, the on-chip antenna integration requires an electrically small antenna, due to wafer cost and devices size constrains, and operating on a substrate that was not initially intended for that purpose.

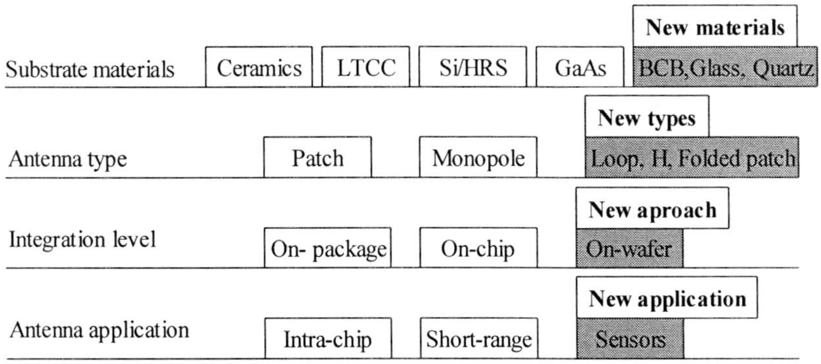

Figure 2. Summary of solutions for on-chip antenna integration.

Substrates

The antenna integration can be achieved with the direct placement of a radiator on the low-ohmic silicon. However, this approach requires a layer of thick oxide to reduce the losses or a proton implantation and still the obtained efficiencies are very low (~10 %).

Due to the high-losses observed on low-ohmic silicon, one way to improve the properties of integrated antennas may be to look for new materials with lower losses, but compatible with silicon processing. Up to now, several different materials have been used to implement on-chip antennas. GaAs has been used very often as antenna substrate due to its lower losses and high dielectric constant, which allows small and efficient antennas. However, since cost and integrability is pushing RF circuitry to be implemented in CMOS, the use of high-resistivity silicon (HRS) was also been subject for several works. The use of HRS increases the antenna efficiency, keeps the antenna dimensions small, and allows easy integration with electronics. The main drawback is the increased wafers cost. On the other hand, efficiency can also be increased using bulk micromachining techniques. A different approach to obtain good efficiency values is to use non-standard materials for antenna substrate. Using this approach, it is possible to use materials like BCB, quartz, glass, or artificial substrates. Another attractive option is to use ceramics for the antenna substrate. It has a high dielectric constant, which allow further size reduction and can be found with low losses. The attempts made so far to use ceramics are based on the LTCC technology or in the use of a ceramic package to place the antenna.

Antenna Types

A good substrate material is essential to obtain an antenna with good radiation properties. However, finding a good substrate is only part of the solution. Also critical, is the antenna type selection. It must be small to fit in a small chip area, it should be easy to design, fabricate and characterize, and it should interfere as low as possible with the remaining circuitry. Notwithstanding all the requirements, mainly two types of antennas have been suggested for integration – patch and dipole or monopole antennas. Those antennas have been chosen mainly due to their simplicity to design and to interface with the RF front-end. Nevertheless, even if in a small number, slots, loops, and other types have also been proposed. Another antenna very popular for miniaturization is the slot antenna.

Antenna Integration

Also important, is how the antenna is integrated with RF front-end: on-package, on-chip, or on-wafer. Integration on-package is a very straightforward concept where the chip package is used for antenna housing. Despite the great advantage of using, for free, a dummy package for antenna placement, it also shows some disadvantages. The use of package as antenna substrate requires a good control of the electrical properties of the package material, the package becomes more complex since its necessary to provide feeding to the antenna, and the advantage of integration gets lost a bit since what we get is more an antenna connected on-chip then an antenna integrated on-chip. On-chip integration is the easiest, simplest, and used more often since the antenna is simply designed to operate on the same substrate as the circuitry. However, despite its simplicity, this solution has to deal with the usual low efficiencies obtained. Moreover, noise coupling to and from RF circuitry, interference with RF passive components, such as inductors, is still lacking analysis. Also very important, this approach consumes very expensive chip area.

On-chip Antennas

The integration of antennas requires the availability of simple, small, and efficient antennas that can be designed and fabricated with techniques and materials compatible with IC processing. Since the substrate materials available for integrated circuits fabrication were not chosen to design antennas on it, first it's necessary to characterize those substrates and find the most suitable one for this purpose.

Substrate Characterization

Several different methods can be used to extract the electrical intrinsic properties of a material. From the most widely used techniques to obtain those properties in the microwave region, the transmission line technique is the simplest method for electromagnetic characterization in wideband frequencies [15, 16]. The S-parameters measurements of a planar test cell can be used to obtain the desired parameters, where either a microstrip or a coplanar waveguide (CPW) can be used as test cell.

In this work, the CPW was used, with its parameters chosen to allow only the dominant quasi-TEM mode to be present. For on-wafer measurements the CPW, without bottom ground, is the easiest structure to feed and the probe station tips are able to touch directly the CPW lines. The S-parameters are then easily measured with a vector network analyzer.

Extraction Method

The electrical properties were obtained from the S-parameters measurements of a planar transmission line test-cell. The coplanar waveguide was used because of the possibility to define a planar shape that can propagate a dominant mode (quasi-TEM). In the case of dominant mode, the coplanar characteristic impedance is quasi-constant in a broad frequency range, for a large variety of substrates and a cell structure obeying $h > W + 2S$ [15]. This cell has also the advantage of avoiding the use of vias to ground.

The CPW cell geometry used for S-parameter measurements is shown in Fig. 3.

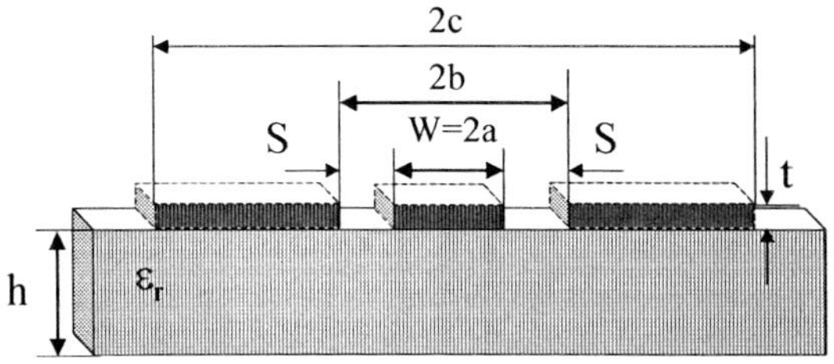

Figure 3. CPW cell used for S-parameters measurement.

The effective dielectric constant for this type of CPW can be obtained from [15]:

$$\varepsilon_{reff} = -\left(\frac{-\ln T}{\omega d\sqrt{\varepsilon_0 \mu_0}}\right)^2 \tag{1}$$

where ω is the angular frequency, ε_0 and μ_0 are the free space permittivity and permeability, d is the coplanar line length and T is the first transmission coefficient. The transmission coefficient can be obtained from the measured scattering parameters using the following equation [16]:

$$T = \frac{S_{11} + S_{21} - \Gamma}{1 - (S_{11} + S_{21})\Gamma} \tag{2}$$

with

$$\Gamma = X \pm \sqrt{X^2 - 1} \tag{3}$$

and

$$X = \frac{S_{11}^2 - S_{21}^2 + 1}{2S_{11}} \tag{4}$$

The electrical permittivity is obtained from the equations characterizing a coplanar waveguide with finite-width ground planes [17]. In this way, the effective electrical permittivity is given by:

$$\varepsilon_{reff} = 1 + \frac{1}{2}(\varepsilon_r - 1)\frac{K(k)}{K(k')}\frac{K(k_1')}{K(k_1)} \tag{5}$$

where ε_r is the relative permittivity of the substrate, K is the complete elliptical integral of the first kind and $k' = \sqrt{1-k^2}$ [18]. The arguments k and k' are dependent on the line geometry and are given by [19]:

$$k = \frac{c}{b}\sqrt{\frac{b^2 - a^2}{c^2 - a^2}} \tag{6}$$

and

$$k_1 = \frac{\sinh(\pi c/2h)}{\sinh(\pi b/2h)}\sqrt{\frac{\sinh^2(\pi b/2h) - \sinh^2(\pi a/2h)}{\sinh^2(\pi c/2h) - \sinh^2(\pi a/2h)}} \tag{7}$$

The characteristic impedance of the coplanar cell can also be computed from the measured S-parameters [20]:

$$Z_c^2 = Z_0^2\frac{(1+S_{11})^2 - S_{21}^2}{(1-S_{11})^2 - S_{21}^2} \tag{8}$$

where Z_0 is the reference impedance (50 Ω).

To compute the attenuation, its necessary to obtain the propagation constant $\gamma = \alpha + j\beta$ for the CPW cell. This can be computed by means of [21]:

$$e^{-\gamma l} = \left\{ \frac{1 - S_{11}^2 + S_{21}^2}{2S_{21}} \pm K \right\}^{-1} \tag{9}$$

where

$$K = \left\{ \frac{(S_{11}^2 - S_{21}^2 + 1)^2 - (2S_{11}^2)^2}{(2S_{21}^2)^2} \right\}^{\frac{1}{2}} \tag{10}$$

Attenuation in microwave lines occurs due to radiation, metal and substrate losses. Assuming that radiation losses are very small, it's possible to obtain the dielectric loss tangent from the value of the total attenuation and conductor losses.

The attenuation due to conductor losses in the center strip conductor and ground planes of a CPW is given by [22]:

$$\alpha_c \approx \frac{R_{sm} b^2}{16 Z_0 K^2(k)(b^2 - a^2)} \cdot \left\{ \frac{1}{a} \ln\left(\frac{2a}{\Delta} \frac{(b-a)}{(b+a)} \right) + \frac{1}{b} \ln\left(\frac{2b}{\Delta} \frac{(b-a)}{(b+a)} \right) \right\} \tag{11}$$

with

$$R_{sm} = \omega \mu_c t \operatorname{Im}\left(\frac{\cot(k_c t) + \csc(k_c t)}{k_c t} \right) \tag{12}$$

where k_c is the wave number, ω the angular frequency and μ_c the permeability of the conductor.

From the knowledge of the total and metal losses we can obtain the dielectric loss tangent from the attenuation constant due to the dielectric losses [18]:

$$\alpha_d = \frac{\pi}{\lambda_0} \frac{\varepsilon_r}{\sqrt{\varepsilon_{eff}}} q \tan\delta \tag{13}$$

where q is the filling factor that depends on the geometry.

Next, the description of the test cells required to measure the S-parameters is presented.

CPW Cells Design

Because the substrate losses are relatively small, lines with 5-mm length were used to increase the calculations accuracy. The metal areas were fabricated with a 2 μm layer of aluminium on top of each wafer. A sample of the fabricated lines is shown in Fig. 4. Since the exact electrical permittivity value at the desired frequencies was not known, several CPW cells were designed in order to obtain a suitable configuration for the material properties extraction. It is recommended the use of some mismatch in order to obtain a good accuracy [15]. In this way, the CPW cells were designed with different *W/S* ratios in order to obtain different characteristic impedances. Namely, lines with the following dimensions were used:

(W = 75 μm, S = 15 μm), (W = 50 μm, S = 35 μm), (W = 75 μm, S = 50 μm) and (W = 100 μm, S = 60 μm).

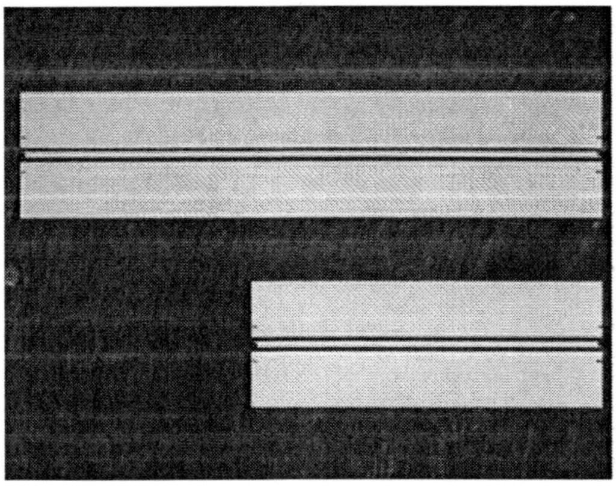

Figure 4. Sample of the fabricated coplanar waveguides used for S-parameter measurement.

Materials Properties

The measured S-parameters of a coplanar waveguide (CPW) propagating the dominant mode were used to obtain the electrical permittivity and the dielectric loss tangent of three different glass wafers: non-alkaline Schott AF45, Corning Pyrex #7740 and Hoya SD-2. These properties were obtained up to 10 GHz. A vector network analyzer and a probe station were used to perform the on-wafer measurements of the two-port network S-parameters. It was calibrated by means of TRL method, providing a measuring reference plane at the edge of the coplanar lines.

Glass Wafers Characterization

The electrical permittivity is presented in Fig. 5 for the three glass wafers under. As can be seen from that figure, for high frequencies, and as expected, the electrical properties show only a slight variation with frequency. Also, we can observe an abrupt change on the measured characteristics at low frequencies. This happens because at those frequencies the assumptions behind the theoretical formulation are not anymore valid. At the frequencies of interest (5-6 GHz), the dielectric constants for SD-2, borofloat and AF45 are 4.7, 5.9 and 6.1, respectively.

The losses are difficult to obtain since they are relatively small for such line lengths. The results can change significantly from one measurement to another, if not enough attention is paid when the contact between probes and lines is established. To compute loss values, the remaining data after discarding the inaccurate measurements were again submitted to a new selection. Only the ones giving the lower values for the losses were used. It was considered that the higher losses were due to imperfect contacts between probes and CPW cells.

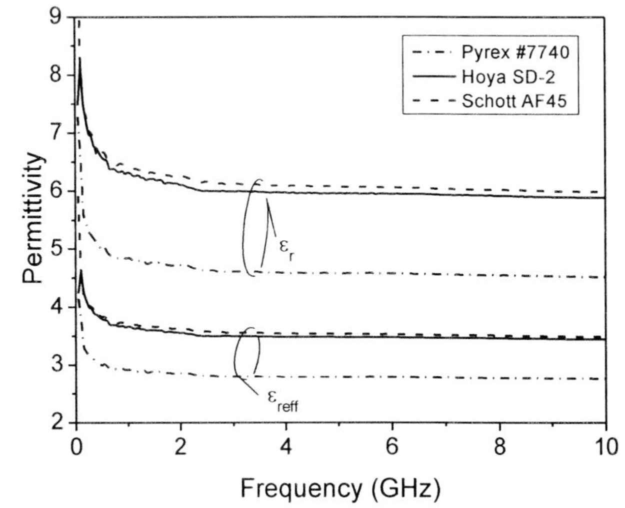

Figure 5. Measured permittivity for three glass wafers.

In Fig. 6, the total attenuation measured in a CPW cell is plotted together with the attenuation due to conductor loss, computed from equation 11. Assuming the radiation losses being very small, the difference between total losses and conductor losses give us the substrate losses.

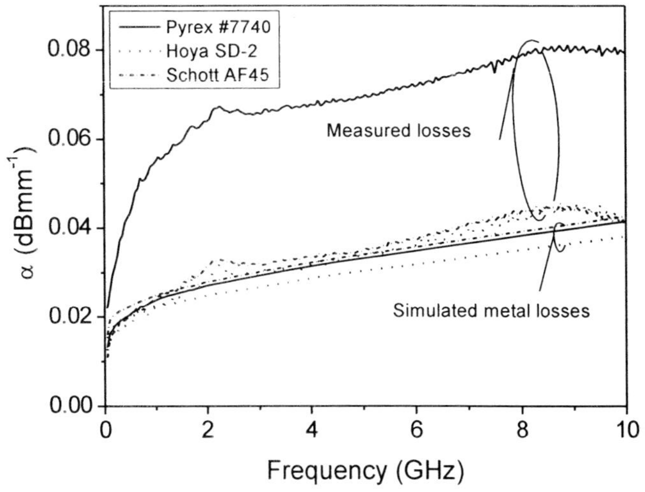

Figure 6. Computed and measured attenuation of a CPW cell with $W = 75$ μm and $S = 50$ μm.

From the above figure we can see that the SD-2 and AF45 substrates suffer from almost the same losses, but the #7740 substrate suffer from increased losses. When designing RF and microwave elements, the structures on the Pyrex and AF-45 wafers should be similar but not the device losses.

The data from Fig. 6 was used to compute the loss tangents. The obtained results are plotted in Fig. 7. As expected from total losses, the SD-2 and AF45 wafers show similar values, and Pyrex wafer presents a higher value for the loss tangent. When possible, AF45 should be used instead of Pyrex #7740.

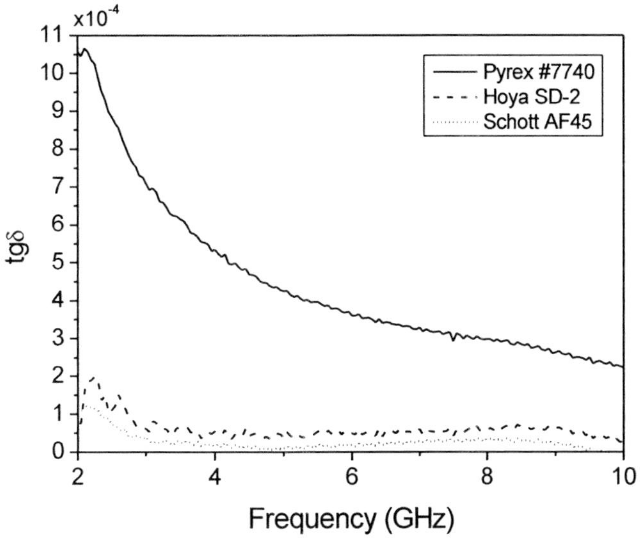

Figure 7. Measured loss tangent (W=75 μm, S=50 μm).

High-resistivity Polycrystalline Silicon

The electrical permittivity of HRPS was also obtained from the measured S-parameters and is displayed in Fig. 8. This plot shows the results obtained from the three different CPW cells. As can be observed in that figure, for high frequencies, and as expected, the electrical properties show only a slight variation with frequency. Also, we can observe an abrupt change on the measured characteristics at low frequencies. This happens because at those frequencies the assumptions behind the theoretical formulation are not anymore valid. At the frequencies of interest (5-6 GHz), the obtained dielectric constant is $\varepsilon_r \approx 11.5$.

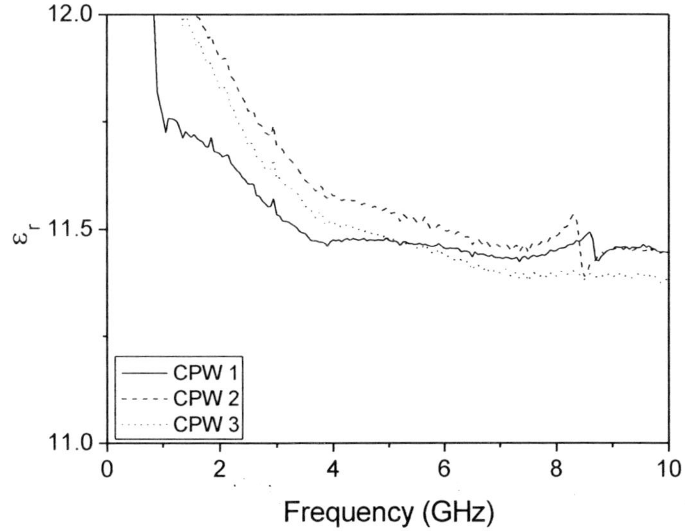

Figure 8. Extracted electrical permittivity for three different CPW cells.

Together with dielectric constant, loss tangent is also a fundamental parameter as it represents the performance achievable with the designed passives. The results obtained for loss tangent are plotted in Fig. 9. As expected, all the CPW lines show similar values since the substrate is always the same.

The method used to obtain the data in Fig. 8 is based on the measured total losses coming from the CPW lines. Then, assuming no radiation losses, it is possible to identify the metal losses and substrate losses. In this way, it is important to refer that the obtained values for loss tangent are heavily dependent on the models used to describe the losses in these kinds of transmission lines.

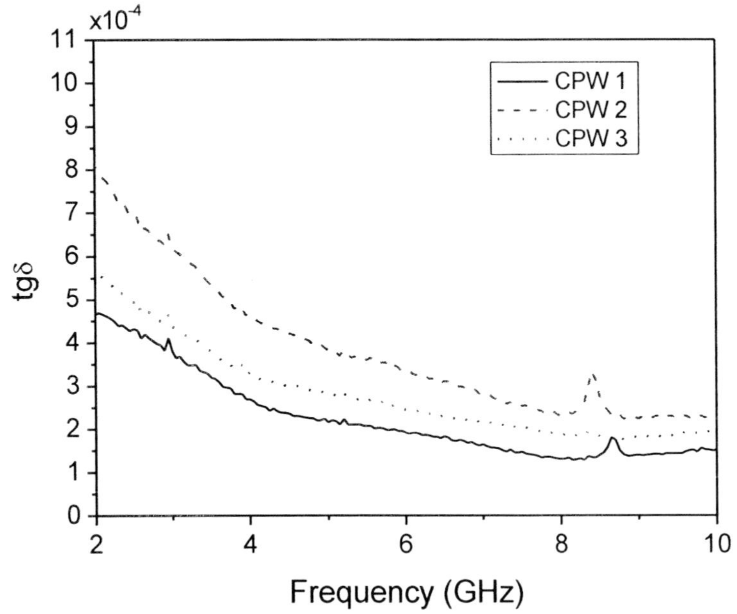

Figure 9. Extracted loss tangent for three different CPW cells.

Patch Antennas

One of the simplest structures that can be used as an integrated antenna is a patch antenna. It is planar and can be fabricated using only two metal layers, one to form the ground plane and other to form the radiating element.

The next antenna design was performed with 3D models built using a 3D FEM simulator tool. The layout was then obtained with the standard IC layout tools. From the set of available materials compatible with IC fabrication, standard silicon was not chosen for use as substrate due to its low resistivity. The option was to use HRS together with insulating layers, to keep the losses as low as possible, and HRPS. Since the antenna dimension is related with its electrical length, which is inversely proportional to the operating frequency, we have chosen to start the antenna design for operation in the 5-6 GHz ISM band. This is an available free band where the antennas can be relatively small and still achieving a good range.

Patch Antennas Using HRS

To obtain the smallest antennas it was used the material with the highest electrical permittivity. Fig. 10 shows the cross-section of the stacked materials with the configuration used in the fabrication.

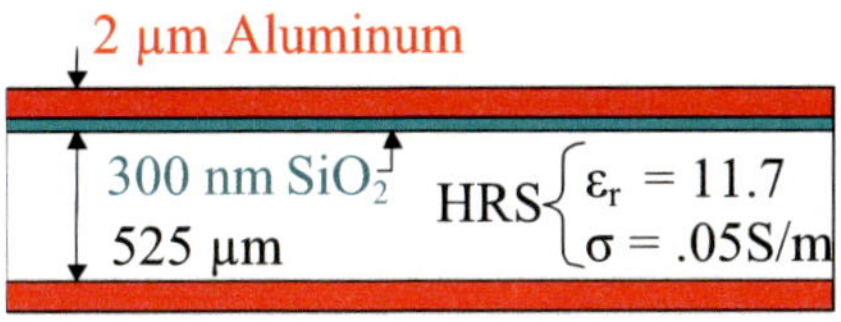

Figure 10. Cross-section view of the realized patch antenna on HRS substrate.

The patch antenna design was supported with a model built using a high frequency structure simulator based on finite elements modeling (FEM) (Fig. 10). A tool with 3D modeling capabilities was necessary due the fact that, for small ground planes, the antenna behavior depends on the ground size.

The two critical steps in designing the patch antenna were the definition of the patch dimensions and the feeding configuration. The patch dimensions have direct influence on the operating frequency and on the antenna gain. The difficulty to predict accurately the patch dimensions is related to the fringing fields together with the small size of the ground plane used. The starting value used for the antenna length, *L*, was half wavelength in the substrate, which is known to give a close value for the operating frequency. This value was then trimmed by simulation.

The antenna feeding should be designed carefully since it must provide a correct impedance matching. At high-signal frequencies it is necessary to design a feeding line with specific characteristic impedance. Also, that line must be connected in a point of the antenna where the input impedance is the same than the feed-line characteristic impedance. The patch antenna was fed with a microstrip line connected to a point inside the patch where the input impedance matches 50 Ω. This connection was achieved with an inset, which had to be properly adjusted with the help of the antenna model.

The model, as well the projected antenna dimensions, is presented in Fig. 11.

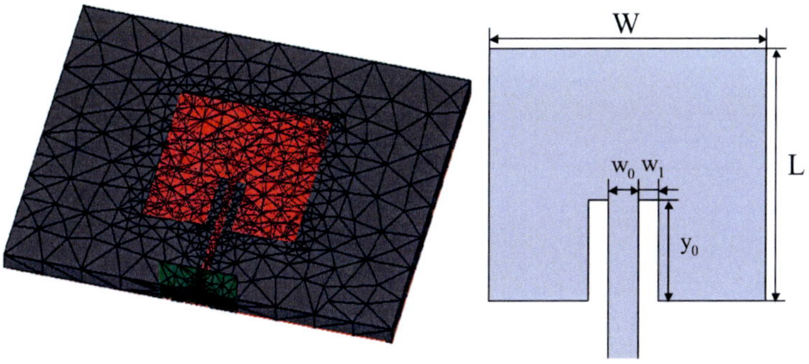

(a) (b)

Figure 11. FEM meshed model (a), and layout (b) describing the studied square patch antenna ($L = 7.7$, $W = 7.6$, $y_0 = 3.1$, $w_0 = 0.36$, $w_1 = 0.32$, in mm).

Fig. 12 shows several different patch antennas standing on the top of a HRS wafer. The HRS substrate shown has a dielectric permittivity of 11.7, conductivity in the range of 0.02-0.05 S/m, and the wafer thickness is 525 ± 25 μm. A 300 nm layer of thermal silicon dioxide layer between the silicon substrate and the metal patch was used for insulation. This layer has an ε_r of 3.9 and for design purposes it is assumed to be an insulator. The metal patch and ground plane were obtained using a 2-μm layer of aluminum. Instead, copper could be used to further reduce the metal losses.

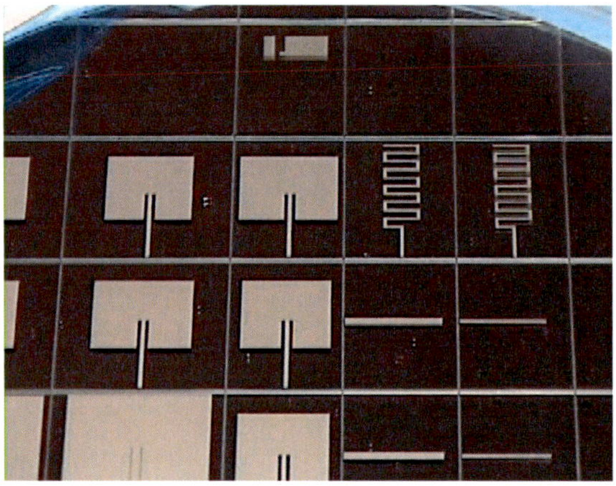

Figure 12. Patch antennas fabricated on a HRS wafer.

To measure the antenna characteristics, it was interfaced with a 3.5 mm coaxial connector. This was achieved by placing the die containing the patch antenna on top of a PCB board with the coaxial connector soldered underneath. One fabricated prototype, ready for measurements, is shown in Fig. 13.

Figure 13. A 7.7x7.6 mm^2 patch antenna realized on a HRS substrate ready for reflection measurements.

All the return loss measurements were performed using an HP vector network analyzer, which was previously calibrated with one-port calibration. The antenna operating frequency, bandwidth, and efficiency were obtained from those return loss measurements.

The simulated and measured values for the patch antenna using HRS substrate are plotted in Fig. 14. The simulated data shows good agreement with the measurements. The obtained operating frequency was 5.705 GHz, providing a -10 dB return-loss bandwidth of 90 MHz.

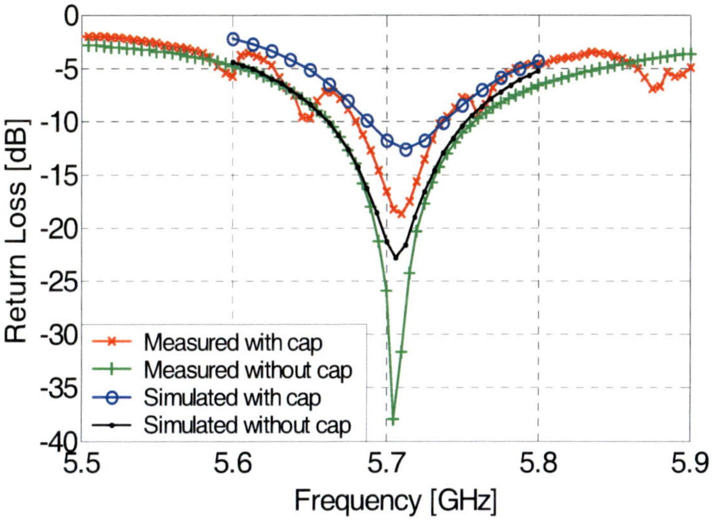

Figure 14. Measured and simulated return loss versus frequency used to obtain the operating frequency, bandwidth and efficiency.

The antenna efficiency was measured using the Wheeler cap method. This method is based on the measurements of the antenna input return loss when it is radiating or not radiating. The last condition is usually met with a metallic cap enclosing the antenna under test. With those measurements the efficiency can be easily computed. The antenna efficiency was also obtained from simulations and compared with the measured values. Using the data from measurements, it was obtained an efficiency of 18.6 %, which is in good agreement with the value computed by the 3D model, that was 19.6 %.

The suitability of an antenna to be integrated depends also on its gain and radiation pattern. The antenna should contribute to a good wireless link range and should not interfere with the already on-chip sensors and electronics. This requires the antenna to radiate mainly in the upward direction, with the backside lobe as weaker as possible. The radiation to the backside can be reduced if the ground plane is big enough. However, all the antenna components should be as small as possible, including the ground plane. The far-field gain patterns measurements were obtained using an anechoic chamber facility. The results are plotted in Fig. 15.

As it was expected, the patch antenna exhibits a linear polarization characteristic, and as it was desired the power is mainly radiated upwards. Nevertheless, it would be desirable to further decrease the power level at the back of the antenna to keep the interference with backside components as low as possible. This drawback results from the small size of the ground plane. The maximum gain registered was ~0.3 dB. This small gain is essentially due to low efficiency of the antenna, since the substrate suffers from high losses.

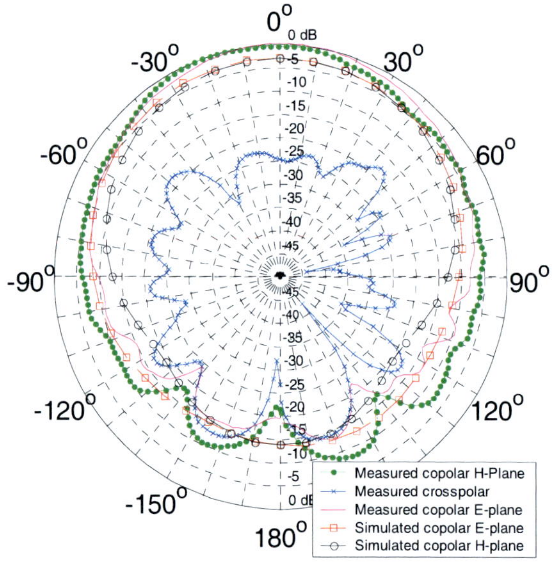

Figure 15. Measured and simulated co-polar and X-polar far-field gain patterns obtained at 5.705 GHz.

After model validation by the measured data, the influence of a few material tolerances were analyzed by simulation. Substrate thickness, substrate conductivity, and oxide thickness were studied. It was observed that varying the substrate thickness from 500 μm to 550 μm and the oxide thickness from 1 μm to 10 μm, the operating frequency changed from about 5.7 GHz to 5.85 GHz. If the substrate conductivity increases from 0.02 S/m to 0.05 S/m, the efficiency decreases from 30.1 % down to 19.6 %.

Patch Antennas on HRPS

The fabrication of patch antennas on HRPS followed a similar process than the antennas fabricated on HRS. The patch antenna was also designed for fabrication on top of an HRPS wafer, without any insulating layer between the metal patch and the substrate. The antenna was designed to operate in the 5-6 GHz ISM band, which yields antenna dimensions of 7.7x7.6 mm^2. The patch metal layer was made with 2 μm of sputtered aluminum and the feeding was realized through a microstrip line.

The measured values used in the Wheeler cap method are plotted in Fig. 16. The figure shows measured values when the antenna is radiating and when it is not. Using the data from measurements, it was obtained an efficiency of 25.6 %, which is in good agreement with the value computed by the 3D model, that was 28.6 %.

It was also verified that this antenna has an operating frequency of 6.25 GHz, with a -10 dB return loss bandwidth of ~200 MHz.

For comparison, a similar patch antenna was fabricated on a Pyrex #7740 wafer. Similarly, 2 μm of sputtered aluminum were used to obtain the metal patch layer. Such antenna has a measured operating frequency of 5.995 GHZ and the -10 dB return loss bandwidth is ≈100 MHz. From the measurements we obtained an efficiency of 51%, which is higher than the obtained with HRPS. The drawback is the increase in the antenna size.

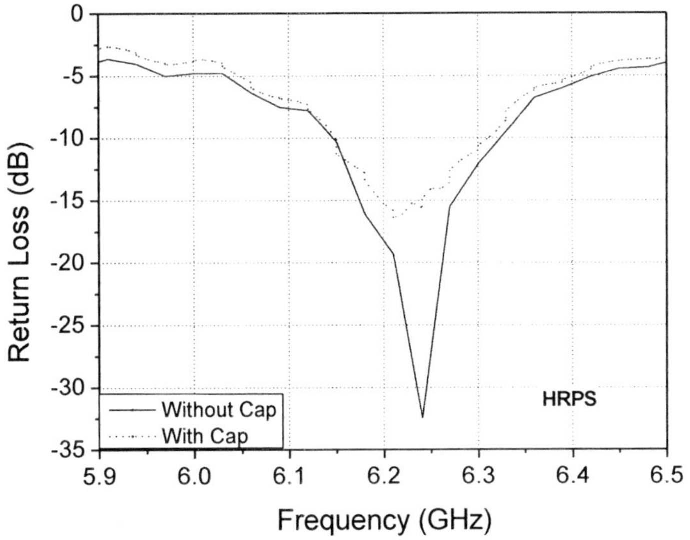

Figure 16. Measured return loss versus frequency used to obtain the operating frequency, bandwidth and efficiency.

MEMS Micro-Antennas

What is the challenge to design wireless communications for biomedical applications? At a first glance, if we make a quick survey in the available databases, it seems that there is no problem since it is not easy to find many references to this topic. However, in previous sections, it was already shown the difficulty in the design of integrated antennas. In this section, the fundamental limits will be revisited and solutions will be discussed to achieve smaller antennas.

Electrically Small Antennas

The physical dimension of the host microsystem limits the available room for antenna integration. In this way, the antenna must be as small as possible, including the ground plane dimensions. This has several implications in the antenna performance and in the neighbour circuits.

Fig. 17 shows the antenna integration using wafer-level chip-scale packaging (WLCSP), together with radio-frequency passives. To save space, the antenna is placed on top of the active circuits. Nevertheless, since the antenna ground plane is very small, not exceeding the antenna dimensions, the system performance is affected.

The antenna is very close to the circuits, allowing a possible coupling between it and any passive devices (inductors or transmission lines). The perfect solution is to obtain an antenna small enough to grant extra space for an antenna ground plane larger then the antenna itself.

Several small and planar antenna types have been proposed for wireless communications [23], but none of them was designed to fulfill all the restrictions and requirements set by on-chip integration. Those restrictions include the properties of available substrate materials and the way they can be processed. Many of the previously proposed solutions to integrate antennas on-chip have been based on the design of planar antennas using silicon as substrate. Since the low-ohmic silicon substrate suffers from high losses, high-resistivity silicon or bulk micromachining have to be used in order to increase the antenna efficiency. Nevertheless, the afore-mentioned solutions have the drawback of increased cost, and the micromachining solution have also the penalty of large area used for antenna implementation. In this way, a preferable solution to decrease the antenna losses may be to use a combination of a low-loss material with silicon. The new material can be used as antenna substrate and any required high-quality factor passives [24], and the silicon will be used to implement the necessary circuitry.

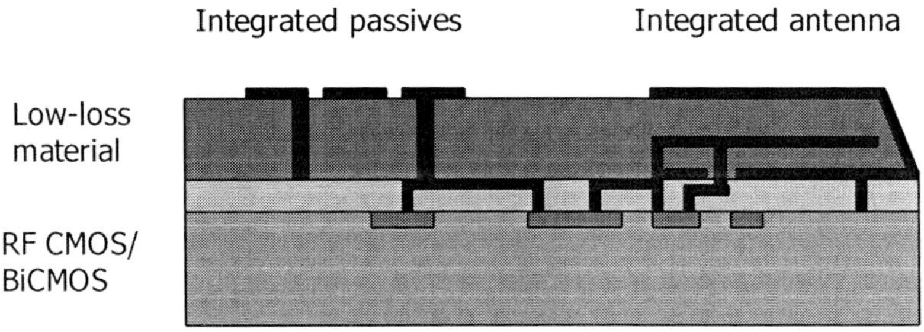

Figure 17. Envisioned application of WLCSP for antenna and other passives integration.

The combination of materials may be achieved with the use of WLCSP techniques, like adhesive wafer bonding and through-wafer electrical via formation, which allows the use of silicon together with different silicon-compatible substrates [25]. However, typical silicon-compatible substrates (e.g. glass, BCB, polyimide, SU-8) have lower dielectric constant compared to silicon. In this way, the use of such materials reduces the losses at the expense of a size increase in the integrated antenna. Therefore, the use of an advanced antenna design may be required to overcome this drawback, providing a small and effective radiator.

In our previous work [26] and other related work [27], the use of shorted-folded patch antennas was considered as a solution to obtain a small antenna. Notwithstanding the obtained success, the dimensions of the developed antenna are still rather large.

Fundamental Limits

A common question asked when the antenna miniaturization is required is: "What is the theoretical limit for antenna size reduction?". The first work trying to answer this question dates back from 1947 [28]. In that work, Wheeler investigated the fundamental limits of electrically small antennas. An electrically small antenna was defined to have overall dimensions smaller then $\frac{\lambda}{2\pi}$. Sometimes, this is also referred under the relation $ka < 1$, where $k = \frac{2\pi}{\lambda}$, where λ is the free space wavelength and a the radius of the smaller sphere enclosing the antenna, also called the radian-sphere. Fig. 18 shows the concept widely adopted to characterize an electrically small antenna.

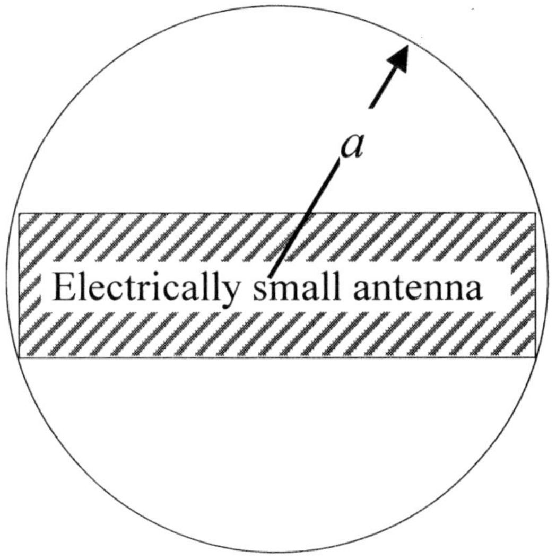

Figure 18. Electrically small antenna enclosed by the radian-sphere.

In his work, Wheeler analysed which could be the highest efficiency of an electrically small antenna, considering it as a capacitive or inductive load. Some time later, Chu published his work about electrically small antennas [19]. In that work, spherical wave functions were used to describe the antenna field, gain, and quality factor. One research topic was to find the maximum gain achievable by an antenna of moderate complexity. This is equivalent to find the minimum quality factor for the antenna. Other research topic was to find the maximum ratio gain/quality factor.

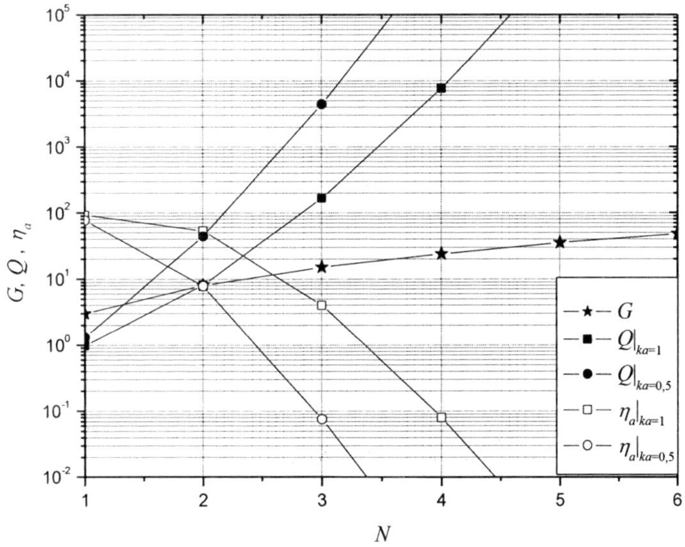

Figure 19. Tradeoffs in antenna parameters as a function of antenna complexity, N.

From there on, Chu results have been widely used as a reference in antenna size reduction. Several authors have since then studied the behaviour of electrically small antennas [30-40]. From those studies it is possible to conclude that, theoretically, it is possible to obtain an antenna with an extremely high gain. However, there are other important parameters in an antenna, namely efficiency and quality factor (bandwidth).

Fig. 19 shows the tradeoffs in antenna parameters as the antenna complexity, N, increases, for two antenna dimensions. It can be observed that the gain, G, and quality factor, Q, increases with the antenna complexity, and the efficiency, η_a, is reduced. It can also be observed that the larger antenna has a smaller quality factor and a larger efficiency.

Other interesting analysis is to observe the behaviour of antenna quality factor, efficiency, and gain for an antenna operating in the fundamental mode. Fig. 20 shows the results of such analysis, where it can be observed that all antenna parameters are improved as the antenna dimension increases. It can also be observed the fast degradation of antenna efficiency for $ka < 0.4$.

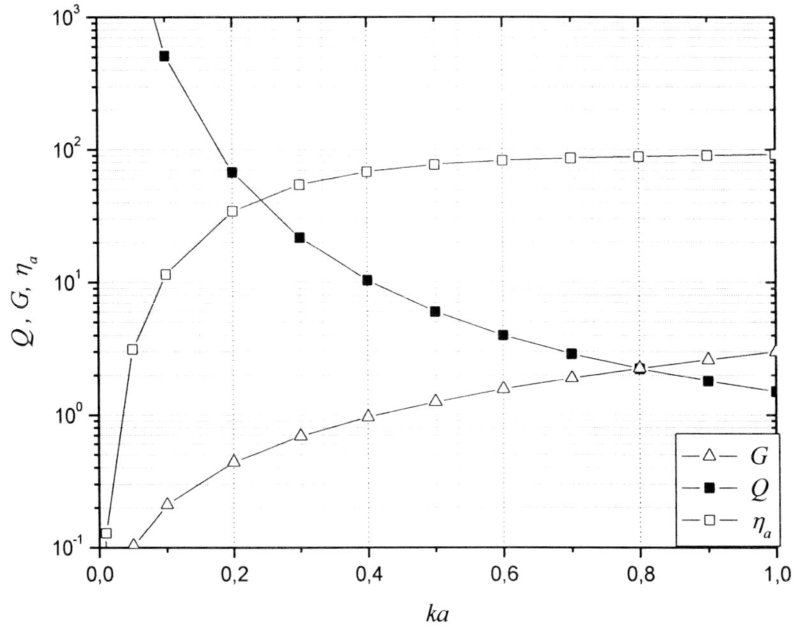

Figure 20. Influence of antenna dimensions on gain, quality factor, and efficiency.

Wafer Level Packaging

The properties of a small antenna are dependent on how it fills it radian-sphere [40]. This means a dipole antenna will have worst properties then a patch antenna, since the patch antenna uses more space inside the radian-sphere. In this way, a very attractive option is the antenna integration using stacked wafers, where the antenna uses not only two dimensions, but also a third dimension, at a reduced cost.

The technical solution for this integration approach is the application of wafer-level chip-scale packaging (WLCSP) techniques. This approach represents a truly added value as at a limited cost 3D passive structures can be realized without increasing the chip active dimensions. Also, WLCSP allows combination of a different substrate (e.g. HRS, glass) together with low-ohmic silicon. The use of new wafer materials may reduce the losses with the potential of allowing the integration of other passive devices. Moreover, the possibility to fabricate 3D structures allows the implementation of more advanced antenna structures, where size and efficiency restrictions may be met more easily. Fig. 21 shows two options on how to use the WLCSP concept to integrate antennas and/or other passive devices.

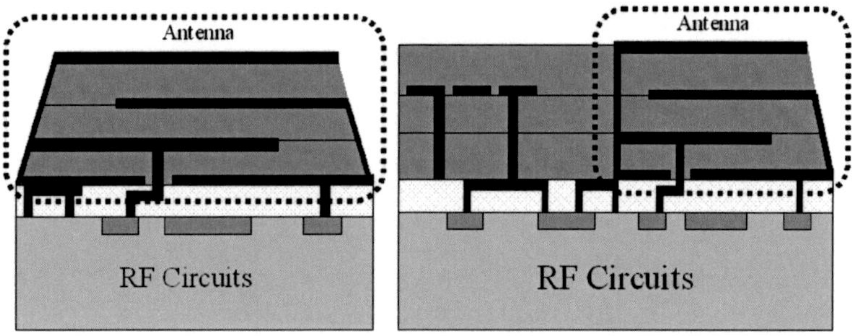

Figure 21. Integration of a small size antenna with RF circuitry.

In the next section, we will introduce the use of WLCSP for integration of planar antennas built using silicon compatible substrates. The suitability of folded patch antennas, built on HRS and glass, are investigated as candidates for on-wafer integration. The 3D antenna models were built and the measured antenna parameters were compared with the values obtained by simulation. Despite its higher fabrication complexity, the folded patch antenna has an increased performance when compared with the patch antenna, and at the same time allows reduction of the used chip area. In this way, the use of glass substrate enables a small on-wafer antenna and RF electronics direct coupling. This offers potential of low cost, low profile and simplified assembly.

Folded-patch Antenna

The possibility to integrate on-chip antennas for biomedical devices is highly dependent on the achievable antenna dimensions and efficiency. Reduction of dimensions together with efficiency improvement can be obtained through proper device geometry. Since the patch antennas are rather large for on-chip integration, the use of a shorted-folded patch antenna on glass was considered.

Antenna Modelling

The proposed, on-chip integrated, folded short-patch antenna (FSPA) is shown in Fig. 22. It consists of three horizontal metal sheets that are electrically connected by two vertical metal walls.

All structure is embedded in a dielectric substrate having certain electrical permittivity and dielectric losses. These two parameters together with the antenna geometry and its actual dimensions will determine its radiation characteristics and overall performance.

For the best performance, the metal sheets should have minimum resistivity and the dielectric should be a low-loss material, which allows high efficiency. Also, to achieve small antenna dimensions, a substrate with high electrical permittivity is desirable. High antenna efficiency requires thicker substrates (>300 μm) and therefore high aspect ratio vias in glass are required.

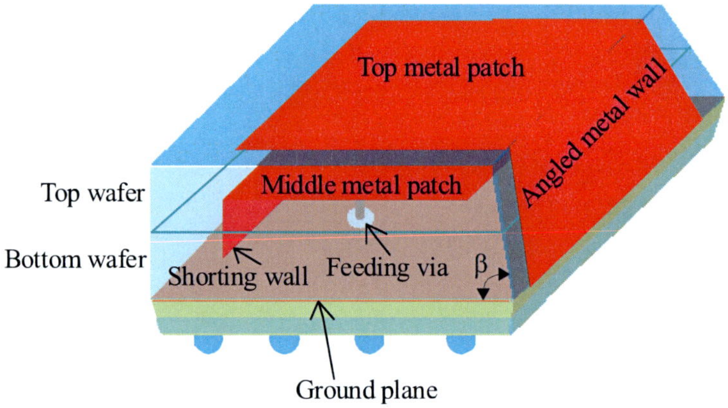

Figure 22. Proposed shorted-folded patch antenna.

At frequencies above 1 GHz, glass becomes a very attractive option. Its main advantages are low losses, reasonable ε_r, availability in a form of wafers with any required thickness and diameter, and last but not least low cost. There is also sufficient experience in processing of glass wafers from MEMS and WLP applications.

The antenna was designed to operate at 5.7 GHz, a frequency chosen to be inside the 5-6 GHz ISM band. All the simulation analysis was performed with an antenna model that was built using the High Frequency Structure Simulator from Ansoft, a 3D tool based on finite element modeling. This simulation tool was intensively used previously in our patch antenna design, where good match of modeling and experimental results were achieved. The developed antenna model is displayed in Fig. 23.

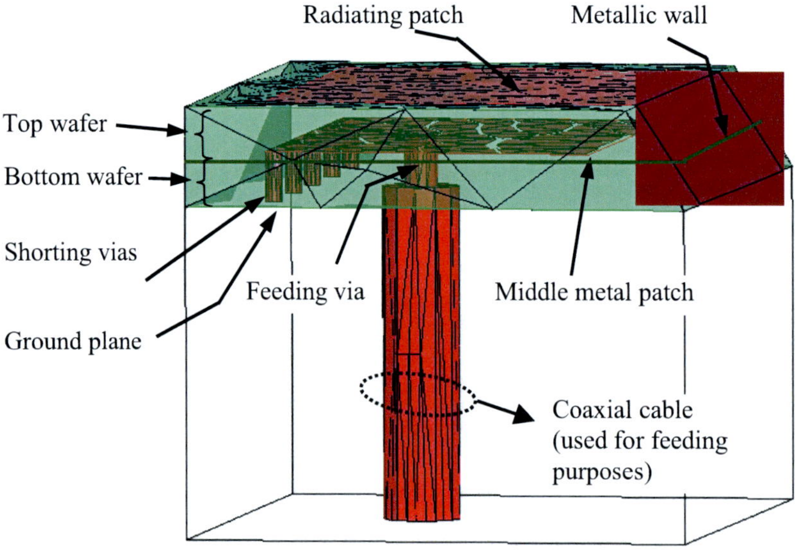

Figure 23. 3D FEM model of the fabricated folded shorted-patch antenna.

From the fabrication point of view, it is preferable to keep the overall antenna thickness small. To achieve this, thin or thinned wafers can be used. However, the wafer thickness is a relevant parameter that influences the antenna performance, since it corresponds to the antenna substrate thickness. To study this effect, all the FSPA dimensions were kept constant, except the wafer thickness, *h*, which was as a parameter. The obtained results for the return loss are presented in Fig. 24.

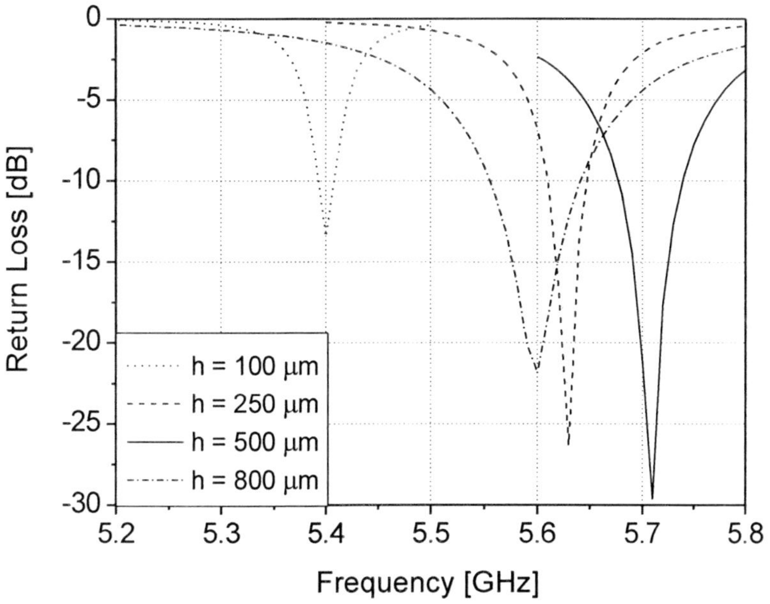

Figure 24. Return loss of the FSPA for different substrate thickness values.

It can be observed that decreasing the wafer thickness from 500 μm down to 100 μm leads to a reduction of the antenna operating frequency. This is a desirable effect since, for the same operating frequency, the antenna can be smaller. However, a thinner substrate, and thus a smaller volume of the antenna dielectric, will cause reduction of the antenna radiation efficiency as well as reduction of the antenna bandwidth. If the wafer thickness is increased to 800 the operating frequency starts to decrease again, instead of a monotonously increase with thickness increment. In opposition to what we could be induced by observation of Fig. 24, the expected behavior would be a decrease in the operating frequency as the wafer thickness is increased since the electrical length of the antenna becomes larger. However, the gap between the middle patch and the ground plane has a significant effect in the antenna operating frequency. In this particular design, that gap starts to be the dominating effect in setting the operating frequency when the wafer thickness is more then 500 μm, which explains the frequency decrease with wafer thickness decrease.

Together with the reduction in the antenna dimension and increased efficiency, despite the added cost of fabrication complexity, with this type of antenna the following material combinations can be formed: glass/glass, HRPS/HRPS, or glass/HRPS. An option to avoid the difficult task of doing through-wafer vias in glass could be to replace the glass wafers by HRPS. This will inevitably increase a bit the dielectric losses, but at the same time the antenna dimensions could be reduced. Other option is to substitute only the bottom glass wafer by HRPS. In this way, the vias in glass are not required and the overall losses are expected to be smaller.

All the proposed options were analyzed based on 3D FEM modelling built using. Considering a 10 kΩ-cm substrate, the predicted results are summarized in Table 1. For various substrate options, the dimensions of the antenna model were kept constant as possible, only some minor adjustments were implemented to adjust the operating frequency and/or to achieve impedance matching.

Table 1. Summary for the different stack options.

	glass/glass	HRPS/HRPS	HRPS/glass
F_c	5.66 GHz	5.66 GHz	5.64 GHz
BW	60 MHz	57 MHz	63 MHz
Eff.	66 %	64 %	65 %
L_l	3.2 mm	1.8 mm	3.2 mm
L	2.6 mm	1.65 mm	2.3 mm

Parameters shown: F_c – operating frequency, BW – bandwidth, Eff. - efficiency, L_l - top patch length, L - middle patch length.

As can be seen from Table 1, the antenna built on a stack of two glass wafers has the highest efficiency and the largest dimensions. The antenna on a stack of two HRPS wafers is the smallest and has the lowest efficiency. When the glass/HRPS stack is used, a compromise can be obtained. The losses are slightly increased and the dimensions don't change significantly. The antenna dimensions are strongly dependent on the projected efficiency and bandwidth and also on the substrate thickness but, if we use HRPS/HRPS, they can fit inside an area of 3x3 mm^2. The achievable –10 dB return loss bandwidth is around 50 MHz (+/- 10 MHz).

The use of a 10 kΩ-cm HRPS wafer makes on-chip antenna integration possible with an antenna efficiency and electrical performance similar to the one obtained with glass, but with the benefit of a smaller size (12.4x11.7 vs. 7.7x7.6 mm^2 for patch antenna and 4x4 vs. 3x3 mm^2 for folded-patch antenna at 5.7 GHz) since the dielectric constant is two times higher for HRPS. Next to that, the inherent problems associated with glass substrate processing (e.g. difficulty to form high-aspect ratio vias) are avoided.

Antenna Fabrication

Despite the potential of HRPS, it was decided to evaluate the suitability of glass wafers for antenna applications because of its popularity in biomedical diagnostic devices (e.g. lab-on-chip) that may also benefit from on-chip antenna.

The folded-patch antenna fabrication sequence is schematically shown in Fig. 25. Two AF-45, 100 mm diameter and 500 μm thick glass wafers are used as the starting material.

Firstly, 200 μm diameter through-wafer vias were formed in the bottom glass substrate. A laser system at Philips CFT with a 30:1 reduction mask was used for ablation of the feeding and shorting vias. Because the selected glass substrate material exhibits sufficient light absorption in the UV region only, a 193 nm excimer laser was required. The initial tests with 248 nm laser were not successful due to formation of cracks. The vias were then metallized by sputtering of 4μm Al layers from both sides of the wafer. Due to the low aspect ratio of the vias (2.5:1) and the fact that the sidewalls are not perfectly vertical, it was possible to form a continuous conductive layer within the vias. The middle antenna patch was then patterned using electroplated photoresist and plasma etching (see Fig. 26).

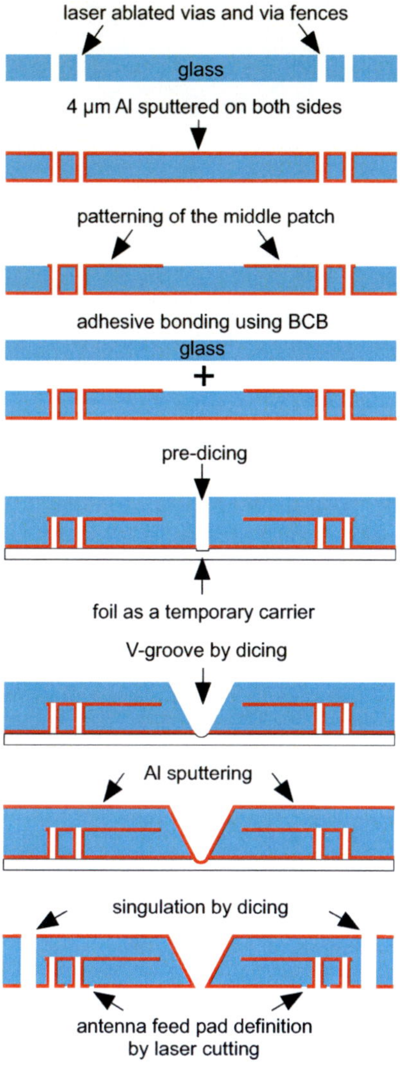

Figure 25. Schematic fabrication sequence used for antenna fabrication.

Then, the second glass wafer was adhesively bonded using ~5-7 μm thick BCB layer as the adhesive. This wafer stack was attached to a temporary carrier (foil or Si wafer) to allow formation of the slanted antenna sidewall by V-blade dicing. The sidewall shaping was performed in two steps. First, a vertical trench through the wafer stack was formed using a 400 μm wide dicing blade. In the second step, a V-shaped dicing blade (60 deg. angle) was applied to shape the antenna sidewalls. The accuracy of this step is critical for antenna electrical properties and care has to be taken to achieve alignment with the middle antenna patch. The achievable accuracy of blade positioning is close to 2 μm, which is more than sufficient.

Figure 26. Photograph showing detail of the bottom glass substrate with the 200 μm diameter ablated vias and the patterned middle Al antenna patch.

The antenna fabrication then continues by Al-layer sputtering to metallize the second glass wafer including the V-shaped trenches. Finally, a standard dicing with vertical sidewalls is applied to define the remaining three antenna sidewalls and thus the lateral dimensions of the final antenna. The fabricated antenna prototype, compared to a 1 eurocent coin, is shown in Fig. 27.

Figure 27. Photograph of the folded shorted-patch antenna prototype realized on a stack of two AF-45 glass substrates.

Antenna Results

For fabrication simplicity, laser cutting on the to-be-measured samples defined the antenna-feeding pad. For the measurement purposes, the antenna was attached to a PCB with a 50 Ω microstrip line, and the antenna feeding pad was connected using multiple bond wire connections (see Fig. 28).

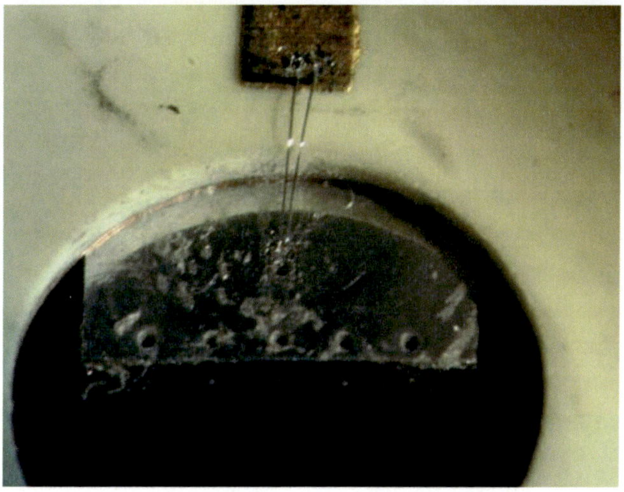

Figure 28. Close-up of antenna backside feeding point connected to a PCB, using multiple bond wires.

A prototype was fabricated with the technique described before, where the laser ablation was used to obtain the through-wafer vias. To measure the fabricated antenna, it was necessary to attach it to a PCB board, where a 50 Ω microstrip line was designed to interface the antenna with a coaxial connector that provides the connection to the vector network analyzer. The connection between the microstrip line and the antenna feeding via was made by wire bonding. The return loss measurements were performed with an E8358A vector network analyzer. Fig. 29 shows the measured and simulated values for a fabricated prototype.

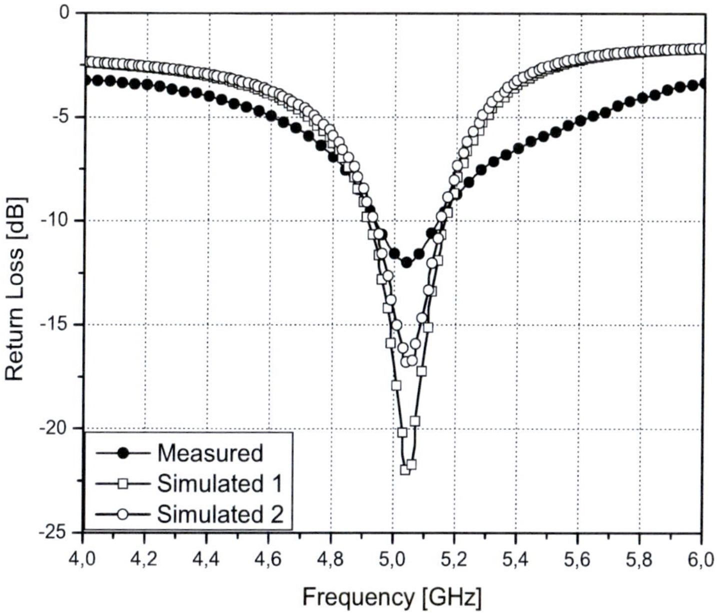

Figure 29. Measured and simulated return loss of the fabricated antenna prototype.

The simulated values on this plot were obtained after antenna fabrication, where relevant process dependent parameters were updated to better match the values obtained during antenna processing. Also, it was included the effect of the ground plane that was used to mount the antenna for simulations. The difference between the projected and measured bandwidth are mainly due to the thickness of the metal inside the shorting and feeding vias, as we were able to verify by simulation. If the metal thickness of vias metallization changes from, e.g., 2 to 0.02 μm it was obtained by simulation that the bandwidth changes from ~50 to ~200 MHz. The small shift in the desired operating frequency from 5.1 to 5.05 GHz is due to a small increase in the dimensions of the top metal patch. This top metal patch was first designed to be defined by patterning but to reduce costs, was defined by the singulation step (Fig. 25), resulting in a larger patch.

The simulated values on Fig. 29 were obtained after antenna fabrication, where relevant process dependent parameters were updated to better match the values obtained during antenna processing. Also, it was included the effect of the ground plane that was used to mount the antenna for simulations. After that adjustment, it was possible to obtain a good agreement between the measured and the simulated data. The final antenna dimensions are 4x4x1 mm^3, the operating frequency is 5.03 GHz with a bandwidth of ~200 MHz.

Cantilever Antenna

From the previous antenna design sections, it is clear that the FSPA shows advantage in occupied area requiring, however, a more complex fabrication process. The dimensions of the FSPA is suitable for small biomedical devices like the required for spinal cord stimulation or for implants used in neural signal recording. The main drawback is that it operates only inside the 5-6 GHz ISM band. This results in high losses in signal propagation inside the human body, requiring placement of the biomedical devices at the body surface. This is not a satisfactory solution for, e.g., human hear implants, or implants required to be inside the spinal cord.

To communicate with those devices its necessary to find a new solution, which will be discussed next.

MEMS Magnetic Sensors

Micro-Electro-Mechanical Systems (MEMS) are an available option for RF communications systems, since they can offer, simultaneously, devices with improved performance and integration capability in a silicon chip, side by side with semiconductor circuits, since they use IC-compatible materials.

The basic principle of micromachined cantilevers offers an interesting possibility to measure a variety of physical parameters [41]. They can be applied as magnetometers for measuring the magnetization [42, 43] and as viscosity sensors [44]. These devices are excited at the resonant frequency to achieve high sensitivity.

Up to now, MEMS have been used in antenna applications, not as an antenna itself, but to obtain non-conventional front-ends with improved, or new characteristics However, it is well known that some MEMS structures can be used as magnetic flux sensors, allowing the detection of time varying fields [45].

When used as a sensor, a MEMS structure requires the use of a sensing mechanism and the most widely used is the capacitive method. The moving structure, and a fixed plate, forms a parallel plate capacitor, where the structure movement is translated into a capacity change. On the other hand, when it has to be used as an actuator, the electrostatic actuation is widely used as the actuating mechanism due to its simplicity. The two main structures used are comb drives [46] and parallel plates [47]. While in the comb drive, which is based on area-varying capacitors, the displacement varies linearly with the gap, in actuators that relies on gap-width varying capacitors (parallel-plate) the pull-in phenomenon has to be considered [47]. Pull-in causes the displacement range due to electrostatic force to be limited to one-third of the gap between the electrodes, in case of a motion perpendicular to the capacitor plate orientation. This effect also limits the dynamic range of capacitive accelerometers operating in the feedback mode. Charge drive (current drive with a series capacitance), rather than direct voltage drive can be used to circumvent pull-in, however, at the expense of attainable maximum force for given device dimensions.

A U-shaped cantilever, proposed to detect a time-varying magnetic field, is presented in Fig. 30.

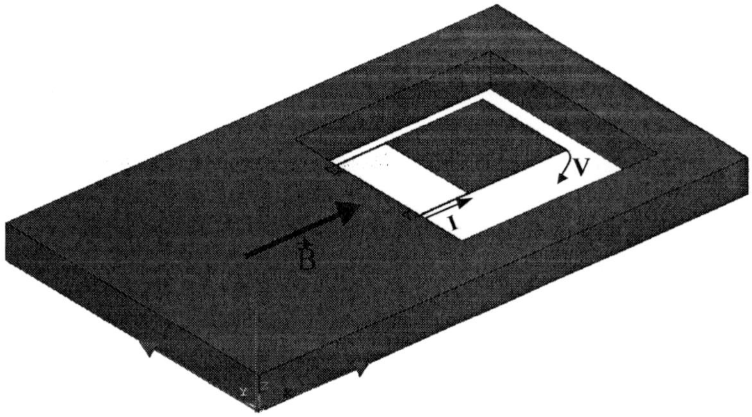

Figure 30. Cantilever used to detect a time-varying magnetic field.

To measure magnetic fields with cantilever structures, the Lorentz-force is utilized on a current carrying lead [50-52]. A cantilever of this type measures only the magnetic flux density in the direction parallel to the arms of the cantilever, i. e., x-axis of Fig. 30. The Lorentz-force acting on a lead is used to bend a micromachined cantilever. Deflections, which are small compared to the length of the cantilever, are a directly proportional measure of the applied force. To reach an as high as possible sensitivity it is advisable to utilize a resonant mechanism where the cantilever is excited by an AC current with a frequency equal to an eigenfrequency of the elastic structure. Due to the high quality factors of Si structures, which are at least several hundred, this is an efficient way to enhance the sensitivity.

Wafer Level Packaging

Fig. 31 shows how WLCSP can be used as an advantage to integrate the proposed antenna structure. It consists of three stacked wafers, where the bottom wafer is used to place the reading and controlling electronics, the middle wafer is used to implement the U-shaped cantilever, and the bottom wafer encapsulates the device, enabling a very small microsystem with integrated antenna.

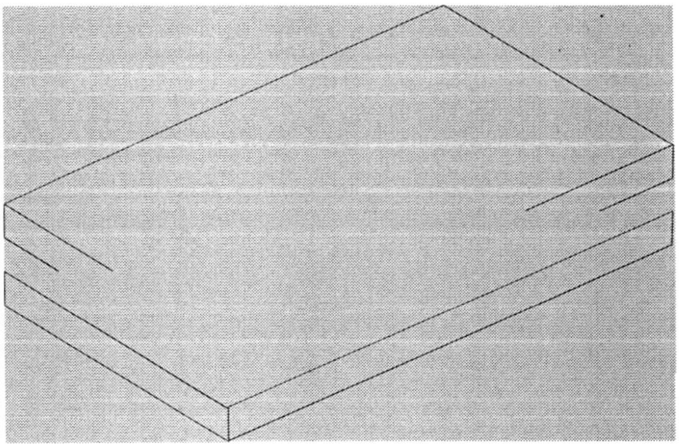

Figure 31. Use of WLCSP to integrate the proposed MEMS antenna.

Additionally, if it is added an energy-harvesting scheme, the microdevice is ready for use. If batteries are required, an extra step will be necessary to implement the required interface.

The starting MEMS structure is a cantilever placed to operate as a magnetic sensor, using the Lorentz force (see Fig. 30), where the electromagnetic field can be sensed using an optical, capacitive, or piezoelectric sensing solution. The most attractive options are capacitive and piezoelectric. These solutions can be easily integrated with the MEMS structure and have the potential for low power consumption (except the optical solution). Since the desirable displacement depends on structure dimensions and material properties, electrostatic actuation can be used as the actuation mechanism for MEMS micro-antennas. However, if large displacements are required or if the MEMS structure area becomes too small for capacitive detection, the use of a piezoelectric material can be the solution since it can act both as sensor and actuator. Moreover, the operation is only voltage based, leading to low power driving operation. Furthermore, it produces a voltage in response to a deflection leading to simple readout electronics.

Piezoelectric is a promising mechanism for realizing RF MEMS structures with low driving power and a wide continuous tuning range. A few papers have reported on piezoelectric actuators [48], which adopt ferroelectric PZT and a bulk MEMS process with wet etched holes through a Si DIE. PZT contains a high vapour pressure oxide of PbO, and requires repeated annealing at more than 600° C. These materials and processes make it difficult to be employed as a CMOS compatible process. To overcome those drawbacks, it was proposed piezoelectric actuator, which uses CMOS compatible AlN and Al as piezoelectric and electrode materials, and surface micromachining processes [49]. The deflection was proportional to the voltage and was in the range 0 – 10 um for a voltage range 0-5 V.

To radiate, the operation of the MEMS structure should be reversed and a field will be produced by an electrostatic or piezoelectric actuation.

Antenna Modelling

The research focus is now in MEMS structures for non-conventional front-ends, where the MEMS structure itself will be operating as an antenna. MEMS will be explored as a new solution to obtain structures that can sense and generate an electromagnetic field. Thus, instead of having the need to design very advanced antenna structures to achieve antenna size reduction, the standard MEMS devices, e.g. cantilevers, will be used to save system space and improve system integration.

The proposed MEMS structures will be engineered to have the desired electrical and geometrical properties, as well the requirements to be used in a post-process module compatible with integrated circuit (IC) fabrication.

Fig. 31 shows the model being used to analyse the receiving properties for a cantilever operating as an antenna.

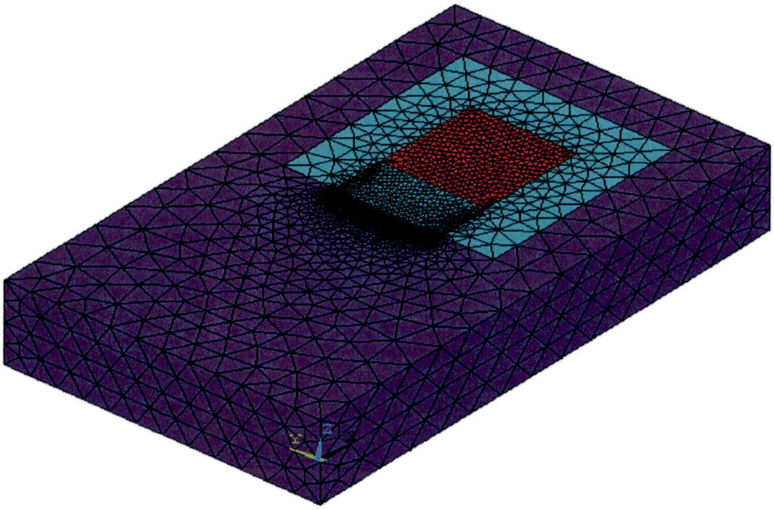

Figure 32. Model of structure used to sense the electromagnetic field.

The main goal is the design, fabrication and characterization of an electrically very small antenna using a MEMS structure. The main target will be implantable and invasive biomedical devices requiring low frequency wireless communications. However, the structures under research could also become a solution for underwater and underground communications. The research will allow to fully understand the bi-directional radiating properties and to optimise the radiation properties from an oscillating beam or cantilever. The driving properties of a piezoelectric material will be explored to obtain structures with large and bi-directional displacement. All the research will focus on low-power, low-size devices, as well the requirements to be used in a post-process module compatible with integrated circuit (IC) fabrication.

Antenna Results

To check the ability to operate as an antenna itself, some preliminary tests were conducted where it was verified that this MEMS structure could also have the potential to operate as a bi-directional wireless link, together with the potential to be smaller than the conventional antennas.

To explore the possibility to use a magnetic sensor, it was used a scaled model of the proposed structure. A commercially available magnetic sensor was connected to a signal acquisition board that was connected to a personal computer. A current was injected into a scaled structure of Fig. 32 and the signal was recorded with the magnetic sensor. When the transmitting structure was oscillating at 100 KHz, it was possible to easily detect that signal with the magnetic sensor.

The received signal was compared with the signal received by a conventional dipole antenna and it was verified that the signal received by the magnetic sensor was easily detected. Thus, instead of having the need to design very advanced antenna structures to achieve antenna size reduction, the standard MEMS devices, e.g. cantilevers, can be used to save system space and improve system integration, when compared to the actual solutions.

Conclusion

This chapter described the design, fabrication and test of chip-size antennas for short-range wireless microsystems. These antennas allow the fabrication of a microsystem with integrated wireless communications. The antenna integration is based on wafer-level packaging techniques, which enables the integration of new materials with the standard silicon processing steps, as well the fabrication of complex three-dimensional structures. Wafer-level packaging techniques, like the adhesive wafer bonding and through-wafer vias, were used to overcome the challenge of obtaining a fully integrated microsystem, including the antenna, in an economically acceptable way.

The use of a folded patch was first explored as a solution for significant antenna area reduction, enabling the antenna integration with wafer-level packaging techniques and giving rise to a good spatial cover. The chip-size antenna works in the 5-6 GHz ISM band, with the small size making it well suited for low data-rate and short-range applications. Two types of antennas were fabricated: patch and folded patch. The patch antenna on glass (11,7x12,4 mm^2), operating at 5,995 GHz, has a gain of 3 dB, 100 MHz of bandwidth, and an efficiency of 51%. On the other hand, the patch antenna on silicon (8x8 mm^2), designed to operate at 5,705 GHz, has a gain of 0,3 dB, a bandwidth of 90 MHz and an efficiency of 19%. Finally, a folded patch antenna was designed on glass (4x4 mm^2) to operate at 5,1 GHz. It revealed a measured gain of approximately -8 dB, a bandwidth of ~200 MHz and a radiation efficiency of ~32%.

MEMS are then explored as a new solution to obtain structures that can sense and generate an electromagnetic field. Thus, instead of having the need to design very advanced antenna structures to achieve antenna size reduction, the standard MEMS devices, e.g. cantilevers, will be used to save system space and improve system integration.

The original contributions are: novel electrically very small antenna using MEMS structures, a model to describe the bi-directional operation of that structure; a radiating element that can be integrated on the same microsystem as the RF circuits, without affecting circuit requirements, and different material compositions in order to maximize the performance for the proposed applications.

References

[1] S. Remke, et al., "Wireless Capsule Endoscopy and Push-Enteroscopy in Chronic Gastrointestinal Bleeding: a Prospective Controlled Trial," *Gastrointestinal Endoscopy*, Vol. 55, No 5, 2002.

[2] J. T., Carlo, et al., "The Utility of Capsule Endoscopy and its Role for Diagnosing Pathology in the Gastrointestinal Tract," *The American Journal of Surgery*, Vol. 190, pp. 886-890, 2005.

[3] W.A. Qureshi, "Current and Future Applications of the Capsule Camera," *Nature Reviews*, Vol. 3, pp. 447-450, 2004.

[4] D. Fleischer, "Capsule Imaging," Clinical Gastrology and Hepatology, Vol. 3, 2005.

[5] P. Hunter Peckham, et al., "Functional Electrical Stimulation for Neuromuscular Applications," *Annu. Rev. Biomed.* Eng. 2005. 7:327–60, 35 pp., March 23, 2005.

[6] Muller N., "What Americans Understand How they Affected by Bladder Control Problems: Highlights of Recent Nationwide Consumer Research," Urologic Nursing. pp. 109-115, 2005.

[7] W. R Lenderking, JF Nackley, RB Anderson, MA Testa. "A review of the quality-of-life aspects of urinary urge incontinence; importance of patients' perspective and explanatory lifestyle," *J Am Geriatr Soc.*, pp. 683-692, 1998.

[8] Robert A Gaunt, Arthur Prochazka, "Control of urinary bladder function with devices: successes and failures," *Progress in Brain Research*, Elsevier, Vol. 152, pp. 163-194, 2005.

[9] R. Macintosh, Lumbar Puncture and Spinal Analgesia, Blackwells Books, Oxford, 1951.

[10]Pedram Mohseni, et al., "Wireless Multichannel Biopotential Recording Using an Integrated FM Telemetry Circuit," *IEEE Trans. Neural. Syst. Rehab. Eng.*, vol. 13, no. 3, pp. 263-270, Sept. 2005.

[11]P.M. Mendes, et al. "Integrated Chip-Size Antennas for Wireless Microsystems: Fabrication and Design Considerations," J. Sens. Act. A Physical, Vol. 125, No. 2, pp. 217-222, 10 Jan. 2006.

[12]Anders J. Johansson, "Wireless Communication with Medical Implants: Antennas and Propagation," PhD Thesis, Lund, Sweden , June 2004

[13]Ronald Kitchen, *RF Radiation Safety Handbook*, Butterworth-Heinemann, 1993.

[14]European Telecommunications Standards Institute, ETSI EN 301 839-1 Electromagnetic compatibility and Radio spectrum Matters (ERM);Radio equipment in the frequency range 402 MHz to 405 MHz for Ultra Low Power Active Medical Implants and Accessories;Part 1: Technical characteristics, including electromagnetic compatibility requirements, and test methods., 2002.

[15]Juan Hinojosa Jimenez, "Contribuition a l'Elaboration d'une Nouvelle Methode de Caracterisation Electromagnetique de Materiaux a Partir de Lignes Plaquees – Applications a l'Etude de Nouveaux Materiaux", Universite des Sciences et Technologie de Lille, Lille, France, These a Docteur de l'Universite, May 1995.

[16]P.M. Mendes, A. Polyakov, M. Bartek, J.N. Burghartz, J. H. Correia, "Extraction of Glass-Wafers Electrical Properties Based on S-Parameters Measurements of Coplanar Waveguides," ConfTele, June 2003, Portugal.

[17]Abdel-Hakim Boughriet, Christian Legrand, Alain Chapoton, "Noniterative Stable Transmission/Reflection Method for Low-Loss Material Complex Permittivity Determination", IEEE Trans. Microwave Theory Tech., Vol. 45, nº 1, pp. 52-56, January 1997.

[18]Rainee N. Simons, *Coplanar Waveguide Circuits, Components and Systems*, John Wiley & Sons, 2001.

[19]Robert E. Collin, *Foundations for Microwave Engineering*, 2nd edition, McGraw-Hill, 1992.

[20]C. Veyres, V. F. Hanna, "Extension of the Application of conformal Mapping Techniques to Coplanar Lines with Finite Dimensions", Int. J. Electron., Vol.48, nº. 1, Jan. 1980.

[21]William R. Eisenstadt, Yungseon Eo, "S-Parameter-Based IC Interconnect Transmission Line Characterization", *IEEE Trans. Components*, Hybrids, and Manufact. Techn., Vol. 15, nº 4, pp. 483-489, August 1992.

[22]C. L. Holloway, E. F. Kuester, "A Quasi-Closed Form Expression for the Conductor Loss of CPW Lines, with an Investigation of Edge Shape Effects", *IEEE Trans. Microwave Theory Tech.*, Vol. 43, nº 12, pp. 2695-2701, December 1995.

[23]Kin-Lu Wong, *Planar Antennas for Wireless Communications*, John Wiley & Sons, Inc., 2003.

[24]A. Polyakov, P.M. Mendes, S.M. Sinaga, M. Bartek, B. Rejaei, J.H. Correia, J.N. Burghartz, "Processability and Electrical Characteristics of Glass Substrates for RF Wafer-Level Chip-Scale Packages", *in Proc. 53rd ECTC*, New Orleans, USA, 2003.

[25]P.M. Mendes, S. M. Sinaga, A. Polyakov, M. Bartek, J.N. Burghartz, J. H. Correia, "Wafer-Level Integration of On-Chip Antennas and RF Passives Using High-Resistivity Polysilicon Substrate Technology", *In Proc. 54th ECTC,* Las Vegas, USA, 2004, pp. 1879 – 1884.

[26]P. M. Mendes, A. Polyakov, M. Bartek, J. N. Burghartz, J. H. Correia, "Integrated 5.7 GHz Chip-Size Antenna for Wireless Sensor Networks," Transducers'03, Boston, USA, June 8-12, 2003, pp. 49-52.

[27]R.L. Li, G. DeJean, E. Tsai, E. Tentzeris, J. Laskar, "Novel small folded shorted-patch antennas", *in Proc. Antennas and Propagation Soc. Int. Symp.*, vol. 4, 2002, pp. 26-29.

[28]H. A. Wheeler, "Fundamental limitations of small antennas", *Proc. IEE*, Vol. 35, pp 1479-1484, Dezembro 1947.

[29]L. J Chu, "Physical Limitations of Omnidirectional Antennas", Technical report nº 64, *Research Laboratory of Electronics*, MIT, Maio 1948.

[30]R.F. Harrington, "Effect of Antenna Size on Gain, Bandwidth and Efficiency," Journal of research of national bureau of standards, D-radio Propagation, Vol. 64D, pp. 1-12, Janeiro 1960.

[31]R. E. Collin, S.Rothschild, "Evaluation of Antenna Q," *IEEE Trans. Antennas Propagat.*, Vol. AP-12, pp. 23-27, Janeiro 1964.

[32]R. L. Fante, "Quality Factor of General Ideal Antennas", *IEEE Trans. Antennas Propagat.*, Vol. AP-17, pp. 151-155, Março 1969.

[33]R. C. Hansen, " Fundamental Limitations in Antennas", *Proc. of the IEEE,* Vol. 69, pp. 170-182, Fevereiro 1981.

[34]J. S. McLean, "A Re-examination of the Fundamental Limits on the Radiation Q of Electrically Small Antennas", *IEEE Trans. Antennas Propagat.*, Vol. AP44, pp. 672-675, Maio 1996.

[35]R .L. Fante, "Maximum possible Gain for an Arbitrary Ideal Antenna with Specified Quality ·Factor",. *IEEE Trans. Antennas Propagat.*, Vol. AP40, pp. 1586-1588, Dezembro 1992.

[36]J. C. Sten, A. Hujanem, "Notes on the Quality Factor an Bandwidth", *Springer-Verlag, Electrical Engineering*, Vol. 84, pp. 189-195, 2002.

[37]G. A. Thiele, Phil L. Detweiler, Robert P. Penno, "On the Lower Bound of the Radiation Q for Electrically Small Antennas," *IEEE Trans. Antennas Propagat.*, Vol. AP51, pp. 1263-1269, Junho 2003.

[38]G. S. Smith, "Efficiency of Electrically Small Antennas Combined with Matching Networks", *IEEE Trans. Antennas Propagat.*, Vol. AP25, pp. 369-373, Maio 1977.

[39]Wen Geyi, "Physical Limitations of Antenna," *IEEE Trans. Antennas Propagat.*, Vol. AP51, pp. 2116-2123, Agosto 2003.

[40]H. A. Wheeler, "Small Antennas", *IEEE Trans. Antennas Propagat.*, Vol. AP23, pp. 462-469, Julho 1975.

[41]D. Lange, O. Brand, and H. Baltes, *CMOS Cantilever Sensor Systems: Atomic Force Microscopy and Gas Sensing Applications.*, Springer, 2002.

[42]J. S. Brooks, M. J. Naughton, Y. P. Ma, P. M. Chaikin, and R. V. Chamberlin, "Small sample magnetometers for simultaneous magnetic and resistive measurements at low temperatures and high magnetic fields," *Rev. Sci. Instr.*, 58(1):117–121, 1987.

[43]M. J. Naughton, J. P. Ulmet, A. Narjis, S. Askenazy, M. V. Chaparala, and A. P. Hope, "Cantilever magnetometry in pulsed magnetic fields," *Rev. Sci. Instr.*, 68(11):4061–4065, 1997.

[44]A. Agoston, F. Keplinger, and B. Jakoby, "A novel MEMS based viscosity sensor," In Proceedings of the Eurosensors 2004 - XVIII, pages B4.4 1–4, Rom, Italy, September 13–15 2004.

[45]T. Kabashima, et al., "A Study of the Cantilever Beam in Time Varying Magnetic Field," *IEEE Trans. on Magnetics*, Vol. 26, No. 2, pp. 563-566, March 1990.

[46]W. C. Tang, et al., "Laterally driven polysilicon microstructures," Sens. Actuators, vol. A20, pp. 25–32, 1990.

[47]L. A. Rocha, et al., "Analysis and analytical modeling of static pull-in with application to mems-based voltage reference and process monitoring," *J. Microelectromech. Syst.*, vol. 13, pp. 342–354, 2004.

[48]H.C. Lee, et al., "Piezoelectrically Actuated RF MEMS DC Contact Switches With Low Voltage Operation", IEEE Micro. Wirel. Comp. letters, Vol.15, No.4, pp.202-204, 2005.

[49]T. Kawakubo, et al., "Piezoelectric RF MEMS Tunable Capacitor with 3V Operation using CMOS Compatible Materials and Process," *IEEE IEDM Technical Digest.*, pp. 294 – 297, 2005.

[50]V. Beroulle, Y. Bertrand, L. Latorre, and P. Nouet, "Monolithic piezoresistive CMOS magnetic field sensors," Sens. Actuators A, 103:23–32, 2003.

[51]F. Keplinger, S. Kvasnica, A. Jachimowicz, F. Kohl, J. Steurer, and H. Hauser, "Lorentz force based magnetic field sensor with optical readout," *Sens. Actuators A,* 110 (1–3):112–118, 2004.

[52] L. Latorre and P. Nouet, “A complete methodology for electro-mechanical characterization of a CMOS compatible MEMS technology,” *IEEE Trans. Electron.*, E82-C(4):582–588, 1999.

In: Wireless Communications Research Trends
Editor: Tong S. Lee, pp. 267-285
ISBN: 978-1-60021-674-9

Chapter 8

A TRANSPORT LAYER SECURITY PROTOCOL FOR HYBRID NETWORKS

*Nikos Komninos**
Algorithms & Security Group, Athens Information Technology,
GR-19002 Peania (Attiki), Greece

Abstract

One of the key enablers for business applications in future mobile communication systems is the ability to set up secure channels across the Internet and mobile networks. In this paper, a hybrid transport layer security protocol (HTLS) is described, which sets-up secure channels across different networks, such as the Internet, Bluetooth, and UMTS, using a single protocol. HTLS's sub-protocols and its unique features are explained versos the features of well known security protocols, such as TLS and WTLS, at the OSI transport layer. A comparison of the implementation results is also presented.

Index Terms: transport layer security protocols, mobile and wired networks, hybrid protocol

1. Introduction

As the use of electronic communications plays an ever-increasing role in business activities, security in communications through networking environments is becoming an important issue. One method is to build secure channels between two parties at the OSI transport layer for end-to-end security. The most widely used security protocols in the Internet and in mobile networks are the Transport Layer Security (TLS) protocol [1], based on the Secure Sockets Layer (SSL) protocol [2], and the Wireless Transport Layer Security (WTLS) protocol [3]. However, the Internet and future mobile systems may integrate satellite, radio, infrared and other means of communication to provide a whole host of upgraded services added to those available today (i.e. voice, text and data), by utilizing the mobile network capability for high-speed information transfer.

* E-mail address: nkom@ait.edu.gr

The basic blocks of transport layer security in wireless, mobile and wired networks are illustrated in Figure 1. A TLS handshake is established between the web server and the wireless application protocol (WAP) gateway and a WTLS handshake is initialized between the gateway and the mobile device. The encrypted content is sent from the web server to the gateway, which then translates it and sends the final message to the mobile phone [5].

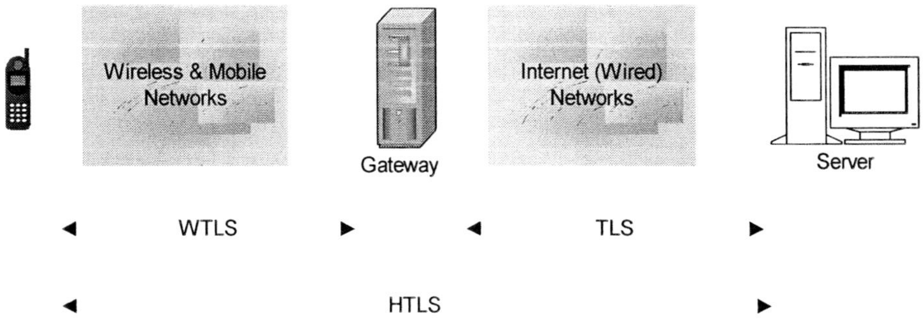

Figure 1. Present and Future Transport Layer Security in Wired and Wireless Mobile Networks.

However, with the merging of mobile and Internet networks, it is essential to develop a hybrid transport layer security protocol (HTLS), which will set up a single secure channel among different networks and will have the following novel characteristics:

- it will provide new services that ensure user privacy, peer authentication and data integrity;
- it will support multiple authentication with hybrid cryptosystems;
- it will support encrypted audio and video services;
- it will achieve end-to-end encryption synchronization;
- it will enable group connections for encrypted audio and video conferences and
- it will overcome known security holes, such as chosen plaintext attacks, data recovery attacks, datagram truncation attacks, and message forgery attacks that have been identified in current transport security protocols [5, 7, 8, 13, 18, 20, 21].

HTLS can also be used among the web server, the gateway and the mobile client (Figure 1). The web server and the gateway first start communicating with the handshake protocol; they select the cryptographic algorithms, authenticate each other and use public key encryption techniques to distribute keys. During the first handshake between the web server and the gateway, a second handshake can take place between the gateway and the mobile client to extend the secure session. At the end of the handshakes, both sessions will share the same group identifier, which will identify a single secure session.

HTLS incorporates new features such as hybrid authentication schemes and group connection support, an optimized handshake protocol suitable for wired and wireless environments, a secure real time audio and video transmission and encryption synchronization support. It has been designed for both high and low bandwidth networks, and single and multiple sets of cipher suites.

In section 2, the security protocols TLS and WTLS are briefly described. Their description exploits the fundamental structure of secure protocols at the transport layer.

HTLS's sub-protocols and its unique features are given in section 3 and implementation results are presented in section 4. In section 4, TLS, WTLS and HTLS handshakes are also analyzed with respect to their response time. The security analysis of HTLS is discussed in section 5. Finally, section 6 concludes with the main characteristics of HTLS.

2. Current Protocol Standards at Transport Layer

Originally developed by Netscape, SSL has been universally accepted on the World Wide Web (WWW) for authenticated and encrypted communications between two peers. SSL was designed to support a range of choices for specific algorithms used in cryptography, digests, and signatures [2]. This allows algorithm selection for specific servers to be made based on legal, export or other concerns and also enables the protocol to take advantage of new algorithms [9]. In 1996, an effort was first made by the Internet Engineering Task Force (IETF) to standardize SSL as an IETF standard under the name of TLS protocol [1]. In the rest of the paper, SSL will be referred to as the Transport Layer Security Protocol, or simply TLS.

TLS, which supports only reliable transport layer protocols, is composed of two main components: the ***record protocol*** used for data transfer and the ***handshake protocol*** used to establish secure sessions between peers. The TLS ***record*** and ***handshake protocols*** provide data transmission and connection security with two basic properties: the transmission and connection is private and reliable. It uses asymmetric, or public key, cryptography (e.g. RSA, DSA, etc.) to authenticate and distribute the session key to the peers. It also uses symmetric, or private key, cryptography (e.g. AES, SAFER SK-128 etc.) and hash functions for data encryption and data integrity, respectively [1, 8].

However, in current wireless telecommunication networks, although the traffic in the air is encrypted, a complete end-to-end security solution is not provided [5]. When WAP was developed to fulfil value added mobile services, WTLS was designed to provide end-to-end security for WAP applications [4]. The WTLS is based on the TLS security layer used in the Internet. A number of modifications and changes were needed because of the nature of wireless environments. Wireless networks in general require support for both reliable and unreliable transport layer protocols. The mobile terminals have limited processing power and memory, and as a result, optimized versions of encryption algorithms are used. WTLS mainly attempts to reduce the overheads associated by establishing a secure connection between two applications. It provides services similar to TLS that ensure privacy, authentication and data integrity [4, 5, 20].

According to the WAP specifications [3], WTLS is composed of the ***record protocol***. The WTLS ***record protocol*** takes messages to be transmitted, optionally compresses the data, applies MACs, encrypts, and transmits the data. Received data is decrypted, verified and decompressed, then delivered to higher-level layers. Four record sub-protocols exist according to the specifications:

- the ***handshake protocol*** produces the cryptographic parameters;
- the ***change cipher spec protocol*** deals with ciphering strategies;

- the ***alert protocol*** is used to inform the peers that the secure connection is ending. It also determines the level of error in the secure connection and provides a description of the error to the peers in the termination of the connection;
- the ***application data protocol*** is used to exchange data between peers.

In the rest of the paper, peers will be referred to as clients[1] or servers[2], depending on which of the two is initiating communication.

In the WTLS handshake protocol, the client sends a *hello* message to which the server must respond with a *hello* message, or else a fatal error will occur and the secure connection will fail [3, 19, 20]. The *client hello* and *server hello* are used to establish security enhancement capabilities between client and server, such as protocol version, key exchange suite, cipher suite, compression method, key refresh, and sequence number mode. Additionally, two random values are generated and exchanged between the client and the server.

Following the *hello* messages the server will send its certificate to be authenticated. Additionally, a *server key exchange* message may be sent if it is required. The server may request a certificate from the client if that is necessary to the key exchange method selected. Then, the client responds with the requested information to the server. At this point, the client and the server may begin to exchange application data as illustrated in Figure 2.

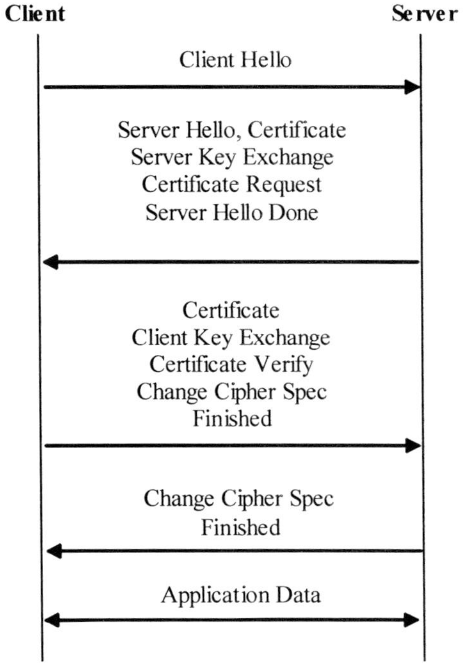

Figure 2. WTLS Handshake Protocol.

---

[2] The ***client*** is a device (or application) that initiates a request for a connection to a server [WAP, 2003]

[3] The ***server*** is a device (or application) that passively waits for connection requests from one or more clients [WAP, 2003]

WTLS also defines an abbreviated handshake [3, 20], where only the *hello* and *finished* messages are exchanged. However, both client and server have a shared session key for data transmission. Another variation of the protocol defines an optimized handshake where the server retrieves the client's certificates using a trusted third party. The information provided by the certificates allows both parties to agree upon a shared secret using the Diffie-Hellman key exchange method. More information about WTLS handshakes can be found in [5, 3, 6].

3. Hybrid Transport Layer Security Protocol

Similarly to TLS and WTLS, HTLS is a layered protocol that accepts data to be transmitted and applies the selected compression and encryption algorithms to the data. It is composed of the HTLS **record protocol**, which is subdivided into six protocols; the ***handshake protocol***, the ***cipher protocol***, the ***alert protocol***, the ***group control protocol***, the ***encryption synchronization protocol*** and the ***application data protocol***.

3.1. HTLS Record Protocol

The HTLS *record* protocol is responsible for taking messages to be transmitted, compressing the data, applying message authentication code (MAC) algorithm, encrypting and transmitting the data. The security parameters of the record protocol are shown in Table 1.

Table 1. HTLS Record Protocol Security Parameters

Parameter	Description
Connection End	Whether the end application is considered a client or a server in the secure session.
Symmetric Key Algorithm	An algorithm to be used for data encryption. It can be either stream or block cipher. The encryption mode is also defined here.
Asymmetric Key Algorithm	An algorithm to be used mainly for authentication & key distribution.
MAC Algorithm	An algorithm to be used for message authentication.
Compression Algorithm	The algorithm to compress data prior to encryption.
Sequence Numbers	Whether sequence numbers are used or not.
Client Random	A 16 byte value provided by the client.
Server Random	A 16 byte value provided by the server.
Key Refresh	Defines how often the encryption key is updated.
Secret Key	A physically planned secret key.

In the record protocol, the end applications, which share a secret key and public key pair, are identified as either client or server. Data can be encrypted using either single or multiple encryptions when symmetric and asymmetric key algorithms are chosen. MAC and compression algorithms are defined and random bytes are assigned to both client and server. The HTLS record protocol parameters are received by the handshake protocol.

Record fragmentation, compression, decompression and data protection follow the same principle as in WTLS [3]. However, the HTLS record protocol supports multiple encryptions at this level. Messages can be encrypted with symmetric and asymmetric key encryption when there is high sensitivity of data. In such circumstances, the message is first encrypted with a symmetric key algorithm and then with an asymmetric key algorithm, similar to the certificate generation/verification process [11, 14].

Multiple authentication schemes are enabled through challenge response techniques, which are based on symmetric or asymmetric key encryption [11, 14]. However, if only one encryption algorithm is used, the symmetric key and asymmetric key parameters of Table 1 will be NULL.

3.2. HTLS Handshake Protocol

The HTLS ***handshake protocol*** is composed of four sub protocols, which are used to allow both parties to agree upon security parameters for the record layer, to authenticate each other and to report error conditions to each other. It is responsible for negotiating a secure session, which consists of the parameters shown in Table 2.

Table 2. HTLS Handshake Protocol Security Parameters

Parameter	Description
Session Identifier	A byte sequence chosen by the server to identify an active or resumable secure session.
Protocol Version	HTLS protocol version number.
Authentication Mode	A flag which specifies if asymmetric or symmetric cryptography will be used at the beginning of transmission.
Session Performance	A byte sequence chosen by both client and server to identify a poor or acceptable or high performance.
Certificate	Client / Server authentication.
Compression Method	The algorithm to compress data prior to encryption.
MAC Algorithm	An algorithm to be used for message authentication.
Symmetric Key Algorithm	An algorithm to be used for the data encryption phase.
Asymmetric Key Algorithm	An algorithm to be used for the authentication and key distribution phase.
Key Refresh	Defines how often the encryption key is updated.
Session Spec	Flags to identify new session id, resumable, extended session.
Sequence Number	Specifies if sequence numbers are used (on / off).

In the HTLS handshake protocol the protocol version, authentication type, session performance and session identifier (ID) are chosen at the beginning of the transmission. Then, certificates, compression and encryption algorithms, are exchanged between the client and server using the session ID. Furthermore, the encryption key is updated later on in the session, as defined by the key refresh parameter, and sequence numbers are enabled or disabled based on the connection type. Finally, the session is agreed to be resumable and/or extended for more than one secure connection.

The client sends an encrypted start message to which the server must respond with an equivalent message. The client start message is encrypted using the shared secret key. The server start message is also encrypted with the shared secret key and/or client's public key. The client start and server start messages are used to establish security enhancement capabilities, such as protocol version, authentication type, channel/network performance, MAC, asymmetric and symmetric key algorithms, compression method, key refresh, server authentication and session key generation.

Following the start messages they exchange their certificates to be authenticated, and choose the encryption algorithms to encrypt traffic. Finally, they agree on the connection type (i.e. connectionless or connection oriented) and assign new session settings for the secure connection. At this point, the client and the server may begin to exchange application data as illustrated in Figure 3.

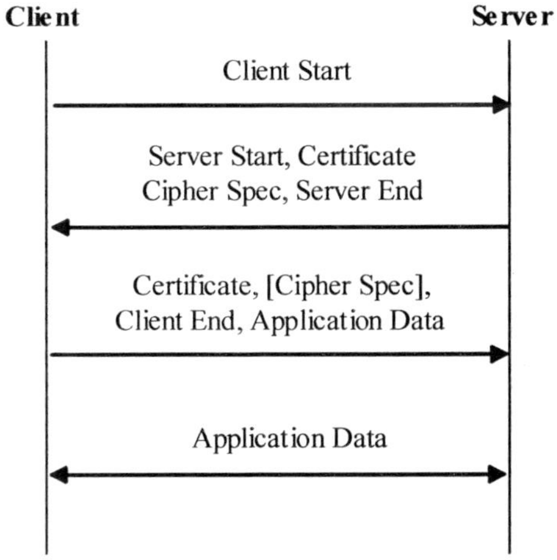

Figure 3. HTLS Handshake Protocol.

The *start* messages are used by the client and the server to agree on a protocol version, session identification, authentication type and exchange information such as random numbers, public and secret keys, and performance parameters. The ***client start message*** is always encrypted with the shared secret key. It consists of nine parameters:

- *protocol version,*
- *session ID,*
- *client environment,*
- *client performance,*
- *authentication type,*
- *client's public key,*
- *secret key number,*
- *random number* and
- *timestamp.*

The *protocol version* parameter identifies which version of the HTLS protocol will be used in the secure session. The *session ID* is assigned an identification number for each secure session. The *client environment* parameter is used to indicate if the client is located in a wired or wireless network where performance boundaries are different. The *client performance* parameter is used to evaluate the quality of service the network requires to support different data types. For example, when the available bandwidth is not enough for the quality of service required in a real time secure audio or video transmission, then the secure session is not established.

The *authentication type* indicates whether asymmetric and/or symmetric cryptography will be used in the authentication and key distribution phase. The *client's public key* parameter is NULL, if symmetric cryptography is used. Otherwise, if asymmetric cryptography is applied, it contains the user's public key. In the case where symmetric cryptography is used in the authentication and key distribution phase the *secret key number* parameter indicates which key will be used if multiple keys are supported for the secure session. The *random bytes* field contains random bits, which will be needed for session key generation. The *timestamp* parameter is used to give a limited lifetime of transmission to prevent replay attacks [14, 15, 20].

The ***server start message*** is sent encrypted using the client's public key and/or shared secret key depending on the authentication type declared in the client start. In a similar way to the client start message, the server start is composed of eight parameters:

- *protocol version,*
- *server environment,*
- *server performance,*
- *authentication type,*
- *server's public key,*
- *secret key number,*
- *random number* and
- *timestamp.*

Following the start message, the server sends its certificate and cipher spec messages. The ***certificate message***, which is used for client and server authentication, consists of seven parameters:

- *certificate type,*
- *validity period,*
- *sign cipher,*
- *sign authority,*
- *authority public key,*
- *user ID* and
- *certificate.*

The *certificate type* defines which type of certificate will be used in this session. Several certificates standards are supported by HTLS (i.e. X509, X968, [11, 14]). In addition to the standards, several parameters are included in the certificate message. The *validity period* parameter defines when the certificate expires. The *sign cipher* identifies the algorithm used to generate the signature. Furthermore, the *sign authority* parameter declares the certification authority assigned to sign the certificate. In order to identify the validity of the certificate, the public key of the certification authority, the user identity and the certificate itself are required. The *authority public key, user ID and certificate parameters* include this information.

The ***cipher spec message***, which is part of the cipher protocol, includes the following six parameters:

- *asymmetric key algorithm,*
- *symmetric key algorithm,*
- *encryption mode,*
- *MAC algorithm,*
- *initialization vector* and
- *key refresh.*

The asymmetric key algorithm parameter defines which cipher is used, in the case when multiple algorithms are supported to secure the distribution of the session key. Likewise, the symmetric key algorithm parameter indicates which cipher is used to secure traffic between client and server. The encryption mode parameter defines the mode of the cipher which is used to enable initialization vectors (IVs) to create entropy. Entropy is needed to protect the symmetric key algorithm that is used in the cipher block chaining mode (CBC). Moreover, the IV parameter contains the initialized input of the block algorithms. The IV used in the algorithms need not be secret and can be transferred in plaintext. Finally, the key refresh option defines how often the encryption key will be updated in this particular secure session.

Finally, the ***end message*** that is the last message sent by the server in response to the client start message, notifies the client for the session properties. It consists of five parameters:

- *session ID,*
- *session resume,*
- *session extend,*
- *session type* and
- *session automatic repeat request (ARQ).*

A new identification number is assigned for the secure session using the *session ID* parameter. If the session is resumable then the session resume flag is set. Similarly, if the session can be extended, the session extend flag is set. Furthermore, the session type parameter identifies continuous and non-continuous data types. Depending on the data type a session can be connection-oriented (reliable) or connectionless (unreliable). Then session ARQ is agreed depending on the data type chosen for this session. The session ARQ parameter enables or disables sequence numbers which are used when real time encrypted audio or video is transmitted in the specified session.

Once the server is authenticated, the client sends its certificate, optionally the cipher spec message and then the client end message. The end message is sent at the end of the handshake to verify that the key exchange and authentication processes were successful. When either the client or server receives the end message, they must verify that the contents are correct. Once this is done, application data can be exchanged over the secure connection.

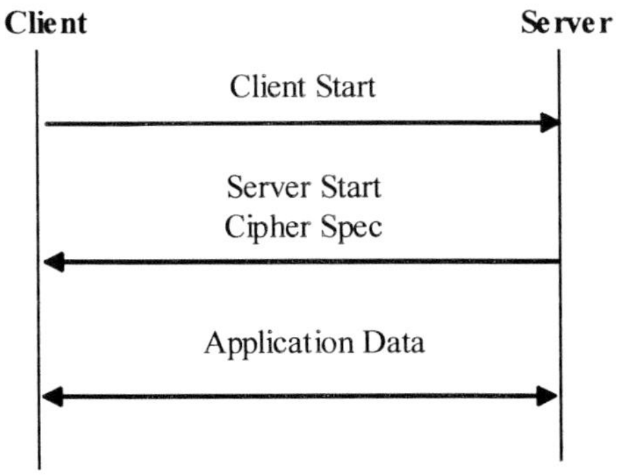

Figure 4. HTLS Session Resume.

As shown in Figure 4, when the client wishes to resume a secure session, it sends a ***client start*** message using the session ID. The server then checks its secure session cache for a match. If a match is found and the session is resumable then the server could re-establish the secure session by sending the ***server start*** message with the same session ID. At this point, the server sends a ***cipher spec*** message. Once the re-establishment is complete, the client and server can start the exchange application data.

3.3. HTLS Application Data Protocol

When a secure connection has been established, the ***application data protocol*** will be responsible for delivering correct information to the end applications and for identifying changes to it. The main advantage of the HTLS application data protocol over TLS and WTLS protocols is that HTLS performs an integrity check on the actual information sent or received, even when a secure connection has been established. This is achieved by hashing the actual information and sending the hash value to the end applications. It consists of a single message which has the following parameters:

- *data* and
- *data integrity check.*

The data parameter contains the actual information that is transmitted in a secure session whereas the data integrity check includes the hash value of the data itself.

3.4. HTLS Alert Protocol

When information is exchanged either in the handshake or in the application data protocol errors can occur. The ***alert protocol***, which is supported by the HTLS record protocol, is used to convey the severity of messages and gives a description of the alert. It consists of two parameters:

- *alert type* and
- *alert description.*

The HTLS alert protocol comprises two types of alerts; *fatal* and *critical*.

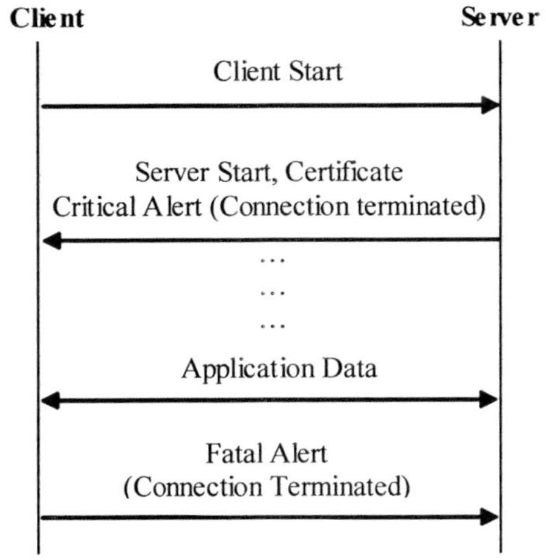

Figure 5. Alert Messages during Handshake and Application Data Protocols.

As illustrated in Figure 5, alert messages with a level of fatal or critical, result in the immediate termination of the secure connection. In the case of fatal alert, the session identifier must be invalidated, thus preventing the failed secure session from being used to establish new secure connections whereas, in critical alerts, the session identifier may be preserved to be used for new secure connections.

In WTLS an alert message can be either sent with or without compression and encryption. In HTLS, however, alert messages **must** be sent compressed and encrypted. This is mainly because NULL encryption algorithms are not supported by HTLS.

3.5. HTLS Group Control Protocol

The ***group control protocol*** enables HTLS to support real time secure audio or video transmission among three or more parties. When a secure connection has been established the group control protocol can be used to identify possible group sessions (i.e. simultaneous sessions) that can be established later on.

In such circumstances, the environment and performance variables, which are found in client and server start messages (see section 3.2), are decision parameters for group connections. For example, audio and video transmission requires a connectionless service environment where clients and servers are allowed to exchange non-confirmed messages. In addition, the client and server performances should be adequate to offer high quality of service in the serving network [12].

Then, when a third party wishes to join a secure session, the *start*, *group* and *end* messages are exchanged between the client and server, as shown in Figure 6.

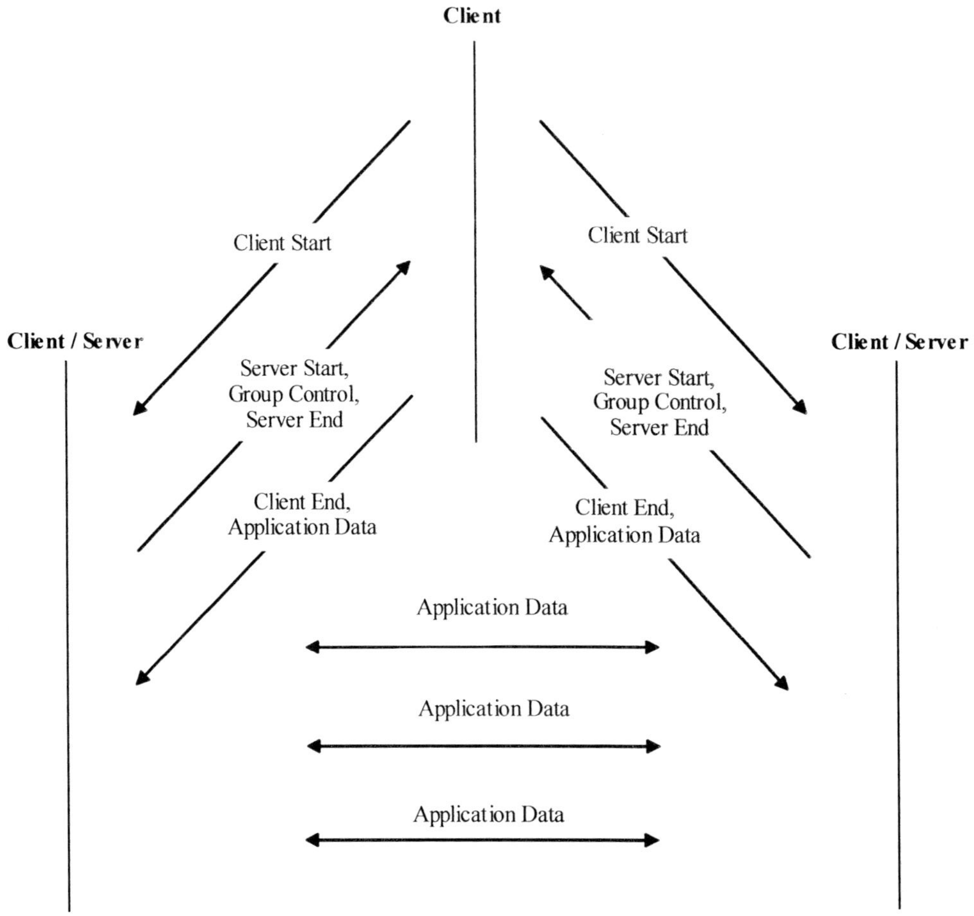

Figure 6. Third Party Joint Session with Group Control Protocol

The group message includes the following parameters:

- *group ID*,
- *join session*,
- *leave session*,
- *suspend session*,
- *resume session* and
- *key exchange*.

At the beginning, all messages are encrypted with the shared secret key. The group connection is initiated when the *group id* requested by a third party is the same as the one used in the current session, and the join flag is set. The *joint session* is established when the performance and the session type parameters agree with those used in the two party session. For example, if bandwidth is available for a group connection, a joint session can be established and cryptographic parameters can be produced by the handshake protocol.

A secure joint session can also be suspended through the *suspend session* flag when, for example, a data circuit in the underlying bearer network is closed. Likewise, a secure joint session can be resumed with the *resume session* flag when the data circuit in the bearer network is opened again. A third party can also leave the group using the *leave session* parameter, which enables changes to the cryptographic parameters.

The group message follows the Burmester-Desmedt conference keying approach by using the *key exchange* parameter. This parameter contains the individual Diffie-Hellman's exponentials $z_i=\alpha^{r_i}$ that form a conference key $K = a^{r_0r_1+r_1r_2+r_2r_3+\ldots+r_{t-1}r_0}$ [11].

3.6. HTLS Encryption Synchronisation Protocol

Several encryption modes (i.e. CBC, OFB, CFB [14]) on block ciphers require encryption synchronization when encrypted data is transmitted over a noisy channel. Moreover, when a third party joins a session, encryption synchronization is required. Initially, the server is synchronized to an incoming encrypted data stream. During transmission, the server must maintain synchronization to the incoming encrypted data stream after the initial synchronization. Encryption synchronization is achieved by the ***encryption synchronization protocol***. It consists of a single message, *synchronization* which is sent periodically to the server, as illustrated in Figure 7.

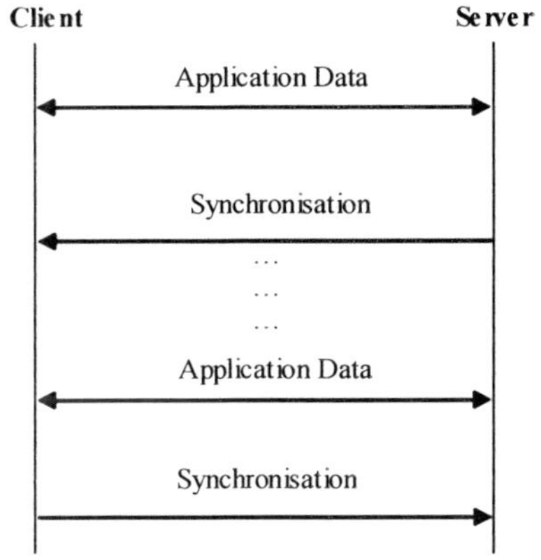

Figure 7. Encryption Synchronization in the Application Data Protocol

The message is transmitted encrypted with the encryption status of the record layer. It includes the:

- *session ID*,
- *group ID* if any,
- updated *IV* and
- the encryption algorithm number which is used, if multiple algorithms are supported.

3.7. HTLS Cipher Protocol

The ***cipher protocol*** consists of a single message, or cipher spec (see section 3.2), which is sent encrypted to the end application either by the client or the server. The cipher spec message can be sent during the handshake or the data transfer phase. When it arrives, the sender waits for a response and the receiver initializes or updates the secure parameters.

4. Implementation Results

HTLS has been implemented using in Figure 1. The testbed consisted of two IBM compatible PCs working as servers with multiple PCs working as clients. Both server models were equipped with Intel CPU running at 450 MHz with 128 MB of RAM. The two client machines were 500 MHz Pentium III using the WIN2000 operating system. The PCs were attached to a local area network to ensure that there were no bottlenecks due to network capacity during any experiment.

For the experiments reported in this session, no modification was made to the workload generated by the Web server, which is designed to mimic the workload on regular Web servers. Although the workload for standard and secure web servers is not the same, we chose to use a standard web server provided by the operating system, as the objective is to compare the performance of TLS, WTLS, and HTLS Handshake protocols in client/server and WAP models.

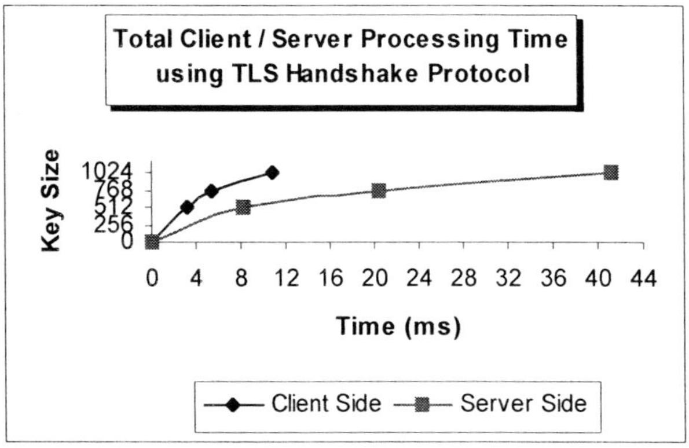

(a)

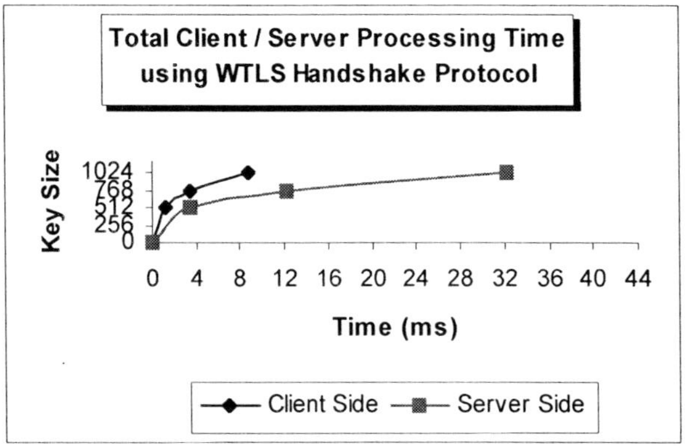

(b)

Figure 8. Timing Analysis of TLS (a) and WTLS (b) in a Client-Server Model

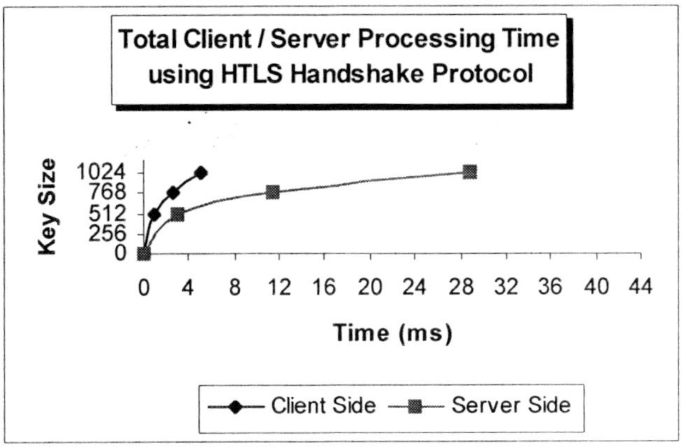

Figure 9. Timing Analysis of HTLS Handshake Protocol in Client-Server Model.

Figure 8a, Figure 8b, and Figure 9 show the overheads involved in setting up a TLS, WTLS, and HTLS client/server session, respectively. There are 3 sets of numbers based on the size of the server's RSA public key i.e. 512, 768, or 1024-bit. As shown in Figure 8a, Figure 8b, and Figure 9 the most expensive operation in the session is the public key decryption calculations on the server side. The verification of the server certificate and the generation and encryption of the session key are the major operations performed on the client side. Note that a TLS session is established in 41.2ms, whereas 32.3ms are required for an optimized WTLS handshake when 1024-bit encryption is used. The HTLS session, on the other hand, is established in 28.8ms, which is even faster than WTLS session with 1024-bit encryption. Hence, in a client-server model HTLS secure session was established 12.4ms and 3.5ms faster than TLS and WTLS respectively. The HTLS secure session performs faster than TLS and WTLS sessions because it uses an optimized handshake, which offers the same security level with the handshakes in TLS and WTLS.

When a connection-oriented transport protocol, such as TCP, is used for data transmission, HTLS performs much better than TLS and slightly better than WTLS in the client-server model. When a connectionless transport protocol is used, however, the HTLS client and the HTLS server also agree upon the network performance required to support different data type transmission. In a normal TLS or WTLS handshake, a connection is established independently of the type of data transmitted and the kind of network performance required to support the session. Therefore, it must be decided whether the network can support transport layer secure protocols. This is the reason why security in the transport layer is optional in most wireless networks, such as WTLS in WAP [8]. However, HTLS can adapt to the environments being used by enabling or disabling secure features. For instance, in HTLS both symmetric and/or asymmetric cryptography can be used for key distribution, whereas, in TLS or WTLS only asymmetric cryptography can be applied.

We have also developed a WAP model to compare the performance of TLS, WTLS and HTLS. The WAP model consists of a mobile client, a WAP gateway, and a web server. In the WAP model, a session is opened from the mobile client to the gateway and a wireless markup language (WML) URL is requested from the web server. The requested URL is then returned to the mobile client and its contents displayed on the micro-browser. The data packets from the mobile client pass along the network in WML format to a WAP gateway. This then reconfigures the essential data and passes it again to a WML format. Conversely, when WML data packets need to reach the mobile client, they have to first pass through a WAP gateway.

In this model we compared the secure session between the mobile client and the web server. In the first implementation two secure sessions were opened between the mobile client and the WAP gateway using WTLS, and between the WAP gateway and the web server using TLS. In the second implementation both secure sessions were established using HTLS. The results obtained show that HTLS performs much better than a combined TLS and WTLS. In Figure 10, we have also included 3 sets of numbers based on the size of the server's public key i.e. 512, 768, or 1024-bit. As shown in Figure 10, the mobile client and the web server establish two secure channels using TLS and WTLS in 93.5ms when a 1024-bit key is used.

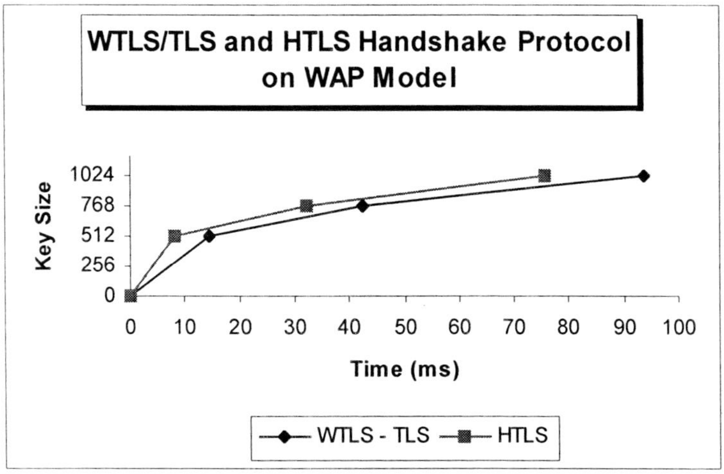

Figure 10. Timing Analysis of TLS, WTLS, and HTLS in WAP Client-Server Model

On the other hand, only 57.6ms were required to set up the secure channels between the mobile client and the web server when HTLS with 1024-bit key encryption was used. Note here that, in the implementation, we have also included the proposed scheme presented in [12], where the web server acts as a client when audio and video conferences are to be established in WAP. Hence, in the WAP model of [12] HTLS was 35.9ms faster than TLS and WTLS.

However, when different data types such as audio and video are supported [12] by WAP, it is necessary that the contents of the data be known to the server. In this case, the WAP gateway can enable or disable some functionality in the different media types. For example, when audio or video is transmitted, a translation of HTML to WML is not required. It is essential to provide the least amount of latency possible for continuous media types.

5. Security in HTLS

The HTLS specification has been adopted by TLS and WTLS specifications with several modifications. These modifications have been designed carefully to avoid the security problems that have been identified in WTLS [6, 7]. The security problems are based on weaknesses in authenticity, privacy and data integrity.

Allowing anonymous connections to be established can be very risky because anonymous authentication increases the chances for intruder-in-the-middle attacks [13, 14]. In order to prevent this problem, both the client and the server should always authenticate each other before a key exchange. Currently, in WTLS the client optionally requests a certificate from the server to be authenticated. In HTLS this feature is not supported since both the client and the server are authenticated before establishing a secure connection.

In HTLS the use of IVs creates the entropy that is needed to protect the symmetric key in CBC mode block ciphers. When the plaintext is encrypted, if IV is used, the brute force method can find the shared secret key if applied. The usage of IV prevents this from happening because the first block in the packet is first XORed with IV. Knowing the content of the original packet does not help in any way, because it is XORed.

Furthermore, because HTLS supports an unreliable transport protocol, datagrams may be lost, duplicated or reordered. The encryption synchronization protocol updates the IV and maintains synchronization. The IV is computed based on random values being sent encrypted to prevent eavesdropping during the handshake.

It has been identified in [6] that the alert messages used in WTLS allow the active attacker to replace an encrypted datagram with an unauthenticated plaintext alert message with the same sequence number without being detected. In HTLS, on the other hand, the data integrity check parameter, which is part of the application data protocol, ensures that each datagram is assigned a unique number.

Finally, a large number of encryption algorithms have been proven to be weak. Allowing connections to be encrypted using weak algorithms raises security issues. HTLS supports a limited number of encryption algorithms that have been identified to be strong (i.e. RSA and AES).

6. Conclusion

In this paper, a hybrid transport layer security protocol was presented which provides end-to-end encryption for combined wireless and wired networks. HTLS follows the same principle as TLS with extra features. It uses most of the WTLS secure parameters and protocols, however, it interconnects different environments based on the channel / network performance, and the media type. HTLS introduces the client / server environment, performance and data type parameters, which enable or disable the HTLS functionality according to the network performance and type of media. HTLS is a layered protocol that allows data to be transmitted and applies the selected compression and encryption algorithms to the data. It is composed of the HTLS record protocol, which is subdivided into six protocols.

The ***handshake protocol*** allows both parties to agree upon security parameters, authenticate themselves, and report error conditions to each other. When one of the parties wishes to exchange new cryptographic parameters in the secure session, he/she uses the ***cipher protocol***, which can be sent either by the client or the server. If an error occurs in the secure session, it is handled by the ***alert protocol***, which is mainly used to convey the severity of messages and gives a description of the alert.

Furthermore, group sessions are supported by HTLS for either audio or video conference with the use of the ***group control protocol***. Encryption synchronization for feedback encryption mode is achieved with the ***encryption synchronization protocol*** and data is delivered to the end applications with the ***application data protocol.***

In the implementation results, it was found that the time required to setup a single HTLS secure session in a client-server model with 1024-bit key, was about 30% and 10% faster than TLS and WTLS respectively. Moreover, when group secure sessions were established using the WAP model of [12], it was also found that HTLS, using 1024 -bit key, was about 40% faster than TLS and WTLS.

References

[1] T. Dierks and C. Allen, "The TLS Protocol", *Request for Comments RFC* (2246), 1999.

[2] A.O. Freier, P. Karlton and P. C. Kocher, "The SSL Protocol Version 3.0", Netscape Corporation, 1996.

[3] Wireless Application Protocol (WAP) Forum, *Wireless Transport Layer Security Specification*, Technical Report, 2003.

[4] Wireless Application Protocol (WAP) Forum, *Wireless Application Protocol Specification*, Technical Report, 2003.

[5] Kwon Eun-Kyeong, Cho Yong-Gu, and Chae Ki-Joon, "Integrated transport layer security: end-to-end security model between WTLS and TLS", *15th International Conference on Information Networking*, 31 Jan.-2 Feb. 2001, Pages: 65 – 71.

[6] Saarinen, Markku and Juhani, "Attacks against the WAP WTLS Protocol", http://www.jyu.fi/~mjos/wtls.pdf, 1999,

[7] A. Yasinsac and J. Childs, "Analyzing Internet Security Protocols", *6th International IEEE Symposium on High Assurance Systems Engineering*, USA, 2001.

[8] G. Apostolopoulos, V. Peris and D. Saha, "Transport Layer Security: How much does it really cost?", *8th Annual Joint Conference of the IEEE Computer and Communications Societies*, USA, 1999.

[9] P. A. Lambert, "The lowdown on lower layer security protocols", *6th Annual IEEE Computer Security Applications Conference*, USA, 1990, Pages: 181-187.

[10]C. J. Mitchell, "Limitations of challenge-response entity authentication", *Electronics Letters*, vol. 25, 1989, Pages: 1195-1196.

[11]A. J. Menezes, P. C. van Oorschot and S.A. Vanstone, *Handbook of Applied Cryptography*, CRC Press LLC, 1996.

[12]N. Komninos and B. Honary, "Modified WAP for Secure Voice and Video Communication", *2nd IEE International Conference on 3G Mobile Communication Technologies, London*, 2001, Pages: 33-37.

[13]R. L. Rivest and A. Shamir, "How to expose an eavesdropper", *Communications of the ACM*, vol. 27, 1984, Pages: 393-395.

[14]B. Schneier, *Applied Cryptography*, John Wiley & Sons, USA, 1996.

[15] H.B Mahmood, "Transport layer security protocol in Telnet", The 9th Asia-Pacific *Conference on Communications*, Vol: 3, 21-24 Sept. 2003, Pages: 1033 – 1037.

[16]Leung, V.C.M., "Proxy services for the mobile Internet", *IEEE International Symposium on Personal Indoor and Mobile Radio Communications*, Vol: 2, 5-8 Sept. 2004, Pages: 1230 – 1235.

[17]M. Portmann and A. Seneviratne, "Selective security for TLS", *Ninth IEEE International Conference on Networks*, 10-12 Oct. 2001, Pages: 216 – 221.

[18]M. Badra, and A. Serhrouchni, , "A new secure session exchange key protocol for wireless communications", *14th IEEE Proceedings on Personal, Indoor and Mobile Radio Communications*, Vol: 3 , 7-10 Sept. 2003, Pages: 2765 – 2769.

[19]A. P. Levi, and E. Savas, "Performance evaluation of public-key cryptosystem operations in WTLS protocol", *Eighth IEEE International Symposium on Computers and Communication*, 30 June-3 July 2003, Pages: 1245 – 1250.

[20]G. Radhamani and K. Ramasamy, , "Security issues in WAP WTLS protocol", *IEEE International Conference on Communications, Circuits and Systems and West Sino Expositions*, Vol: 1 , 29 June-1 July 2002, Pages: 483 – 487.

[21]A. Yasinsac and J. Childs, "Analyzing Internet security protocols", *Sixth IEEE International Symposium on High Assurance Systems Engineering*, 22-24 Oct. 2001, Pages: 149 – 159.

[22] G. Kahraman and S. Bilgen, "Wireless application protocol transport layer performance", *Eighth IEEE International Symposium on Computers and Communication*, Vol: 2, 30 June-3 July 2003, Pages: 1141 - 1149

In: Wireless Communications Research Trends
Editor: Tong S. Lee, pp. 287-299
ISBN 978-1-60021-674-9

Chapter 9

HOW CAN 3G AND 2G SYSTEMS COOPERATE?

S-E. Elayoubi*, L. Sartori† and B. Fourestié‡
France Telecom, Research and Development Division
38-40 rue du Général Leclerc - 92130 Issy Les Moulineaux - France
Phone +33 1 45 29 88 67 - Fax +33 1 45 29 63 07

Abstract

In this chapter, we develop Markovian models to study the dynamics of real time and elastic calls in a cell served by UMTS/HSDPA and GSM/EDGE systems. We first present analytical models for interference and throughputs in GERAN and UTRAN. We then consider different strategies of Joint Radio Resource Management (JRRM) with or without inter-system vertical handovers and show how to calculate the steady-state probabilities and the performance measures (blocking probabilities, mean sojourn times, loads). Our numerical results compare the different JRRM strategies and show that the best performance is obtained with the strategy where several vertical handovers are allowed all over the communication in order to continuously choose the best system.

1. Introduction

The presence of two or more radio network systems in the same area is becoming more and more frequent as fourth generation systems will be composed of heterogeneous networks [1]. Various interworking and handover strategies have been presented in the literature for the interworking between 3G and WLAN systems [2][3][4]. The reason for this large interest in the combination of UMTS and WLAN systems is that they are viewed as complementary as the former provides wide-area coverage with high mobility and medium rates, while the latter provides local coverage and low mobility associated with high data rates.

However, less attention has been given to the cooperation between 2G and 3G systems, although these networks already coexist. In fact, the majority of operators deploying 3G networks have legacy 2G networks and will still exploit these two networks for the next

*E-mail address: salaheddine.elayoubi@orange-ftgroup.com
†E-mail address: luca.sartori@orange-ftgroup.com
‡E-mail address: benoit.fourestie@orange-ftgroup.com

years. It is then interesting for the operators to integrate 2G systems with 3G ones to provide users with a better service. With regard to the data traffic it is important to make interwork EDGE and the High Speed Downlink Packet Access (HSDPA) evolution of 3G. In this work, we model and analyze a network composed of the cooperation of UTRAN and GERAN systems. We consider co-localised base stations as it is likely to reduce deployment costs.

However, we must first have reliable models for the individual systems to study the cooperative system. To our knowledge, there is no complete analytical model in the literature for inter-cell interference and throughputs in EDGE. We then develop a model to evaluate the Signal-to-Interference-plus-Noise Ratio (SINR) and the throughput, based on the number of collisions and the propagation conditions.

Conversely, a large number of models have been developed for HSDPA [6][5]. However, they do not model completely the interaction between dedicated channels (Release 99) and shared ones (HSDPA). Furthermore, they consider a simplistic model where the inter-cell interference is proportional to the (instantaneous) intra-cell interference. We then develop a model that takes into account both dedicated and shared channels and that models accurately the inter-cell interference.

Having modeled the individual systems, we are able to study the cooperative system in order to compare different strategies of Joint Radio Resource Management (JRRM) including joint admission and rate control and vertical handover. We first consider a strategy based on the service at the user arrival and which depends on the state of each system load. In this case, once the selection is made, the user cannot be transfered to the other system even if the network conditions in the chosen system become worse. Secondly, we consider a more complete JRRM policy where the best serving system is chosen upon user arrival and when the loads in the system change, for instance, due to a call departure. This involves vertical (inter-system) handovers. Note that we consider several regions in the cell, as the propagation conditions are different for the two systems, leading to different behaviors for cell-edge and cell-center users in both systems.

For each of the considered JRRM strategies we construct the corresponding Markovian model. We then find several performance measures, namely the global blocking probability, the mean sojourn time and the loads in each system.

The remainder of this paper is organized as follows. In Section II, we develop a model for inter-cell interference and throughputs in GERAN. Section III presents the UTRAN model. Several JRRM strategies for the cooperative network are analyzed in Section IV. These strategies are compared by the numerical results in Section V. Section VI eventually concludes the chapter.

2. GSM/EDGE Model

Radio resources are allocated in GERAN on both time and frequency bases. Given a total set of N_{tot} frequencies we can partition them amongst a number of cells to cover a certain area. This number is called pattern size and denoted by S_{pat}. We provide each base station with N_{TRX} transmitters (TRXs) which, in the case of synthesized frequency hopping, is smaller than the number of frequencies in each cell. Since the pattern size is defined as the maximal number of cells where no frequencies are repeated, the latter must always be smaller than N_{tot}/N_{TRX}.

We assume that N_{tot} does not account for the carriers of the logical channels (e.g. beacons, pilots). Similarly, N_{TRX} is the number of transmitters only dedicated to the traffic channel.

We now assume that each mobile uses randomly all the frequencies in its cell. During a call the frequency hopping consists of changing frequency while passing from one time slot (TS) to another. In case of GSM voice calls only one TS per frame (1 frame=8 TSs, from which 7 TSs carry traffic) is occupied as data transmission and decompression take up at least one frame to be executed by the receiver. In case of EDGE calls we allow the system to use up to 3 TSs per frame as NRT applications data can be stored at the receiver once they have been sent and wait for being decompressed until the network conditions allow to do it.

With regard to 2G voice/data services, N_{GSM} time slots are reserved to GSM voice calls, N_{EDGE} are reserved to elastic EDGE calls and the rest are the so-called "GSM-shared" ones, namely voice calls will occupy them if the traffic is such that the TSs reserved to them are no longer sufficient and will be used by the elastic calls otherwise.

2.1. Load

We partition the entire set of frequencies amongst the pattern cells in such a way that no intra-cell interference is generated. This aim can be achieved by using a distributed allocation procedure based on grouping carriers sufficiently far apart that wave tail interference effects (e.g. co-channel, first and second adjacent channel interference) can be neglected. For the sake of concreteness we adopt here a reuse of 3 model to allocate all the available frequencies among the cells, that is each cell of the pattern is assigned only one third of the N_{tot} frequencies.

To the contrary, all the carriers in the neighboring cells interfere to different extent with the carriers in the target cell. Here we consider only co-channel and first adjacent channel interference. Such a type of interference manifests itself only between frequencies used at the same TSs. The latter occurrence is due to the assumption that the base stations of all the pattern cells are synchronised. Therefore we will be concerned with calculating the collision probability at each frequency in the target cell.

As a consequence of this resource sharing policy, if the cell contains M_v^G and M_e^G RT and NRT calls, respectively, the number of TSs occupied by the RT applications is $N_v^{TS}(M_v^G) = M_v^G$, while the number of TSs occupied by the NRT applications depends on the number of RT as well as NRT calls as follows:

$$N_e^{TS}(M_v^G, M_e^G) := \begin{cases} 7N_{TRX} - N_{GSM} & \text{if } M_v^G \leq N_{GSM} \text{ and } M_e^G > \left\lfloor \frac{N_{EDGE}}{3} \right\rfloor \\ 7N_{TRX} - M_v^G & \text{if } M_v^G > N_{GSM} \text{ and } M_e^G > \left\lfloor \frac{N_{EDGE}}{3} \right\rfloor \\ 3M_e^G & \text{otherwise} \end{cases} \quad (1)$$

We define the load of the cell, say χ^G, as the ratio between the mean number of allocated

TSs per frame and the number of all the available TSs:

$$\chi_0^G(M_v^G, M_e^G) = \frac{N_v^{TS}(M_v^G) + N_e^{TS}(M_v^G, M_e^G)}{7N_{TRX}}. \quad (2)$$

2.2. Collision Probability

Here we calculate the distribution of number of collisions with $S_{pat} - 1$ interfering cells in the first and in the second ring, supposing that we have an homogeneous network with the same (mean) load in all cells. We denote with $\mathbf{X}^{(1)}$ the vector of dimension equal to the number of interfering cells in the first ring and with $\mathbf{X}^{(2)}$ the vector of dimension equal to the number of interfering cells in the second ring. The elements of each $\mathbf{X}^{(i)}$ are 1 or 0 depending on whether a collision occurs with the corresponding cell belonging to ring i.

Lemma 1 *The probability of having a particular configuration of collisions with a frequency in cell* 0 *represented by the pair of vectors* $\left(\mathbf{X}^{(1)}, \mathbf{X}^{(2)}\right)$ *is:*

$$\mathbb{P}\left[\mathbf{X}^{(1)}, \mathbf{X}^{(2)} | \chi^G\right] = g^{\mathbf{X}^{(1)}.\mathbf{X}^{(1)} + \mathbf{X}^{(2)}.\mathbf{X}^{(2)}} (1-g)^{2S_{pat}-2-\mathbf{X}^{(1)}.\mathbf{X}^{(1)} - \mathbf{X}^{(2)}.\mathbf{X}^{(2)}}, \quad (3)$$

where $\mathbf{X}^{(i)} \cdot \mathbf{X}^{(i)}$ *gives the number of collisions with the* i*-th ring cells and* $g = \frac{N_{TRX}}{N_{tot}/3}\chi^G$ *for a reuse of 3 frequency allocation model.*

Proof: Firstly, independent from the reuse scheme, only one frequency of each interfering cell can interfere with a given frequency in the target cell. E.g., if a frequency resue of 3 is used, as in Figure 1, the same frequencies are used in the cells of the second ring, while adjacent frequencies are used in the first interfering ring, and the absence of intra-cell interference implies that only one frequency in each interfering cell is close enough to interfere with the frequency of the target cell. The probability of having a collision between two cells corresponds than to a Bernoulli distribution. The probability g of collision between each cell i and a frequency f is the product of the probability for a TS (the one used by f) to be occupied multiplied by the probability of using one frequency out of $k(i)$ possible ones. Therefore

$$g = \frac{N_i^{TS}}{7N_{TRX}} \frac{N_{TRX}}{k(i)} = \frac{N_{TRX}}{k(i)}\chi^G$$

Taking into account the independence of the frequency allocation between cells, this leads to Eqn. (3).

2.3. Inter-cell Interference

Typically, in GERAN, the SINR perceived by a mobile, numbered k, is given by:

$$SINR_k^G = \frac{P_e/q_{k,0}^G}{I_{inter,k}^G + N_0}, \quad (4)$$

where P_e is the power emitted by the base station in the target cell, $\gamma \in [2, 4]$ is a constant, I_{inter}^G is the inter-cell interference, N_0 is the background noise and $q_{k,l}^G = r_{k,l}^{\gamma} 10^{\xi_{k,l}^G/10}$ is

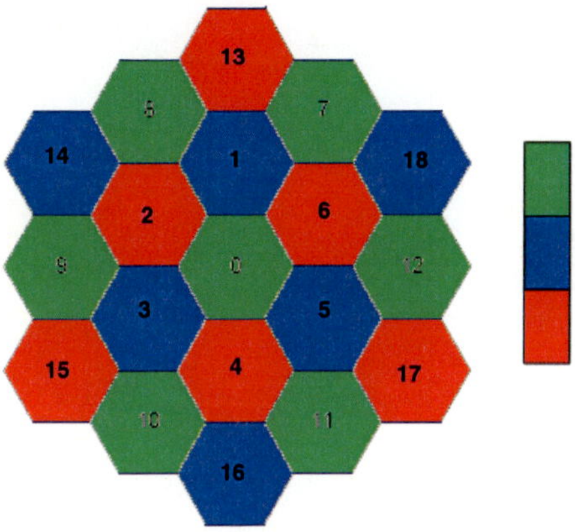

Figure 1. A 2G system with frequency reuse factor of 3.

Table 1. Protection ratio between adjacent frequencies.

$\lvert F_j - F_i \rvert$	$R(i \to j)$	$[R(i \to j)]$dB
0 (co-canal)	$R(i \to j) = R_0 = 1$	$[R_0]$dB = 0
1 (first adjacent)	$R(i \to j) = R_1 = 10^{-1.8}$	$[R_1]$ dB = -18
2 (second adjacent)	$R(i \to j) = R_2 = 10^{-5}$	$[R_2]$ dB = -50
3 (third adjacent)	$R(i \to j) = R_3 = 10^{-58}$	$[R_3]$ dB = -58
≥ 4 (beyond fourth adjacent)	$R(i \to j) = R_{\geq 4} = 0$	$[R_4]$ dB = -200

the path-loss between user k and base station l, with $r_{k,l}$ and $\xi^G_{k,l}$ respectively the distance and the shadowing between them.

Our primary task now is to calculate $I^G_{inter,k}$ in order to evaluate the SINR. Let us notice how a particular pair of vectors $\left(\mathbf{X}^{(1)}, \mathbf{X}^{(2)}\right)$ has an impact on the total interference, which is given by

$$I^G_{inter}(r_0, \mathbf{X}^{(1)}, \mathbf{X}^{(2)}) = \sum_{k=1}^{\mathbf{X}^{(1)}.\mathbf{X}^{(1)}} R_k \frac{P_e}{q^G_{k,l}} + \sum_{k=1}^{\mathbf{X}^{(2)}.\mathbf{X}^{(2)}} \frac{P_e}{q^G_{k,l}}, \tag{5}$$

where R_k the protection ratio of the adjacent channel, given by Table 2.3..

2.4. Throughput of Elastic Calls

For each SINR value in a given interference configuration $\left(\mathbf{X}^{(1)}, \mathbf{X}^{(2)}\right)$, we obtain an average throughput (over the shadowing) at each point r_0 in the target cell, $T^G\left(SINR^G(r_0, \mathbf{X}^{(1)}, \mathbf{X}^{(2)})\right)$, via link level curves with a tolerance of 10^{-2} for the received BLER.

For each configuration of the number of interferes $(\mathbf{X}^{(1)}, \mathbf{X}^{(2)})$, we can obtain a mean throughput $\left\langle T^G(\mathbf{X}^{(1)}, \mathbf{X}^{(2)})\right\rangle$ by averaging over the surface of the cell:

$$\left\langle T^G(\mathbf{X}^{(1)}, \mathbf{X}^{(2)})\right\rangle \simeq \mathbb{E}_{r_0}[T^G(SINR^G(r_0, \mathbf{X}^{(1)}, \mathbf{X}^{(2)}))] \tag{6}$$

Finally, the mean throughput over all the possible configurations of interfering neighboring cells is given by:

$$\bar{T}^G(\chi^G) = \sum_{\mathbf{X}^{(1)},\mathbf{X}^{(2)}} \left\langle T^G(\mathbf{X}^{(1)}, \mathbf{X}^{(2)})\right\rangle \mathbb{P}\left[\mathbf{X}^{(1)}, \mathbf{X}^{(2)} | \chi^G\right], \tag{7}$$

If there are M_e^G elastic calls in the cell, their departure rate is given by:

$$\nu_e^G(\mathbf{M}_v^G, \mathbf{M}_e^G, \chi^G) = \mu_e \bar{T}^G(\chi^G) N_e^{TS}\left(M_v^G, M_e^G\right) \tag{8}$$

where μ_e is the inverse of the mean file size.

3. UMTS/HSDPA Model

In WCDMA, the SINR achieved for user k, situated at distance r_k from its own base station, is given by:

$$SINR_k^U = \frac{P_{k,0}/q_{k,0}^U}{I_{inter,k}^U + I_{intra,k} + N_0} S_k$$

where S_k is the corresponding spreading factor.

To analyze the interference, let us first note that the intra-cell interference originates from the common channels and from other users: $I_{intra}(r_0) = \alpha(P_{tot,0} - P_{k,0})/q_{k,0}^U$, α being the orthogonality factor and $P_{tot,l}$ the total transmitted power by base station l.

For the inter-cell interference, it is given by: $I_{inter,k} = \sum_{l \neq 0} P_{tot,l}/q_{k,l}^U$. Differently from other works (e.g. see [9]) in which the powers in the interfering cells equals the instantaneous power in the target cell, here we consider that such an equality holds only for the average power and it is independent of the instantaneous fluctuations of $P_{tot,0}$. Here the total power of the base station $l \neq 0$ can be written as $P_{tot,l} = \bar{\chi}^U P_{max}$, where $\bar{\chi}^U$ is the average load in a typical cell of the system, defined as the ratio between the used and total powers,and P_{max} is the maximal transmission power. On the other hand, the value $\sum_{l \neq 0} q_{k,0}^U/q_{k,l}^U$ is the well-known F-factor F_k [7] [11]. The SINR is then equal to

$$SINR_k^U = \frac{P_{k,0}}{\alpha(P_{tot,0} - P_{k,0}) + \chi P_{max} F_k + N_0 q_{k,0}^U} S_k$$

leading to the expression:

$$\beta_k = \frac{P_{k,0}}{\alpha P_{tot,0} + \chi P_{max} F_k + N_0 q_{k,0}^U}, \quad \beta_k = \frac{SINR_k^U}{S_k + \alpha.SINR_k^U}$$

If M_v^U voice users, with no HSDPA users, were present in the cell, this leads to the power expression:

$$P_{tot,0} = \frac{P_{Com} + \beta_v \sum_k (\chi P_{max} F_k + N_0 q_{k,0}^U)}{1 - \alpha M_v^U \beta_v}$$

where $\beta_v = \frac{SINR_v^U}{S + \alpha . SINR_v^U}$ and the required SINR for voice calls is equal to 11 dB. If we now suppose that all users have the same radio conditions, characterized by a mean F-factor $\bar{F}$ and a mean path loss $\bar{q}^U$, this expression becomes:

$$P_{tot,0} = \frac{P_{Com} + \beta_v (\chi P_{max} \bar{F} + N_0 \bar{q}^U) M_v^U}{1 - \alpha \beta_v M_v^U} \quad (9)$$

Considering the constraint on the maximal transmission power ($P_{tot,0} \leq P_{max}$), the constraint on the number of users becomes:

$$\beta_v (\chi P_{max} \bar{F} + N_0 \bar{q}^U + \alpha P_{max}) M_v^U \leq P_{max} - P_{Com}$$

the maximal number of voice users that can be accepted is equal to:

$$M_{v,max}^U = \frac{P_{max} - P_{Com}}{\beta_v (\chi P_{max} \bar{F} + N_0 \bar{q}^U + \alpha P_{max})}$$

In the presence of HSDPA users, and supposing that all the available power will be used by the HSDPA calls, we have $P_{tot,0} = P_{max}$, leading to:

$$P_v = \beta_v (\alpha P_{max} + \chi P_{max} \bar{F} + N_0 \bar{q}^U)$$

For an elastic user, the achieved value of β is equal to:

$$\beta_e = \frac{P_e}{(\alpha P_{max} + \chi P_{max} \bar{F} + N_0 \bar{q}^U)} \quad (10)$$

Using the fact that, when an HSDPA user receives transmission, he is receiving all the available power, we obtain:

$$P_e = P_{max} - P_{Com} - P_{SCCH} - P_v M_v^U \quad (11)$$

leading to the SINR of HSDPA users:

$$SINR^H = \frac{S[P_{max} - P_{Com} - P_{SCCH} - \beta_v (\alpha P_{max} + \chi P_{max} \bar{F} + N_0 \bar{q}^U) M_v^U]}{\chi P_{max} \bar{F} + N_0 \bar{q}^U + \alpha (P_{Com} + P_{SCCH}) + \alpha \beta_v (\alpha P_{max} + \chi P_{max} \bar{F} + N_0 \bar{q}^U) M_v^U} \quad (12)$$

Note that, in order to have the system in equilibrium, we must verify that the power constraints for the voice calls can be achieved, this can be expressed as the condition:

$$P_{DSCH} = P_{max} - P_{Com} - P_{SCCH} - P_v M_v > 0$$

leading to the constraint on the number of real-time calls in the system:

$$\beta_v (\chi P_{max} \bar{F} + N_0 \bar{q}^U + \alpha P_{max}) M_v^U < P_{max} - P_{Com} - P_{SCCH} \quad (13)$$

the maximal number of voice users that can be accepted is equal to:

$$M^U_{v,max} = \frac{P_{max} - P_{Com} - P_{SCCH}}{\beta_v(\chi P_{max}\bar{F} + N_0\bar{q}^U + \alpha P_{max})}$$

Knowing the SINR of HSDPA users (12), the choice of the modulation is based on link level curves $t(SINR)$ as described in [8]. The throughput is then:

$$T^U(M^U_v, \chi) = t(\frac{S[P_{max} - P_{Com} - P_{SCCH} - \beta_v(\alpha P_{max} + \chi P_{max}\bar{F} + N_0\bar{q}^U)M^U_v]}{\chi P_{max}\bar{F} + N_0\bar{q}^U + \alpha(P_{Com} + P_{SCCH}) + \alpha\beta_v(\alpha P_{max} + \chi P_{max}\bar{F} + N_0\bar{q}^U)M}$$

and the throughput of one elastic call is:

$$D_e(M^U_v, M^U_e, \chi) = \frac{T^U(M^U_v, \chi)}{M^U_e} \qquad (14)$$

The mean departure rate of HSDPA calls is then:

$$\nu^U_e(M^U_v, M^U_e, \chi^U) = \mu_e T^U(M^U_v, \chi)$$

where μ_e is the inverse mean file size.

Note that there is no physical limit for the number of users in HSDPA. However, in order to guarantee a good QoS, a maximal number of HSDPA users $M^H_{e,max}$ is fixed.

4. Cooperative Model

We analyze the performance of a cooperative network composed of GERAN and UTRAN, based on the previous models for these systems. We will consider two different possible strategies of JRRM:

1. a first strategy that allocates voice calls with priority to GERAN, while elastic calls are directed initially to HSDPA. Although, if a voice call arrives and there is no available room in GERAN, it is directed to DCH in the 3G cell. Equivalently, elastic calls are directed to EDGE if HSDPA is fully loaded. This strategy allows a special form of vertical handovers where voice (data) calls are redirected to GERAN (UTRAN) if there is available space due to call departures;

2. a second strategy that maximizes the throughput of elastic calls by allocating them to the system with the best conditions. Vertical handover here is also allowed in order to keep the best possible conditions all over the communication duration.

4.1. Choice of the Network Based on Service

4.1.1. Markovian Model

We suppose that call durations and file sizes are exponentially distributed. Let A be the set of all the possible states of the cooperative network represented by $\mathbf{S} = (M_v, M_e)$. The first strategy for selecting the system based on the service can be written as:

$$M^G_v = \begin{cases} M_v, & \text{if } M^G_v \leq M^{G,max}_v \\ M^{G,max}_v, & \text{otherwise} \end{cases} \quad \text{and } M^U_e = \begin{cases} M_e, & \text{if } M^U_e \leq M^{U,max}_e \\ M^{U,max}_e, & \text{otherwise} \end{cases} \qquad (15)$$

The system can be described by a Markov chain and the transition matrix $\mathbf{Q}$ associated with it has transition rates:

$$\begin{aligned}
&P(M_v, M_e \to M_v - 1, M_e) = \mu_v M_v \\
&P(M_v, M_e \to M_v + 1, M_e) = \lambda_v \\
&P(M_v, M_e \to M_v, M_e + 1) = \lambda_e \\
&P(M_v, M_e \to M_v, M_e - 1) = \nu_e^U(M_v^U, M_e^U, \bar{\chi}^U) + \nu_e^G(M_v^G, M_e^G, \bar{\chi}^G)
\end{aligned}$$

where μ_v is the mean voice call duration. The steady-state probabilities are obtained by solving the set of equations $\pi\mathbf{Q} = \mathbf{0}$ with the normalization condition $\pi\mathbf{e} = \mathbf{1}$, where $\mathbf{e}$ is a vector of ones of proper dimension.

4.1.2. Performance Measures

Using this model, we calculate several performance measures, namely the blocking probability (by summing the probabilities of the limit cases) and the mean sojourn time (using the Little formula). We also calculate the load in GERAN using eqn. (2):

$$\bar{\chi}^G = \sum_{\mathbf{S} \in A} \chi^G(M_v^G(\mathbf{S}), M_e^G(\mathbf{S}))\pi(\mathbf{S}). \tag{16}$$

The load in UTRAN is defined as the proportion of the total Node B power used for transmission. As the total power that is not allocated to common or dedicated powers will be used for HS-DSCH transmission, the mean HSDPA load becomes:

$$\bar{\chi}^U = \frac{P_{com} + \beta_v(\bar{\chi}^U P_{max}\bar{F} + N_0\bar{q}^U)M_v^U}{1 - \alpha\beta_v M_v^U} \sum_{\mathbf{S} \in D} \pi(\mathbf{S}) + \left(1 - \sum_{\mathbf{S} \in D} \pi(\mathbf{S})\right), \tag{17}$$

where D is the subspace of A where $M_e = 0$.

4.1.3. Iterative Resolution

As the steady-state probabilities in the target cell are calculated using the loads in the interfering cells, we propose the following iterative algorithm in a homogeneous network:

1. set the initial values for the iterations, e.g. with taking loads equal to 0.5;
2. calculate the steady-state probabilities using these initial values, and deduce the loads using (16-17).
3. use the new value of the loads into the transition matrix and repeat the iterations until convergence.

4.2. Throughput Maximization with Vertical Handovers

We consider now the case where a call is accepted in a way that maximizes the throughput of elastic calls in the system. Under this condition inter-system (vertical) handover can also be performed during the communications, whenever a better configuration is possible.

We calculate again, for each state $\mathbf{S} = (M_v, M_e)$, the individual assignment of calls to the different RANs through a policy that imposes to both systems to maximize their throughput:

$$\begin{aligned}\left(M_v^G(M_v, M_e), M_e^G(M_v, M_e)\right) &= argmax_{(x,y)}[\bar{T}^G(\chi^G)N_e^{TS}(x,y) \\ &+ \bar{T}^G(\chi^G)N_v^{TS}(x,y) + \bar{T}^H(M_v - x, \chi^U)]\end{aligned} \quad (18)$$

Due to the nonlinear nature of N_v^{TS} and N_e^{TS} more conditions might be needed to completely define the state of the different RANs. Let us call $A_i(M_v, M_e, \chi^G, \chi^U)$ the set of values of $\left(M_{v,i}^G(M_v, M_e), M_{e,i}^G(M_v, M_e)\right)$ maximizing the throughput, for example maximizing the number of voice calls in GERAN and of elastic calls in UTRAN. An iterative algorithm similar to the one for the first policy is used to calculate the load and the performance measures.

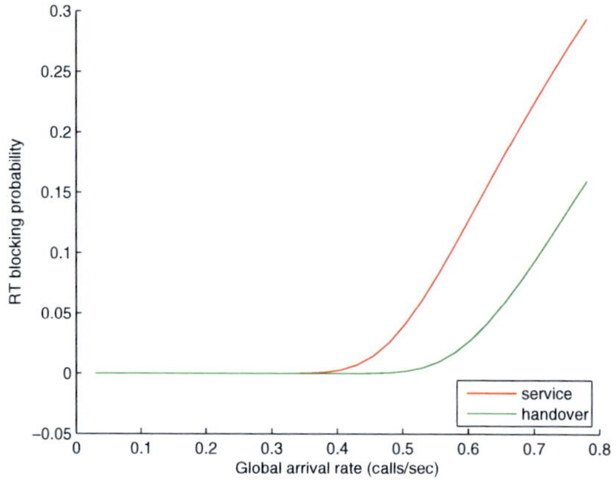

Figure 2. RT blocking probability for 2G/3G in the two strategies.

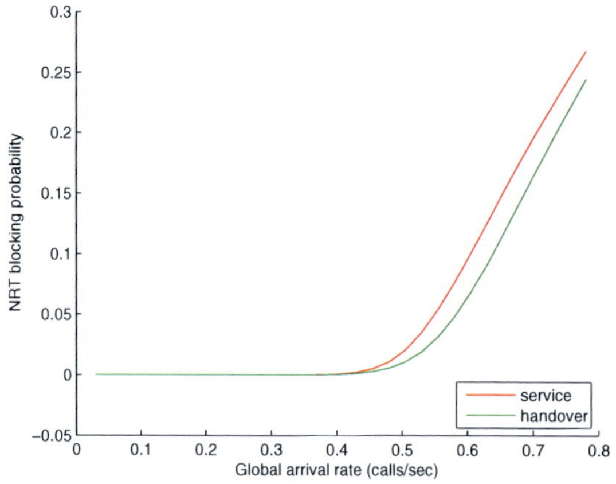

Figure 3. NRT blocking probability for 2G/3G in the two strategies.

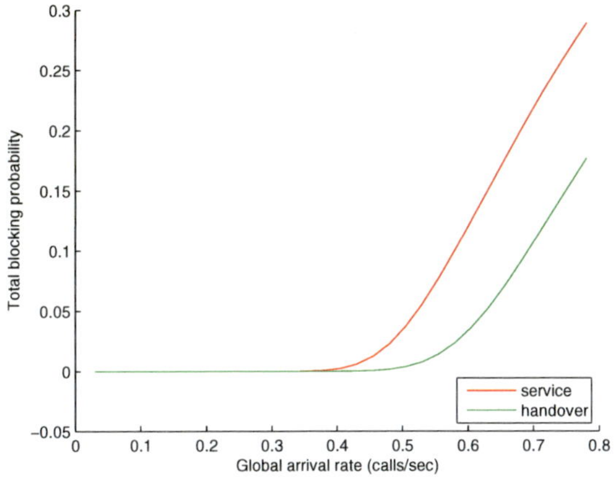

Figure 4. Total blocking probability for 2G/3G in the two strategies.

5. Numerical Applications

In order to evaluate the user QoS, we show here some numerical results in which the blocking probability and the mean sojourn time are compared for the two different strategies. In particular, the service-based strategy compared with the throughput maximization using vertical handovers.

We consider a poissonian flow of both real time (voice) and elastic (file downloading) calls with varying arrival rates. The file size is exponentially distributed with mean equal to $1/\mu = 2$ Mb. Cells are hexagonal with a radius of 500 meters and interference from the first two rings is considered forboth systems.

5.1. 3G System Description

The chip rate W is set to 3.84 Mcps while the spreading factor is equal to 16. The common channels occupy 33% of the power transmitted by the base station (26% if there is no HSDPA active transmission in the cell), and the noise level is equal to -174 dBm/Hz. The values of the F Factor are calculated as in [10][11]. The orthogonality factor α is set equal to 0.45 and the path loss model we used is the well-known Cost-231 Hata model [12].

5.2. 2G System Description

The system we consider contains four transmitters, with 7 TS reserved for communication in each TRX. Only three TS are reserved for EDGE and two others are shared between voice and data users (with a priority for voice users). A frequency reuse of 3 is considered, within a bandwidth of 200 KHZ located at the 900 MHZ frequency. The transmission power of the base station is equal to 43 dBm and the noise level is equal to -174 dBm/Hz. Link adptation is possible for EDGE with a throughput ranging from 3 to 58 Kbits per second per TS.

5.3. Blocking Probabilities

We first plot in Figures 2 and 3 the blocking probabilities for the two strategies, function of the global arrival rate. We can see that the strategy with throughput maximization and vertical handover achieves the best blocking indicating that the capacity is maximized. This is also reflected on the overall blocking probability illustrated in Figure 4.

More in depth, if we aim to calculate an Erlang-like capacity of the network, we set a threshold on blocking probability and calculate the maximal admitted arrival rate. For instance, if the operator requires that blocking must be less than 5%, the arrival rate can be up to 0.65 calls/sec in the strategy with vertical handover, up to 0.52 in the service-based strategy (gain of 25% on the capacity).

5.4. File Download Times

In Figure 5 we plot the mean file download time in the two strategies versus the arrival rate of calls. Obviously, the strategy with vertical handover achieves faster downloads, meaning that the throughput is better. This is due to the intelligent use of resources as users can change their serving system upon availability of resources.

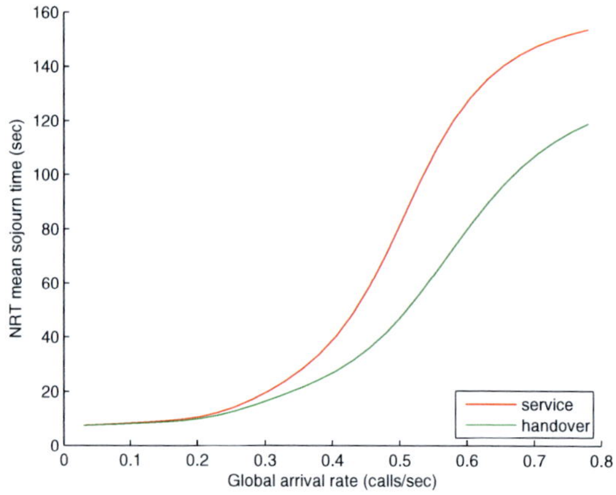

Figure 5. NRT mean sojourn time for 2G/3G in the two strategies.

6. Conclusion

An analytical model of cooperation between 2G and 3G has been presented here. The system selection rules are based either on service or on throughput maximization. Two JRRM policies have been studied depending on whether vertical handovers are implemented or not. We began by developing an analytical model for inter-cell interference and throughputs in 2G and 3G systems and then developed Markovian models that describe the evolution of the network for each of the studied policies.

Our numerical results show that the strategy with vertical handover and throughput maximization gives the best performance (lower blocking rates, and smaller file download times). These results were obtained considering that the handovers were ideal: No delays nor imperfections were introduced. The obtained performance can then be viewed as an upper limit of the performance of the system.

As of future work, we aim at differentiate between users depending on their positions in the system and take into account the vertical handovers due to mobility.

References

[1] Y. Raivio, *4G - Hype or Reality*, IEEE 3G Mobile Communication Technologies, March 2002.

[2] D. Chen, X. Wang, A.K. Elhakeem, Performance Analysis of UMTS Handover with the Help of WLAN, *Second International Conference on Quality of Service in Heterogeneous Wired/Wireless Networks (QShine05)*, 22-24 Aug. 2005.

[3] Q. Song, A. Jamalipour, Network Selection in an Integrated Wireless LAN and UMTS Environment Using Mathematical Modeling and Computing Techniques, *Wireless Communications, IEEE* Volume 12, Issue 3, June 2005 Page(s):42-48.

[4] K. Ahmavaara, H. Haverinen and R. Pichna, Interworking architecture between 3GPP and WLAN systems, *IEEE Communications Magazine* 41 (11) (2003) 74-81.

[5] T. Bonald and A. Proutière, Wireless Downlink Data Channels: User Performance and Cell Dimensioning, *ACM Mobicom'03,* San Diego, September 2003.

[6] M. Assaad and D. Zeghlache, On the Capacity of HSDPA, *IEEE Globecom* 2003, San Francisco, November 2003.

[7] H. Holma and A. Toskala, *WCDMA for UMTS: Radio access for third generation mobile communications*, John Wiley & Sons, 3rd Ed. (2004).

[8] P. Bender, P. Black, M. Grob, R. Padovani, N. Sindhushayana, and A. Viterbi, CDMA/HDR: A Bandwidth-Efficient High-Speed Wireless Data Service for Nomadic Users, *IEEE Communications Magazine*, July 2000.

[9] I. Koukoutsidis, E. Altman, J.M. Kelif, A non-homogeneous QBD approach for the admission and GoS control in a multiservice WCDMA system, *INRIA Research Report* No. RR-5358 (2004).

[10] K. Hiltunten, R. De Bramardi, WCDMA downlink capacity estimation, *IEEE VTC-Spring* (2000) 992-996.

[11] K. Sipila, Z. Honkasalo, J. Laiho-Steffens and A. Wackr, Estimation of Capacity and Required Transmission Power of WCDMA Downlink Based on a Downlink Pole Equation, *IEEE VTC* 2000.

[12] V.S. Abhayawardhana, I.J. Wassell, D. Crosby, M.P. Sellars, M.G. Brown, Comparison of Empirical Propagation Path Loss Models for Fixed Wireless Access Systems, *IEEE VTC* 2005.

In: Wireless Communications Research Trends
Editor: Tong S. Lee, pp. 301-323
ISBN: 978-1-60021-674-9

Chapter 10

IMPACT OF ULTRA WIDE BAND (UWB) ON MACROCELL DOWNLINK OF UMTS, CDMA-450, DCS-1800 AND GSM-900 SYSTEMS

Bazil Taha-Ahmed, Miguel Calvo-Ramón and Leandro Haro-Ariet
Departamento Sistemas, Senales y Radiocomunicaciones,
Universidad Politécnica de Madrid, Spain

1. Introduction

The popularity of wireless networks makes interference and cross-talk between multiple systems inevitable. This chapter describes techniques for quantifying the effect of the UWB system on the second and third generation mobile communications systems.

Ultra-wideband (UWB) radio signals have characteristics that are different from conventional radios. Of special interest is the ability to spread the transmission power over a sufficiently wide bandwidth to make the signal appear as noise to a narrowband receiver, while still being able to transmit very high data rates over short distances. In this context "narrowband" may actually mean 20 MHz Wide. Ultra Wideband was traditionally accepted as impulse radio, but the FCC and ITU-R now define UWB in terms of a transmission from an antenna for which the emitted signal bandwidth exceeds the lesser of 500 MHz or 20% bandwidth. Thus, pulse-based systems—wherein each transmitted pulse instantaneously occupies a UWB bandwidth, or an aggregation of at least 500 MHz worth of narrow band carriers, for example in orthogonal frequency-division multiplexing (OFDM) fashion—can gain access to the UWB spectrum under the rules. Pulse repetition rates may be either low or very high. Pulse-based radars and imaging systems tend to use low repetition rates, typically in the range of 1 to 10 megapulses per second. On the other hand, communications systems favor high repetition rates, typically in the range of 1 to 2 gigapulses per second, thus enabling short-range gigabit-per-second communications systems. Each pulse in a pulse-based UWB system occupies the entire UWB bandwidth, thus reaping the benefits of relative immunity to multipath fading (but not to intersymbol interference), unlike carrier-based systems that are subject to both deep fades and intersymbol interference.

The Federal Communications Commission (FCC) agreed in February 2002 to allocate 7.5 GHz of spectrum for unlicensed use of ultra-wideband (UWB) devices for communication applications in the 3.1–10.6 GHz frequency band, the move represented a victory in a long hard-fought battle that dated back decades. With its origins in the 1960s, when it was called time-domain electromagnetics, UWB came to be known for the operation of sending and receiving extremely short bursts of RF energy. With its outstanding ability for applications that require precision distance or positioning measurements, as well as high-speed wireless connectivity, the largest spectrum allocation ever granted by the FCC is unique because it overlaps other services in the same frequency of operation. Previous spectrum allocations for unlicensed use, such as the Unlicensed National Information Infrastructure (UNII) band have opened up bandwidth dedicated to unlicensed devices based on the assumption that operation is subject to the following two conditions:

(1) This device may not cause harmful interference. Harmful interference is defined as interference that seriously degrades, obstructs or repeatedly interrupts a radio communication service.
(2) This device must accept any interference received, including interference that may cause undesired operation. This means that devices using unlicensed spectrum must be designed to coexist in an uncontrolled environment. Devices utilizing UWB spectrum operate according to similar rules, but they are subject to more stringent requirements because UWB spectrum underlays other existing licensed and unlicensed spectrum allocations. In order to optimize spectrum use and reduce interference to existing services, the FCC's regulations are very conservative and require very low emitted power.

UWB has a number of advantages which make it attractive for consumer communications applications. In particular, UWB systems

- have potentially low complexity and low cost;
- have noise-like signal characteristics;
- are resistant to severe multipath and jamming;
- have very good time domain resolution.

Universal Mobile Telecommunications System (UMTS) is one of the third-generation (3G) mobile phone technologies. It uses W-CDMA as the underlying standard. UMTS supports up to 1920 kbit/s data transfer rates (and not 2 Mbit/s as frequently seen), although at the moment users in the real networks can expect performance up to 384 kbit/s - in Japan upgrades to 3 Mbit/s are in preparation. However, this is still much greater than the 14.4 kbit/s of a single GSM error-corrected circuit switched data channel or multiple 14.4 kbit/s channels in HSCSD.

From the beginning of 2006, UMTS networks in Japan are being upgraded with High Speed Downlink Packet Access (HSDPA), sometimes known as 3.5G. This will make a downlink transfer speed of up to 14.4 Mbit/s possible. Work is also progressing on improving the uplink transfer speed with the High-Speed Uplink Packet Access (HSUPA).

The spectrum for UMTS lies between 1900 MHz to 2025 MHz and 2110 MHz to 2200 MHz. For the satellite service an own sub-band in the UMTS spectrum is reserved (uplink 1980 MHz to 2010 MHz, downlink 2170 MHz to 2200 MHz). The remaining spectrum for terrestrial use is divided between two modes of operation. In the FDD (Frequency Division Duplex) mode there are two equal bands for the uplink (1920 MHz to 1980 MHz) and for the downlink (2110 MHz to 2170 MHz). In the operation mode TDD (Time division duplex) uplink and downlink are not divided by use of different frequency carriers but by using different timeslots on the same carrier. So there is no need for a symmetrical spectrum but the remaining unpaired spectrum can be used.

CDMA-450 is a TIA-EIA-IS-CDMA2000 (CDMA-MC) system deployed in 450 MHz. CDMA-450 provides a larger macrocell size compared to macrocell sizes in other bands (UMTS for example), which translates to fewer macrocell sites. The CDMA-450 utilized band is the 450-470 MHz band.

The Global System for Mobile Communications (GSM) is the most popular standard for mobile phones in the world. GSM service is used by over 2 billion people across more than 210 countries and territories. The ubiquity of the GSM standard makes international roaming very common between mobile phone operators, enabling subscribers to use their phones in many parts of the world. GSM differs significantly from its predecessors in that both signaling and speech channels are Digital call quality, which means that it is considered a *second generation* (2G) mobile phone system. This fact has also meant that data communication was built into the system from very early on. The GSM 900 band provides a GSM uplink in the range 890-915 MHz, a GSM downlink in the range 935-960 MHz. The GSM 900 band is used in all countries in which GSM networks are found.

DCS-1800 is a Digital Communications System based on GSM, working on a radio frequency of 1800 MHz. Also known as GSM-1800 or PCN, this digital network operates in Europe and Asia Pacific. The DCS-1800 band provides for a DCS uplink in the range 1710-1785 MHz, a DCS downlink in the range 1805-1880 MHz.

Hamalainen *et al.* studied the coexistence of the UWB system with GSM900, UMTS/WCDMA, and GPS [1]. They gave the bit error rate (BER) of the above mentioned systems for different pulse length.

Hamalainen *et al.* investigated the coexistence of the UWB system with IEEE802.11a and UMTS in Modified Saleh-Valenzuela Channel [2]. They gave the bit error rate (BER) of the UWB system for different types of modulation (Direct sequence and Time Hopping). Guiliano *et al.* studied the interference between the UMTS and The UWB system [3]. In [4], Hamalainen *et al.* investigated the effect of the in band interference power caused by different kinds of UWB signal at UMTS/WCDMA frequency bands. The UWB interference was given for the UMTS/WCDMA uplink and downlink. In [5], Hamalainen *et al.* studied the effect of the in band interference power caused by three different kinds of UWB signal on GPS L1 and GSM-900 uplink band. In **[6]**, the Impact of the UWB system on the macrocell downlink range of DCS-1800 and GSM-900 systems has been presented.

The aim of this chapter is to present the effect of UWB on UMTS, CDMA-450 , DCS-1800 and GSM-900 on the urban macrocell downlink performance (range and capacity) for a critical distance (distance between the UWB transmitter and the mobile receiver under study) of 1m.

2. Effect of UWB Interference on the UMTS and CDMA-450 Downlink Performance

Fig. 1 shows the studied scenario where the mobile communication base station is at a distance R from the mobile receiver that exists in a building near by an UWB transmitter with small distance d between them. The propagation model for the mobile communication is an outdoor propagation with wall penetration loss of 10 dB. The indoor propagation model (Line of Sight) is used to calculate the UWB interference.

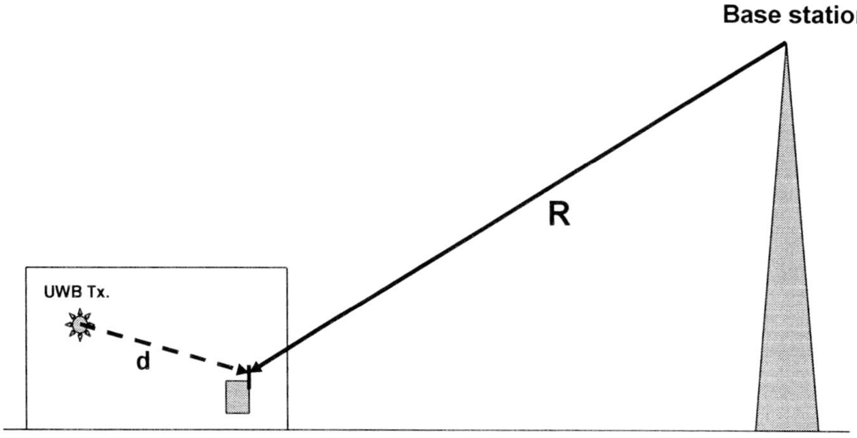

Figure 1. The studied scenario.

To account for UWB interference, an extra source of interference is added linearly to the UMTS and the CDMA-450 intra-system interference. The interference power is calculated by assuming the UWB source to be at different distances from the UMTS receiver (the mobile station). Therefore, the interference power generated by a device UWB, I_{UWB}, is given by (in dBm):

$$I_{UWB} = P_{UWB} - L_{UWB}(d) + G_{UMTS} \tag{1}$$

where

- P_{UWB} is the UWB EIRP in dBm in the UMTS band.
- $L_{UWB}(d)$ is the path-loss between the UWB device and the UMTS receiver which varies with the separation distance, d in m, and
- G_{UMTS} is the UMTS antenna gain .

Given that UWB devices are typically low power, short range devices, then the line-of-sight path-loss model is often most appropriate. Thus the UWB signal propagation loss in dB is calculated as:

$$L_{UWB}(d) \approx 39.03 + 20\log_{10}(d) \tag{2}$$

The effect of the UWB interference is to reduce the UMTS macrocell range or/and the macrocell capacity.

The UMTS signal propagation loss in urban macrocell is given as:

$$L_{UMTS}(dB) \approx L_{o1} + 10s \log_{10}(R_{km}) \quad (3)$$

where

- L_{o1} is the UMTS signal propagation loss at a distance of 1 km.
- s is the UMTS signal propagation exponent (3.5 to 4.5).

Thus

$$10\, s \log_{10}(R_{km}) = L_{UMTS}(dB) - L_{o1} \quad (4)$$

L_{UMTS} depends on the SNR of the UMTS signal, i.e., lower is the noise, higher is the accepted compensated propagation loss. Thus we can rewrite (4) as

$$10\, s \log_{10}(R_{km}) = k - S_{min} - L_{o1} \quad (5)$$

where

- k is constant,
- S_{min} is the UMTS receiver sensitivity.

When the total noise power consists of the UMTS system noise I_{UMTS} only, then:

$$10\, s \log_{10} R_{UMTS,o} = k - S_{min} - L_{o1} \quad (6)$$

where $R_{UMTS,o}$ is the UMTS initial range.

$$R^{s}_{UMTS,o} = 10^{\frac{k - S_{min} - L_{o1}}{10}} \quad (7)$$

With the existence of the UWB noise (I_{UWB}), the UMTS receiver noise will increase. Thus

$$10\, s \log_{10} R_{UMTS} = k - S_{min} - N_r - L_{o1} \quad (8)$$

where R_{UMTS} is the UMTS new range when the UWB affects the UMTS system and N_r is the UMTS receiver noise increment (dB).

$$R^{s}_{UMTS} = 10^{\frac{k-S_{min}-N_r-L_{o1}}{10}} \tag{9}$$

$$\frac{R^{s}_{UMTS,o}}{R^{s}_{UMTS}} = 10^{\frac{k-S_{min}-L_{o1}}{10}-\frac{k-S_{min}-N_r-L_{o1}}{10}} = 10^{\frac{N_r}{10}} \tag{10}$$

$$\left(\frac{R_{UMTS,o}}{R_{UMTS}}\right)^{s} = 10^{\frac{N_r}{10}} \tag{11}$$

$$\left(\frac{R_{UMTS}}{R_{UMTS,o}}\right)^{s} = 1/10^{\frac{N_r}{10}} = 10^{\frac{-N_r}{10}} \tag{12}$$

$$R_{UMTS} = R_{UMTS,o}\left(10^{-N_r/10}\right)^{-s} = R_{UMTS,o}\sqrt[\frac{1}{s}]{10^{\frac{-N_r}{10}}} \tag{13}$$

we can notice that $10^{\frac{N_r}{10}}$ is the UMTS receiver noise increment in natural number, given by:

$$10^{\frac{N_r}{10}} = \frac{I_{UMTS}+I_{UWB}}{I_{UMTS}} \tag{14}$$

Thus, the macrocell range R_{UMTS} with the existence of the UWB interference is given as:

$$R_{UMTS} = R_{UMTS,o}\sqrt[\frac{1}{s}]{\frac{I_{UMTS}}{\left(I_{UMTS}+I_{UWB}\right)}} \tag{15}$$

The normalized range is given as:

$$\frac{R_{UMTS}}{R_{UMYTS,o}} = \sqrt[\frac{1}{s}]{\frac{I_{UMTS}}{\left(I_{UMTS}+I_{UWB}\right)}} \tag{16}$$

The normalized capacity C_n is given as:

$$C_n = \left(\frac{I_{UMTS}}{I_{UMTS}+I_{UWB}}\right) \tag{17}$$

The interference power generated by a device UWB, I_{UWB}, that affects the CDMA-450 receiver is given by (in dBm):

$$I_{UWB} = P_{UWB} - L_{UWB}(d) + G_{CDMA} \tag{18}$$

where

- P_{UWB} is the UWB EIRP in dBm in the CDMA-450 band.
- $L_{UWB}(d)$ is the path-loss between the UWB device and the CDMA-450 receiver which varies with the separation distance, d in m, and
- G_{CDMA} is the CDMA-450 antenna gain .

In the frequency band used by CDMA-450, the UWB signal propagation loss in dB is calculated as:

$$L_{UWB}(d) \approx 25.7 + 20\log_{10}(d) \tag{19}$$

The effect of the UWB interference is to reduce the CDMA-450 macrocell range or/and the macrocell capacity.

The macrocell range R_{CDMA} with the existence of the UWB interference is given as:

$$R_{CDMA} = R_{CDMA,o} \sqrt[\frac{1}{s}]{\frac{I_{CDMA}}{(I_{CDMA} + I_{UWB})}} \tag{20}$$

where

- $R_{CDMA,o}$ is the CDMA-450 macrocell initial range without the UWB interference.

The normalized range is given by:

$$\frac{R_{CDMA}}{R_{CDMA,o}} = \sqrt[\frac{1}{s}]{\frac{I_{CDMA}}{(I_{CDMA} + I_{UWB})}} \tag{21}$$

The normalized capacity of the CDMA-450 system C_n is given as:

$$C_n = \left(\frac{I_{CDMA}}{I_{CDMA} + I_{UWB}} \right) \tag{22}$$

3. Effect of UWB Interference on the DCS-1800 and GSM-900 Downlink Performance

To account for UWB, an extra source of interference is added linearly to the GSM and DCS system interference. The interference power is calculated by assuming the UWB source to be at different distances from the DCS or the GSM mobile receiver. Therefore, the interference power generated by a device UWB, I_{UWB}, is given by (in dBm):

$$I_{UWB} = P_{UWB} - L_{UWB}(d) + G_{Ant} \tag{23}$$

where

- P_{UWB} is the UWB EIRP in dBm in the GSM or the DCS band.
- $L_{UWB}(d)$ is the propagation loss between the UWB device and the GSM or the DCS receiver which varies with the separation distance, d in m, and
- G_{Ant} is the GSM or the DCS antenna gain .

Given that UWB devices are typically low power, short range devices, then the line-of-sight path-loss with log-normal deviation propagation model is the most appropriate. From [7], the UWB signal propagation loss in dB at a distance d can be given as:

$$L_{UWB}(d) \approx \begin{cases} 20\log_{10}\left(\dfrac{4\pi d}{\lambda}\right) & d \le 1\,m \\ 20\log_{10}\left(\dfrac{4\pi d}{\lambda}\right) + N(0,\sigma) & 1\,m < d \le 10\,m \end{cases} \tag{24}$$

where λ is the wavelength and N is normal distribution random variable with zero mean and a standard deviation of σ. The standard deviation σ has a value of 1.25 to 2.00 dB.

The UWB signal propagation loss in dB at the DCS-1800 band is calculated as:

$$L_{UWB}(d) \approx \begin{cases} 37.7 + 20\log_{10}(d) & d \le 1m \\ 37.7 + 20\log_{10}(d) + N(0,\sigma) & 1m < d \le 10m \end{cases} \tag{25}$$

The DCS normalized macrocell range $R_{n,DCS}$ with the existence of the UWB interference is given as:

$$R_{n,DCS} = \frac{R_{DCS}}{R_{DCS,o}} = \sqrt[1/s]{\frac{I_{DCS}}{(I_{DCS} + I_{UWB})}} \tag{26}$$

where

- s is the DCS-1800 signal propagation exponent for the outdoor environment,
- $R_{DCS,o}$ is the DCS-1800 initial range,
- R_{DCS} is the DCS new range when the UWB affects the DCS system,
- I_{DCS} is the DCS receiver noise without the effect of the UWB system,
- I_{UWB} is the UWB extra interference.

I_{DCS} is given as:

$$I_{DCS}(dB) = -114 + 10\log_{10}(BW_{MHZ}) + NF(dB) \tag{27}$$

where

- BW_{MHZ} is the DCS channel bandwidth = 0.2 MHz,
- NF(dB) is the DCS noise figure in (dB).

The UWB signal propagation loss in dB at the GSM-900 band is calculated as:

$$L_{UWB}(d) \approx \begin{cases} 32.0 + 20\log_{10}(d) & d \le 1m \\ 32.0 + 20\log_{10}(d) + N(0,\sigma) & 1m < d \le 10m \end{cases} \tag{28}$$

The GSM normalized macrocell range $R_{n,GSM}$ with the existence of the UWB interference is given as:

$$R_{n,GSM} = \frac{R_{GSM}}{R_{GSM,o}} = \sqrt[\frac{1}{s}]{\frac{I_{GSM}}{(I_{GSM} + I_{UWB})}} \tag{29}$$

where

- $R_{GSM,o}$ is the GSM-900 urban macrocell initial range without the UWB interference,
- R_{GSM} is the GSM-900 urban macrocell range affected by the UWB interference.

From (26) and (29), it can be noticed that the effect of the UWB interference is to reduce the DCS-1800 and the GSM-900 macrocell range.

4. Numerical Results

In the analysis we assume that the UWB data rate is higher than the UMTS and the CDMA-450 chip rate (data rate higher than 3.84 Mbps). Here we address the effect of the UWB system on the downlink of the UMTS and CDMA-450 systems.

In Fig. 2, the UWB interference power on the UMTS downlink (i.e. interference as seen at the mobile) is plotted assuming an average P_{UWB} of -60 dBm/MHz within the UMTS bandwidth. It can be noticed that the UWB interference is high when the distance between the UWB transmitter and the UMTS mobile receiver is lower than 1m.

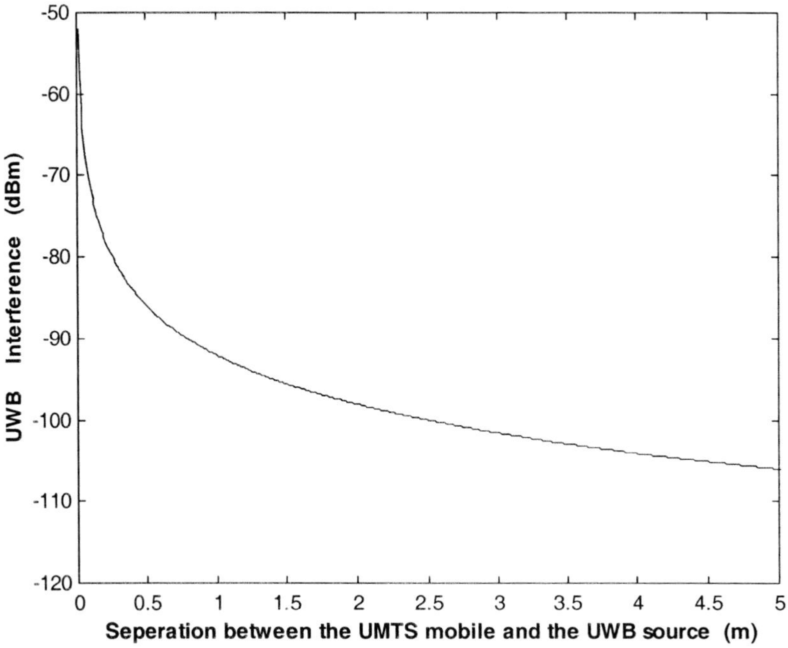

Figure 2. UWB interference as a function of the separation between the UWB transmitter and the UMTS mobile (P_{UWB} = -60 dBm/MHz).

We study the case of voice service (G_p = 25 dB and $(E_b/N_o)_{req}$ = 6 dB) assuming an UMTS interference of -88 dBm (14 dB noise rise). Fig. 3 shows the downlink macrocell normalized range as a function of the separation between the UMTS mobile and the UWB transmitter for three different values of the propagation exponent s. It can be noticed that the UWB signal creates a high interference which reflects a macrocell normalized range reduction of 9.5% when the separation is 1m. For higher separation, the interference is lower and thus the range reduction is also lower.

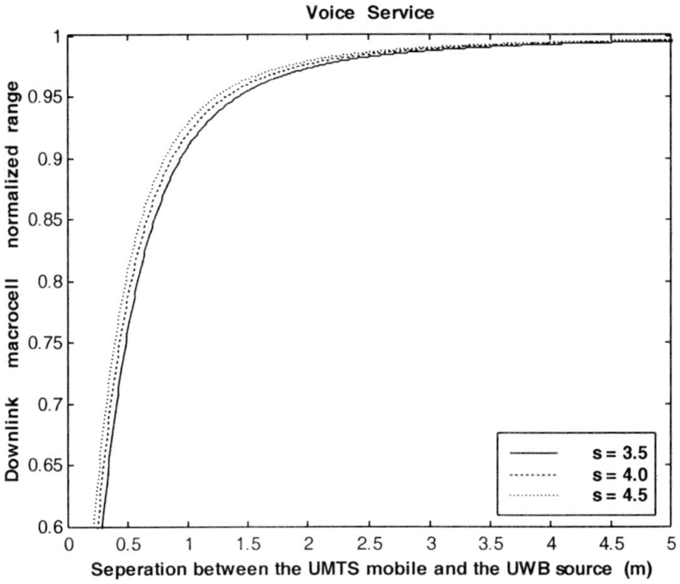

(P_{UWB} = -60 dBm/MHz).

Figure 3. Effect of the UWB interference on the macrocell range as a function of the separation between the UWB transmitter and the UMTS mobile.

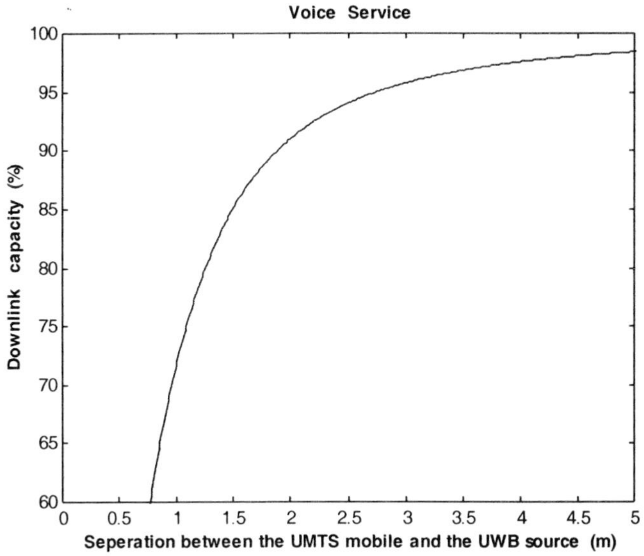

(P_{UWB} = -60 dBm/MHz).

Figure 4. Effect of the UWB interference on the macrocell normalized capacity as a function of the separation between the UWB transmitter and the UMTS mobile.

Fig. 4 portrays the downlink macrocell normalized capacity as a function of the separation between the UMTS mobile and the UWB transmitter. It can be noticed that the UWB signal creates a high interference which provokes an unexeptable macrocell capacity reduction of 29% when the separation is 1m.

Next we study the case data service [G_p = 14.25 dB and $(E_b/N_o)_{req}$ = 4.25 dB] assuming an UMTS total interference of -92.5 dBm (9.5 dB noise rise and thus highly loaded macrocell). Fig. 5 shows the downlink macrocell normalized range as a function of the separation between the UMTS mobile and the UWB transmitter for three different values of the propagation exponent s. It can be noticed that the UWB signal creates a high interference which reflects a high macrocell range reduction of 20% when the separation is 1m.

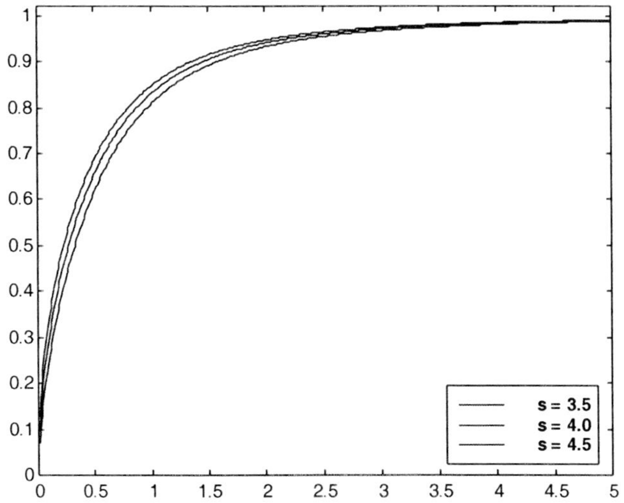

(P_{UWB} = -60 dBm/MHz).

Figure 5. Effect of the UWB interference on the macrocell normalized range as a function of the separation between the UWB transmitter and the UMTS mobile.

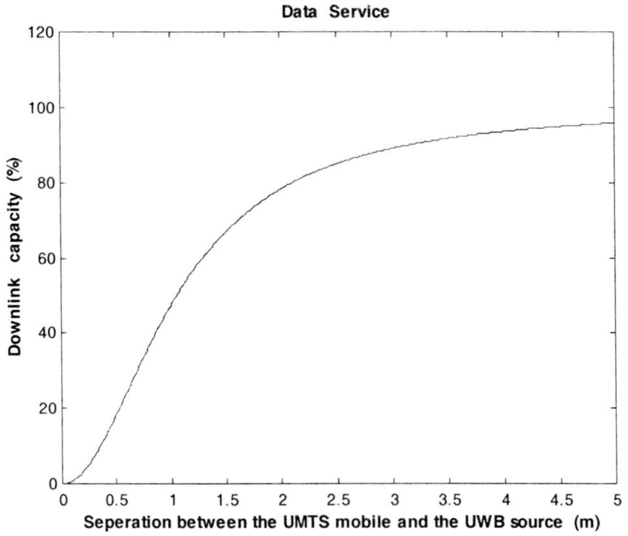

(P_{UWB} = -60 dBm/MHz).

Figure 6. Effect of the UWB interference on the macrocell normalized capacity as a function of the separation between the UWB transmitter and the UMTS mobile.

Fig. 6 depicts the downlink macrocell capacity as a function of the separation between the UMTS mobile and the UWB transmitter. It can be noticed that the UWB signal creates a high interference which gives arise to a high macrocell capacity reduction of 53% when the separation is 1m.

Let us now present the case of the data service case assuming P_{UWB} of -81 dBm/MHz. Fig. 7 portrays the downlink macrocell range as a function of the separation between the UMTS mobile and the UWB transmitter. It can be noticed that the UWB signal creates a high interference which provokes a high macrocell range reduction when the separation is less than 0.25m. For larger separation, the interference is lower. At a distance higher than 1m, the effect of the interference is quasi null.

Fig. 8 shows the downlink macrocell capacity as a function of the separation between the UMTS mobile and the UWB transmitter. Here it can be noticed that the UWB signal creates a low interference which reflects a low macrocell capacity reduction when the separation is equal to or higher than 1m.

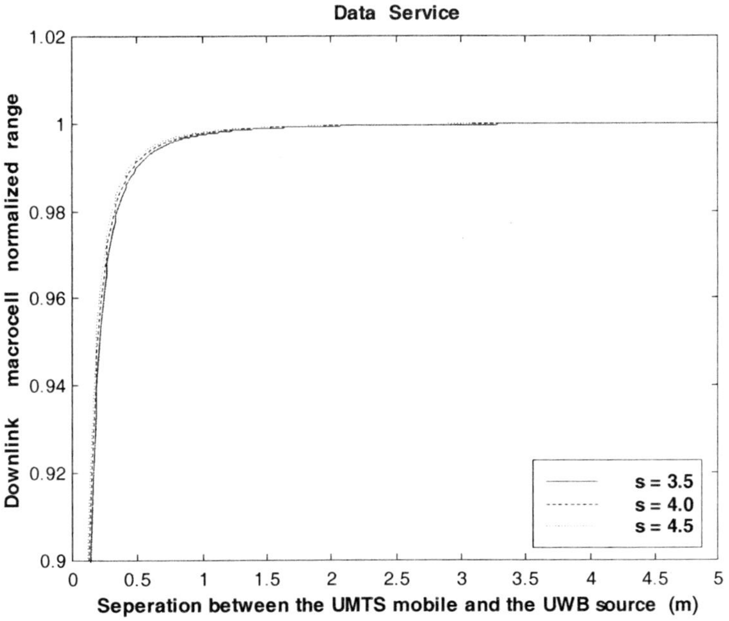

(P_{UWB} = -81 dBm/MHz).

Figure 7. Effect of the UWB interference on the macrocell range as a function of the separation between the UWB transmitter and the UMTS mobile.

Next we study the case data service assuming an UMTS total interference of -94.5 dBm (7.5 dB noise rise and thus medium loaded macrocell) and UWB power density of -83 dBm/MHz. Fig. 9 depicts the downlink macrocell range as a function of the separation between the UMTS mobile and the UWB transmitter for three different values of s. It can be noticed that at a distance equal to or higher than 1.0m, the effect of the interference is quasi null (less than 0.3% range reduction).

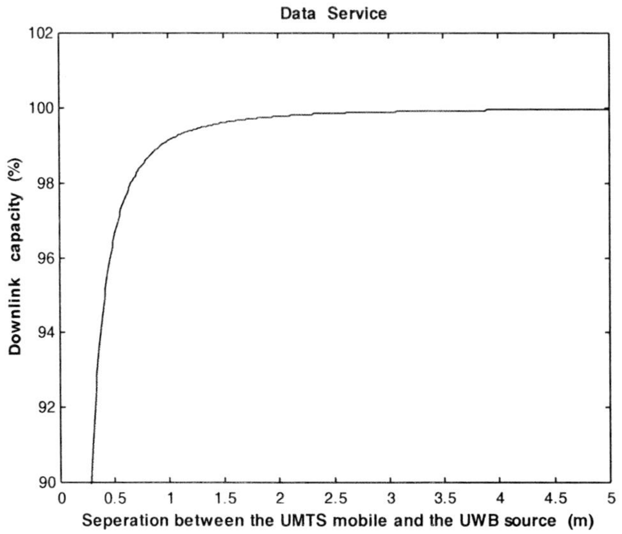

(P_{UWB} = -81 dBm/MHz).

Figure 8. Effect of the UWB interference on the macrocell normalized capacity as a function of the separation between the UWB transmitter and the UMTS mobile.

Fig. 10 shows the downlink macrocell capacity as a function of the separation between the UMTS mobile and the UWB transmitter. It can be noticed that the macrocell normalized capacity reduction is high when the separation is less than 0.3m. For larger separation, the interference is lower. At a distance higher than 1m, the effect of the interference is very little (less than 1%).

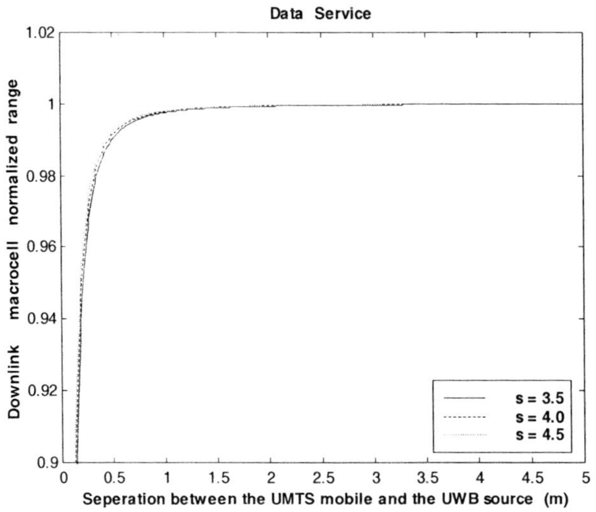

(P_{UWB} = -83 dBm/MHz) and 7.5 dB noise rise.

Figure 9. Effect of the UWB interference on the macrocell range as a function of the separation between the UWB transmitter and the UMTS mobile.

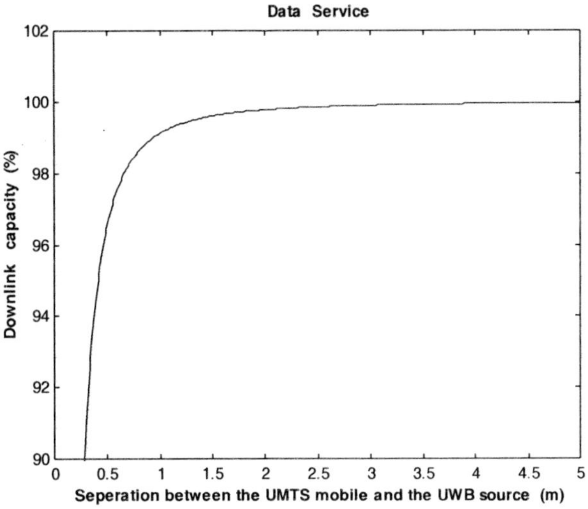

(P_{UWB} = -83 dBm/MHz) and 7.5 dB noise rise.

Figure 10. Effect of the UWB interference on the macrocell normalized capacity as a function of the separation between the UWB transmitter and the UMTS mobile.

Then we study the case of multiple UWB transmitters with one UWB transmitter at each $4 \times 4\,m^2$ area of the indoor environment assuming P_{UWB} of -83 dBm/MHz, 18 UWB transmitters and noise rise of 9.5 dB. Fig. 11 represents the downlink macrocell normalized range as a function of the UMTS mobile distance from the nearest UWB transmitter for three different values of s. It can be noticed that the macrocell normalized range reduction is high when the UMTS receiver is located at 0 to 0.15m. At a distance higher than 0.4 m from the nearest UWB transmitter, the macrocell range reduction is less than 1%.

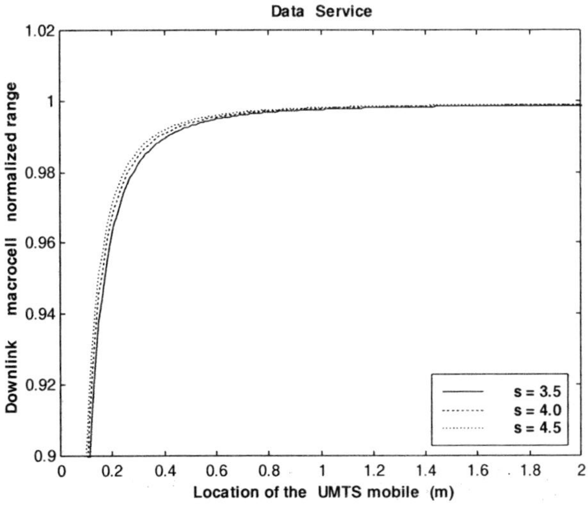

Figure 11. Effect of the UWB interference on the macrocell range as a function of the UMTS mobile location (P_{UWB} = -83 dBm/MHz) for multi UWB transmitters and 9.5 dB noise rise.

Fig. 12 shows the downlink macrocell capacity as a function of the UMTS mobile location. It can be noticed that the macrocell normalized capacity reduction is high when the UMTS receiver is located at a distance less than 0.2m from the nearest UWB transmitter. For a distance greater than 1m from the nearest UWB transmitter, the effect of the UWB transmitters is quasi null (less than 1% capacity reduction).

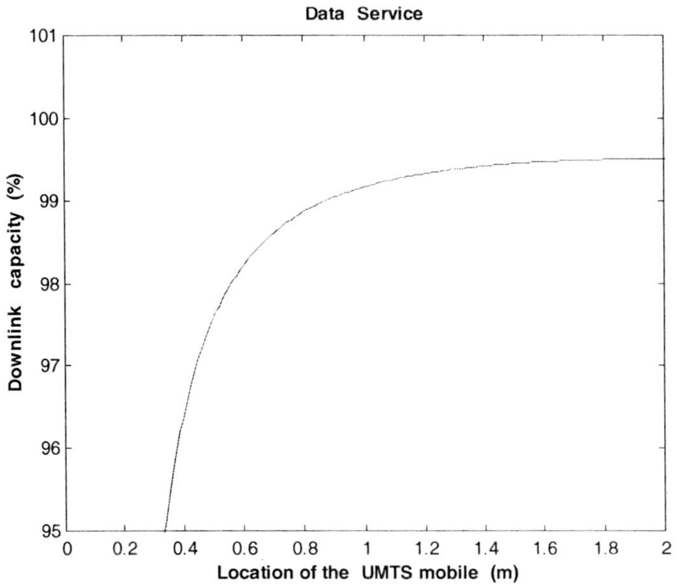

Figure 12. Effect of the UWB interference on the macrocell normalized capacity as a function of the UMTS mobile location (P_{UWB} = -83 dBm/MHz) for multi UWB transmitters and 9.5 dB noise rise.

It can be concluded that, for the case of single UWB transmitter, the UMTS can easily tolerate the UWB interference when the UWB EIRP is -83 dBm/MHz for a distance between the UWB transmitter and the UMTS mobile of 1 m or higher. For the case of multi UWB transmitter, the UMTS can easily tolerate the UWB interference when the UWB EIRP is -85 dBm/MHz. The above mentioned numbers are valid for medium loaded macrocells, i.e., (60-70)% loaded macrocell.

Next we study the case of data service (G_p = 14.2 dB and $(E_b/N_o)_{req}$ = 4.25 dB) of the CDMA-450 3X assuming that the CDMA-450 total interference of -92.5 dBm (9.5 dB noise rise) and UWB power density of -95 dBm/MHz. Fig. 13 shows the CDMA-450 downlink macrocell normalized range as a function of the separation between the CDMA mobile and the UWB transmitter. It can be noticed that the UWB signal creates a low interference when the separation is more than 1m which reflects a normalized range reduction of less than 0.3%.

Fig. 14 shows the CDMA-450 downlink macrocell normalized capacity as a function of the separation between the CDMA-450 mobile and the UWB transmitter. It can be noticed that the effect of the UWB interference is quasi null for a distance greater than 1 m (less than 1% capacity reduction).

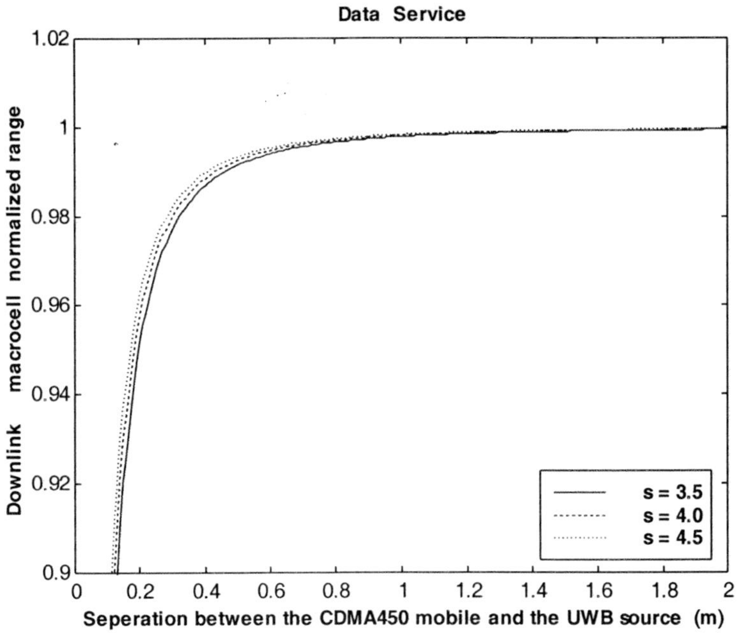

(P_{UWB} = -95 dBm/MHz).

Figure 13. Effect of the UWB interference on the macrocell normalized range as a function of the separation between the UWB transmitter and the CDMA450 mobile.

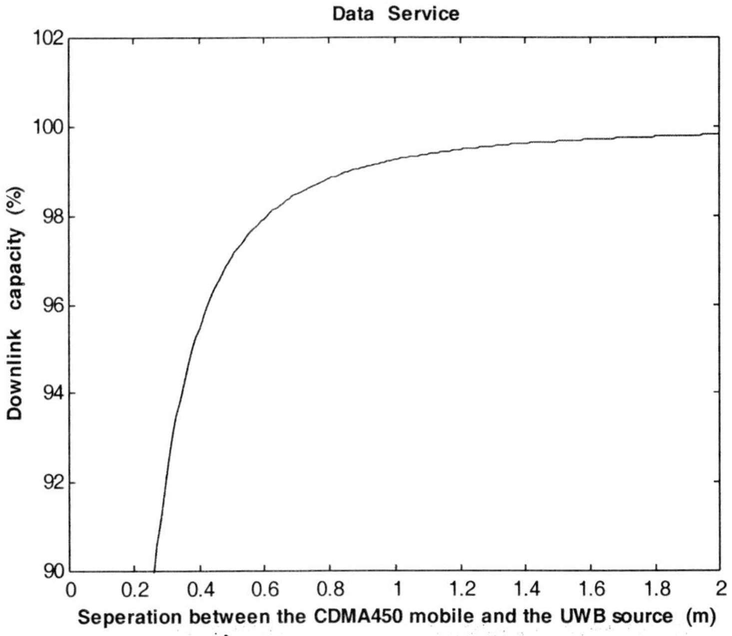

(P_{UWB} = -95 dBm/MHz).

Figure 14: Effect of the UWB interference on the macrocell normalized capacity as a function of the separation between the UWB transmitter and the CDMA450 mobile.

Here we can conclude that the UWB power density that can be tolerated by the CDMA-450 downlink is of the order of -95 dBm/MHz for single UWB. For the case of multi UWB transmitters, the power density that can be tolerated by the downlink of the CDMA-450 system is of the order -98 dBm/MHz. These numbers are valid for high loaded macrocells. For medium loaded macrocells, the accepted UWB power density is of the order -97 dBm/MHz and -100 dBm/MHz for the single UWB transmitter and multi UWB transmitters case respectively.

If we reduce the critical distance to 0.5m, we have to lower the maximum accepted UWB power density by 6 dB. The effect of the UMTS and CDMA-450 signal propagation exponent (s) is very little when its value is changed from 3.5 to 4.5.

Next we presents the effect of the UWB system on the DCS or GSM systems assuming that the DCS or GSM mobile receiver is in an office of 20×18 m and that the propagation exponent s of the DCS and GSM macrocell is 3.5. The three different scenarios considered are:

- The best case for which the propagation loss is 6 dB higher the average case.
- The average case.
- The worst case for which the propagation loss is 6 dB lower than the average case.

Let us present the case of DCS-1800 service assuming that the receiver noise figure is 8 dB, the UWB transmitting antenna gain is 0 dB and that the DCS receiving antenna gain is 0 dB.

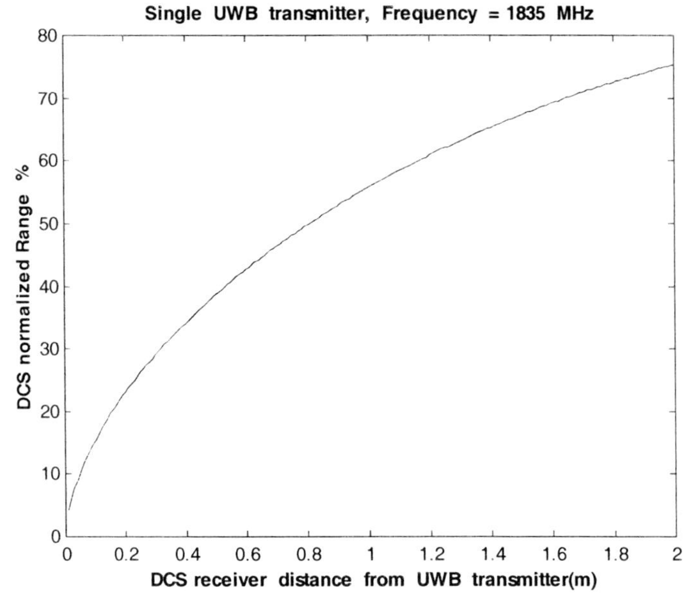

Figure 15. DCS-1800 normalized macrocell range as a function of distances between the UWB transmitter and the DCS-1800 mobile receiver for (P_{UWB} = -60 dBm/MHz).

Fig. 15 shows the DCS normalized downlink range R_n as a function of the distances between the UWB transmitter and the DCS-1800 mobile when the UWB power density is -

60 dBm/MHz assuming the average case. It can be noticed that R_n has a value of 55% which give a rise to an acceptable macrocell range reduction. Thus, the UWB power density recommended by the FCC organization (-51.3 dBm/MHz) is very high to be tolerated by the DCS-1800 system. For this reason, lower UWB power density should be recommended.

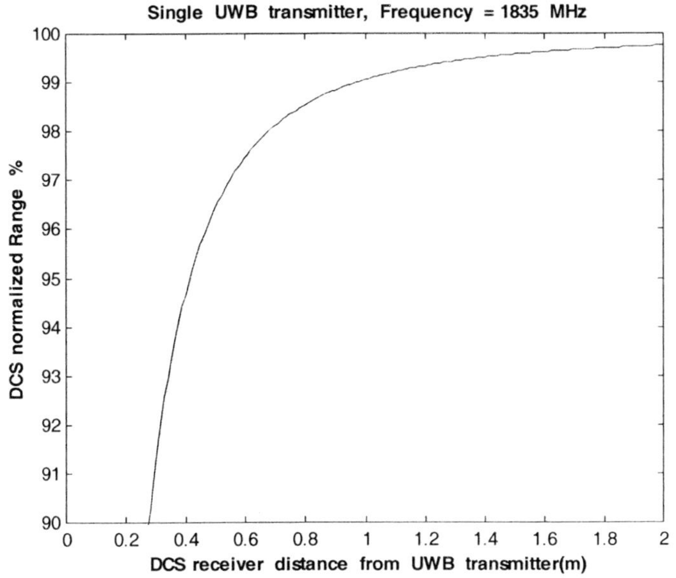

Figure 16. DCS-1800 normalized macrocell range as a function of distances between the UWB transmitter and the DCS-1800 mobile receiver for (P_{UWB} = -83 dBm/MHz).

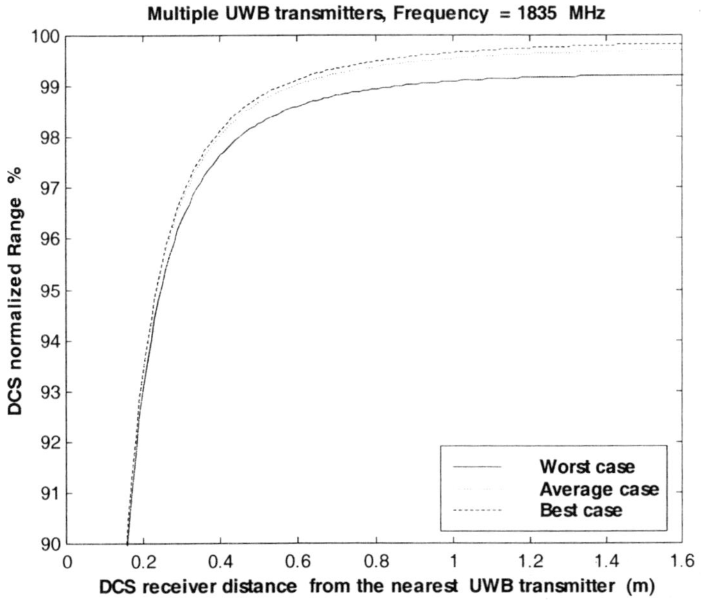

(P_{UWB} = -88 dBm/MHz and 16 UWB transmitters).

Figure 17. DCS-1800 normalized macrocell range as a function of distances between the DCS-1800 mobile receiver and the nearest UWB transmitter for.

Fig. 16 shows the DCS normalized downlink range R_n as a function of the distances between the UWB transmitter and the DCS-1800 mobile when the UWB power density is -83 dBm/MHz assuming the average case. It can be noticed that R_n has a value of 99% which give a rise to acceptable macrocell range reduction (1%).

Let us present the case of multiple UWB transmitters with one UWB transmitter at each $4 \times 4\ m^2$ area of the indoor environment assuming 16 UWB transmitters.

Fig. 17 depicts the DCS normalized downlink range R_n as a function of the distances between the DCS-1800 mobile and the nearest UWB transmitter when the UWB power density is -88.0 dBm/MHz. It can be noticed that for the worst case, R_n has a value of 99% which give a rise to an acceptable macrocell range reduction (1%).

Next we present the case of GSM-900 service assuming that the receiver noise figure is 8 dB, the UWB transmitting antenna gain is 0 dB and that the GSM receiving antenna gain is 0 dB.

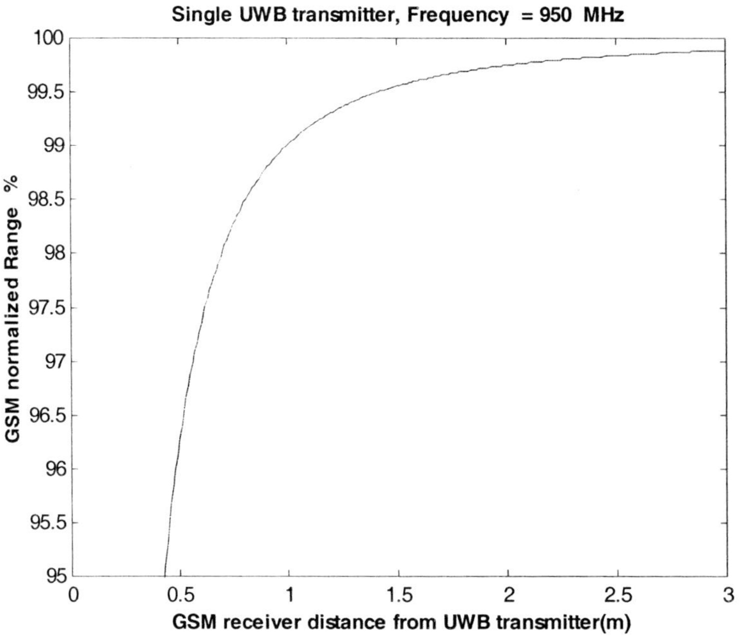

Figure 18. GSM-900 normalized macrocell range as a function of distances between the UWB transmitter and the GSM-900 mobile receiver for (P_{UWB} = -88.5 dBm/MHz).

Fig. 18 presents the GSM normalized downlink range R_n as a function of the distances between the UWB transmitter and the GSM-900 mobile when the UWB power density is -88.5 dBm/MHz assuming the average case. It can be noticed that R_n has a value of 99% which give a rise to an acceptable macrocell range reduction (1%).

Let us present the case of UWB multiple transmitter affecting the GSM-900 system. Fig. 19 shows the GSM normalized downlink range R_n as a function of the distances between the GSM-900 mobile and the nearest UWB transmitter when the UWB power density is -93.5 dBm/MHz. It can be noticed that for the worst case, R_n has a value of 99% which give a rise to an acceptable macrocell range reduction (1%).

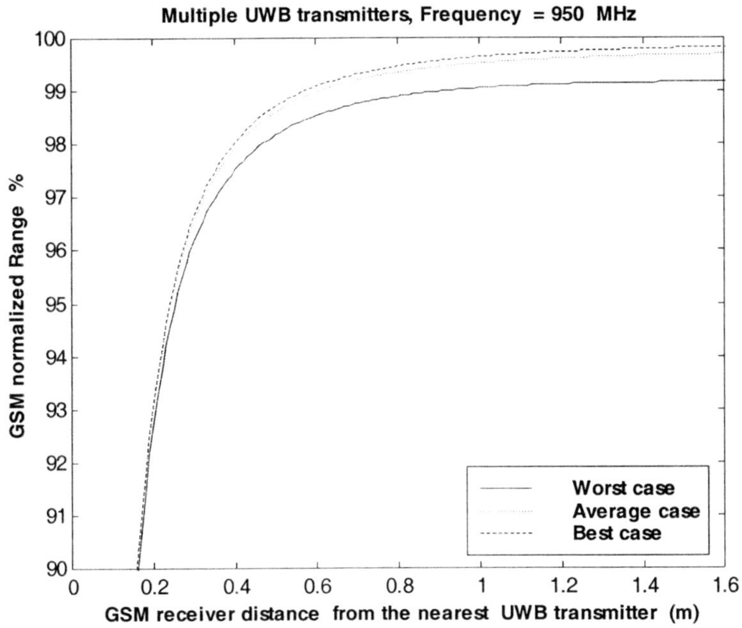

(P_{UWB} = -93.5 dBm/MHz and 16 UWB transmitters).

Figure 19. GSM-900 normalized macrocell range as a function of distances between the GSM-900 mobile receiver and the nearest UWB transmitter for.

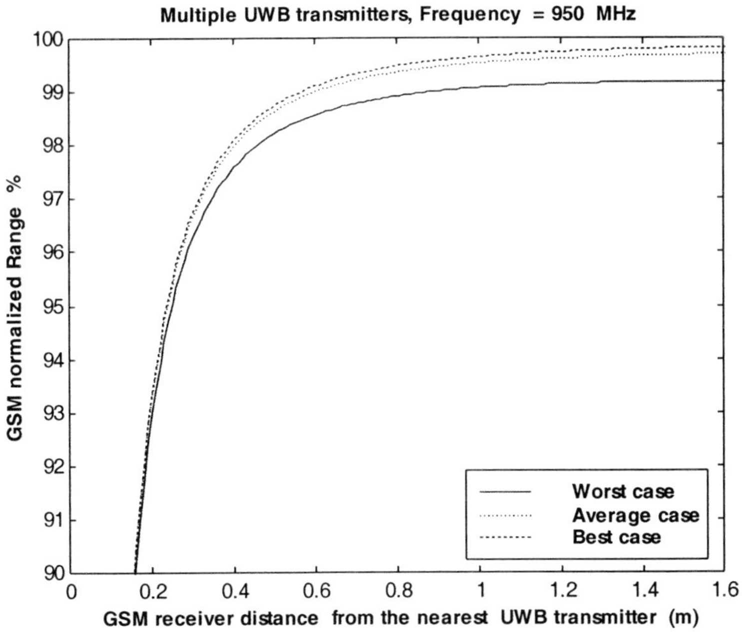

(P_{UWB} = -93 dBm/MHz, 16 UWB transmitters and $\alpha = 4.0$).

Figure 20. GSM-900 normalized macrocell range as a function of distances between the GSM-900 mobile receiver and the nearest UWB transmitter for.

Let us study the case of UWB multiple transmitter affecting the GSM-900 system assuming that the outdoor propagation exponent (s) for the GSM macrocell is 4. Fig. 20 shows the GSM normalized downlink range R_n as a function of the distances between the GSM-900 mobile and the nearest UWB transmitter when the UWB power density is -93.0 dBm/MHz. It can be noticed that for the worst case, R_n has a value of 99% which give a rise to an acceptable macrocell range reduction (1%). Thus, form the results of Figs. 19 and 20, it can be noticed that the effect of the outdoor propagation exponent is very small (The recommended UWB power density when s = 4 is only 0.5 dB higher than the recommended UWB power density when s = 3.5).

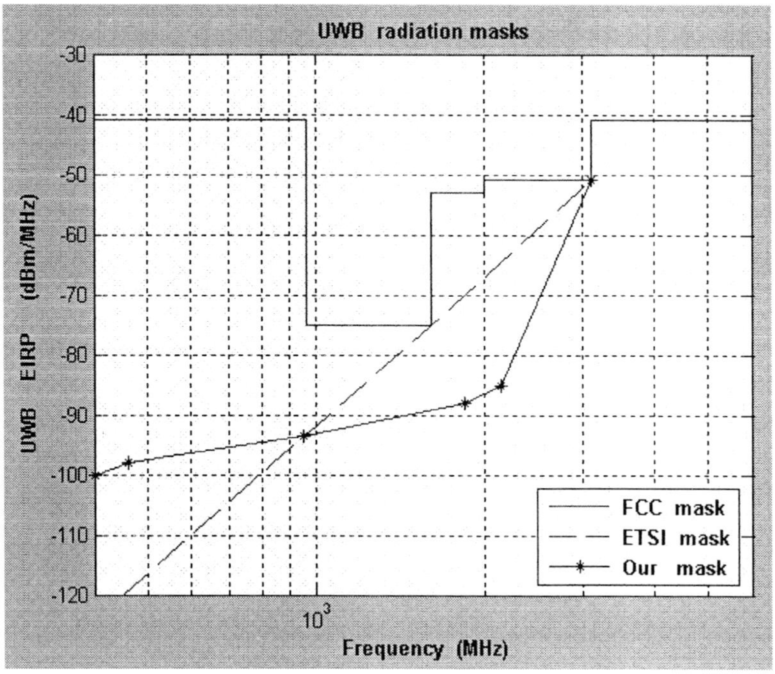

Figure 21. FCC mask, ETSI mask and our recommended UWB radiation mask.

From the above given results we can conclude that the spectrum mask proposed by the FCC for indoor application (-51 dBm/MHz in the UMTS band, -53 dBm/MHz in the DCS-1800 band, -41 dBm/MHz in the GSM-900 band and -41 dBm/MHz for the CDMA450 band) is very high to be tolerated by the four mobile systems. Thus, a new spectrum mask with lower UWB power density should be proposed.

From the results presented by this work, our recommended UWB radiation mask is shown in Fig. 21. Fig. 21, also presentss the FCC mask and the European ETSI mask. For frequencies greater than or equal to 3.1 GHz, the three masks have the same values of UWB accepted power density. For frequencies less than 3.1 GHz, our recommended mask has always lower accepted UWB power density than the UWB power density given by the FCC recommendations. For frequencies lower than 3.1 GHz and greater than 0.95 GHz our recommended mask has always lower accepted UWB power density than the UWB power density given by the ETSI mask. For frequencies lower than 0.95 GHz, our recommended mask has always higher accepted UWB power density than the UWB power density given by the ETSI mask.

References

[1] M. Hamalainen, V. Hovinrn, R. Tesi, J. Iinatti, and M. Latava-aho, " On the UWB System Coexistance with GSM900, UMTS/WCDMA, and GPS", *IEEE Journal on Selected Areas in Communications*, Vol. 20, No. 9, pp. 1712-1721, Dec. 2002.

[2] M. Hamalinen, R. Tesi., J. Iinatti, " UWB co-existence with IEEE802.11a and UMTS in modified Saleh-Valenzuela channel", Ultra Wideband Systems, 2004. Joint with Conference on Ultrawideband Systems and Technologies. 2004 International Workshop on Joint UWBST & IWUWBS., pp:45 – 49, May 18-21, 2004.

[3] R. Giuliano, F. Mazzenga, ,F. Vatalaro, "On the interference between UMTS and UWB systems", pp: 339 – 343, Ultra Wideband Systems and Technologies, 2003 IEEE Conference on , 16-19 Nov. 2003.

[4] M. Hamalinen, V. Hovinen, J. Iinatti, M. Latva-aho: "In-band Interference Power Caused by Different Kinds of UWB Signals at UMTS/WCDMA Frequency Bands", , the 2001 IEEE Radio and Wireless Conference, RAWCON 2001, , pp. 97-100, Waltham-Boston, Massachusetts, USA, Aug. , 2001.

[5] M. Hamalinen, J. Iinatti, V. Hovinen, M. Latva-Aho:" In-band Interference of Three Kind of UWB Signals in GPS L1 Band and GSM900 Uplink Band", the 12th International Symposium on Personal, *Indoor and Mobile Radio Communications*, PIMRC2001, pp. D 76-80, USA, Sep - Oct , 2001.

[6] B. Taha Ahmed, M. Calvo Ramón, L. Haro Ariet " Impact of Ultra Wide Band (UWB) on Macrocell Downlink of DCS-1800 and GSM-900 Systems", *Radioengineering* , Vol. 14, No. 1, pp. 51-55, April 2005.

[7] W. Ciccoganini, A. Durantini, and D. Cassioli, "Time domain propagation measurements of the UWB Indoor Channel Using PN-Sequence in the FCC-Compliant Band 3.6-6 GHz", *IEEE trans. Antennas and Propagation*, Vol. 53, No. 4, pp. 1542-1549, April 2005.

In: Wireless Communications Research Trends
Editor: Tong S. Lee, pp. 325-347
ISBN: 978-1-60021-674-9

Chapter 11

DEPENDABLE PUBLIC WIRELESS LANS WITHOUT HARDWARE SUPPORT

Jenn-Wei Lin[*] and Ming-Feng Yang
Department of Computer Science & Information Engineering,
Fu Jen Catholic University, Taipei, Taiwan

Abstract

This paper presents an efficient fault-tolerant approach for public wireless local access networks (public WLANs). In a public WLAN, multiple access points (APs) are first deployed in the public area to provide wireless communication. For a user in the public WLAN, it must associate with an AP to acquire a wireless communication path before performing data services. If a failure occurs in an AP of the public WLAN, the users under the coverage range of the faulty AP (the affected users) cannot perform data services again. To tolerate the AP failure, previous approaches are based on the hardware redundancy or network planning technique. In this paper, we proposed a new fault-tolerant approach which directs each affected user how to move itself to the coverage range of another AP. For quickly reconnecting the wireless communication, the moving distance is considered in the proposed approach. In addition, the proposed approach also considers the overloading problem to avoid causing significant performance degradation on an AP. Finally, extensive simulations are performed to evaluate the performance overhead of the proposed approach.

Keywords: Fault-tolerant, public wireless local access networks, access point, overloading, simulation.

1. Introduction

With the advent of the IEEE 802.11 standard, wireless LANs (WLANs) are a competitive technology for mobile computing. Currently, WLANs have been extensively deployed in public areas (e.g. airports, hotels, universities, shopping centers, etc). To set up a public WLAN, multiple access points (APs) are first deployed to provide wireless communication in

[*] E-mail address: jwlin@csie.fju.edu.tw

the public area. Before performing data services, each user in a public WLAN must associate with an AP to acquire a wireless communication path. If a failure occurs in an AP, the users under the coverage range of the faulty AP will lose their wireless communication paths. Next, such affected users cannot perform any data services.

The fault-tolerant issue of the AP has been studied in literature [1-2]. The proposed previous approaches can be categorized into the following three techniques.

- Access-point replication: In this technique, each working AP is statically equipped with one redundant AP. If a working AP fails, its corresponding redundant AP is activated as a serving AP to replace the faulty AP.
- Overlapping coverage: The main idea of this technique is to make each place of a public WLAN be covered by at least two APs. If an AP fails, the users under its coverage range are switched to be served by another AP which coverage range is intersected with the faulty AP. In this technique, the fault tolerance is achieved based on the assumption of the overlapping coverage. If an MS is located within a non-overlapping range, its wireless communication cannot be supported if its serving AP fails.
- Link multiplexing: This technique equips multiple wireless network cards for each user. Due to multiple wireless network cards, an active user can have multiple wireless communication paths to distinct APs. If one of the user's serving APs fails, the user still has wireless communication paths to other APs. The data services of the user can be continuously performed.

In this paper, we propose a new approach to tolerating the AP failure in a public wireless LAN. Unlike previous approaches, the proposed approach is not based on the hardware support since it does not need to equip redundant APs or multiple network cards. In addition, the proposed approach is also not required to make each place of a public WLAN be covered by two or more APs. In the proposed approach, if an AP fails, it selects some currently working APs to constitute a backup AP set. Then, the failure-affected users (the users under the coverage range of the faulty AP) are directed to shortly move to coverage ranges of the backup AP set. To avoid severely affecting the performance of an AP, if the loading of a working AP is already large, it will be not selected in the backup AP set. The moving distance and overloading are simultaneously considered in the proposed approach to reduce the fault-tolerant overhead. We also perform extensive simulation experiments to quantify the overhead of the proposed approach.

The rest of this paper is organized as follows. Section 2 gives background knowledge of this paper. Section 3 describes our fault-tolerant approach. Section 4 evaluates the overhead of the proposed approach. Section 5 compares the proposed approach with previous approaches. Simulation results are given in Section 6. Finally, concluding remarks are made in Section 7.

2. Background

This section describes the background knowledge of this paper. First, we give the network model of a public WLAN. Then, we describe how to detect the AP failure. Last, we review related work.

2.1. Network Model

The network model referred to this paper is shown in Fig. 1, which consists of four components: mobile station (MS), access point (AP), distribution system (DS), and simple network management protocol (SNMP) server. The MS is a computing device with wireless network interface, which also represents a user. The AP has a fixed coverage range, i.e., a limited range of operation. An AP with the MSs within its coverage range form a basic service set (BSS). For the MSs within an AP's coverage range, they contend the radio channel of the AP. The IEEE 802.11 standard proposed two radio channel accessing mechanisms: the distributed coordination function (DCF) and point coordination function (PCF) [3].

In a public WLAN, multiple APs are usually deployed for covering all the serving ranges of the public area. A wired network (called a DS in the IEEE 802.11 standard) is required to connect the multiple APs. The IEEE 802.11 standard does not specify any particular technology for the DS. The DS is usually implemented by an Ethernet network. With the DS, the mentioned BSSs are connected to form an extended service set (ESS). In the same ESS, any two MSs within different BSSs can communicate with each other. For further managing all APs of a public WLAN, a SNMP server is also equipped in the network model (Note that SNMP is a standard protocol for managing a wired or wireless network [4]). Most APs also support SNMP [5, 6]. In this paper, the SNMP server is used to monitor the loading status of each AP [7-9], which periodically sends a loading-inquiry message to each AP.

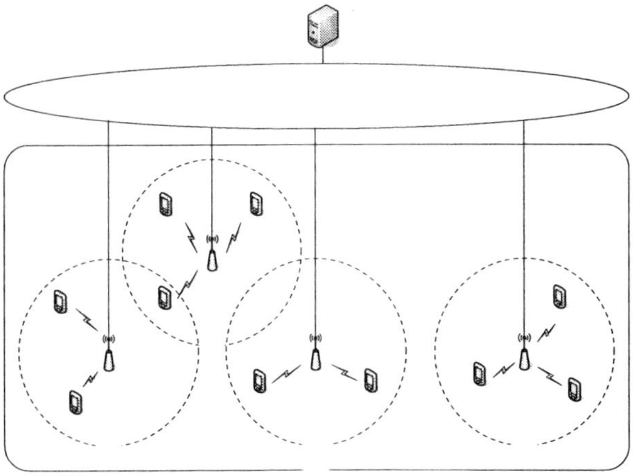

Figure 1. Network model.

2.2. Failure Detection

This paper mainly concerns the fault-tolerant issue of the AP; therefore, failures are assumed to occur in APs only. The SNMP server and the distribution system are assumed to be failure-free; their fault-tolerant issues are not discussed in this paper. Traditionally, the AP failure is detected by monitoring the periodical beacon frame [2]. In [2], it indicated that this detection method is difficult to distinguish between the MS mobility and the AP failure. Therefore, in [2], it proposed an efficient detection method by using the signal strength as the indicator of the AP failure. An AP which is correctly functioning will present a strong signal. The MSs within its coverage range will also receive a strong signal. Conversely, if an AP fails, it cannot present a strong signal to the MSs within its coverage range. Therefore, if a MS detects that there is a sudden drop in the signal from its serving AP or it does not receive a signal from its serving AP for a period of time, the corresponding serving AP is regarded in the faulty state.

2.3. Related Work

As mentioned in section 1, the previous approaches are classified into three basic techniques: access-point replication, overlapping coverage, and link multiplexing.

The approach of [1] belongs to the access-point replication technique. In [1], it concerned the "shadow region" issue. Due to physical obstacles and radio interference, there may be a shadow region in the AP's coverage range. When an MS moves into a shadow region, it will lose the wireless connectivity. To avoid this situation, a redundant AP is placed in the shadow region to serve the MSs moving into this region. Here, the concept of the redundant AP can be also applied on the AP fault-tolerant problem. Each working AP is statically equipped with one redundant AP. If an AP fails, its equipped redundant AP is activated as the backup. However, the redundant AP needs to actively detect the failure of the corresponding AP. In addition, the failure is not allowed to occur in the redundant AP. If the redundant AP also fails, the fault tolerance cannot be achieved.

The approaches of [10, 11-13] belong to the overlapping coverage technique. In the IEEE 802.11 standard, only three radio channels can be used in the ESS of a public WLAN. Due to few useful radio channels and radio interference, it cannot ensure that overlapping coverage can be planned in each place of the public WLAN [1, 2]. In addition, the use of the overlapping coverage technique may introduce the overloading problem [2]. When an AP fails, if all the overlapping APs are in the overloading state, a suitable AP is difficult to be selected for taking over the workload of the faulty AP. Here, the overlapping APs of AP x indicate the APs which radio coverage ranges are overlapped with AP x. Like the access-point replication technique, failures cannot occur in the overlapping APs. For the approaches proposed in [10-13], they mainly handle the load-balance issue of a WLAN, but the concepts of these approaches can be also used to tolerate the AP failure. In [10], it discussed the issue of user congestion in a public WLAN. It proposed two approaches (explicit channel switching and network-directed roaming) to relieve the hop-spot congestion. In the explicit channel switching approach, if two or more APs admit the connection request of a MS, the MS will associate with the AP with the light load, not the AP with a stronger signal. The first approach is relied on the existence of more than one AP being able to signal to the MS. This also means that a MS must locate within the overlapping coverage range. This assumption is not always true. To avoid making this assumption, the network-directed approach directs a MS to move to a distant AP with the light load, which is based on the location of the MS and the capacities of all APs. The concepts of the above two mentioned approaches can be applied to handle the AP failure. Especially, the first approach is very similar to the overlapping coverage technique. However, the second approach cannot be directly applied in the fault tolerance. The reason is that if an AP fails, the MSs under its coverage range cannot acquire the network direction from the faulty AP. In [10], it does not describe how to extend the second approach to handle the fault-tolerant problem. Like the first approach of [10], the approaches of [11-13] also discuss how to distribute the load of the MSs within the overlapping coverage range. Similarly, if the approaches of [11-13] are applied to handle the AP failure, they are regarded as the overlapping coverage technique.

As for the link multiplexing technique, the approach of [2] adopts this technique. It utilizes multiple wireless network cards to provide the AP fault tolerance. In the approach of [2], each MS is equipped with more than one wireless network card. Whenever a MS is formally associated with a working AP, it may also receive signals from neighboring APs. The MS utilizes multiple wireless network cards to connect with the APs signaling to it. In such case, the MS owns multiple wireless communication paths to different APs. Therefore, if one of the MS's serving APs fails, the packets of the MS can be multiplexed over the remaining communication paths. In the approach of [2], it clearly indicated that a software module is required to be installed on each MS to multiplex packets over multiple communication paths. In addition, the approach of [2] also needs the support of overlapping coverage. It indicated that overlapping coverage is easily provided in each place of a public WLAN based on the support of the multiple wireless network cards. However, the approach of [2] does not describe how to plan the overlapping coverage everywhere. When the MS is not located within an overlapping coverage range, it only hears the signal from one AP and only has one communication path to the AP. In such case, if the AP fails, the approach of [2] cannot work correctly.

3. Proposed Approach

This section presents a new fault-tolerant approach for a public WLAN, which can allow multiple APs to fail simultaneously. Unlike the previous approaches, the proposed approach is not dependent on the hardware support and the overlapping coverage.

3.1. Basic Idea

As shown in Fig. 1, there are multiple APs in a public WLAN. If an AP fails, the MSs under its coverage range (the failure-affected MSs) will lose wireless connectivity. From the viewpoint of the public WLAN system, some working (survival) APs still exist in the system. If the failure-affected MSs can move to the coverage ranges of the survival APs, their wireless connectivity can be resumed. The moving distance of a failure-affected MS is not too far. The reason is as follows. The public WLAN is usually deployed in indoor environment (e.g. airport, coffee shop and conference center, and etc.). The responsible area of a public WLAN is not too large, and the distance between two neighboring APs is neighbor long. A failure-affected MS can find a survival AP without moving too far. Based on the above description, the main idea of the proposed approach is triggered as follows. For each failure-affected MS, a direction is given to guide it how to move itself to the coverage range of one survival AP. The basic idea can be also imagined that survival APs are utilized to form a backup AP set. To achieve the idea, the following problems are required to be solved.

- How to make each failure-affected MS select its preferable survival AP as the backup AP to serve it.
- How to avoid selecting an overloading AP as the backup AP.
- How to force each failure-affected MS to move to the coverage range of its corresponding backup AP

To quickly resume the wireless connectivity of each failure-affected MS, the goal of the first problem is to select backup APs based on the viewpoints of failure-affected MSs. For a failure-affected MS, if it is located within the overlapping coverage range of several APs, its preferable backup AP is the overlapping AP with the strongest signal, as shown in Fig. 2(a). In such case, the failure-affected MS can resume its wireless connectivity without moving. Conversely, if a failure-affected MS is not located within the overlapping coverage range, it needs to move to the coverage range of another AP. In such case, the failure-affected MS wants to move as short as possible. Under this consideration, the APs neighboring the faulty AP are the possible backup APs, as shown in Fig. 2(b). The failure-affected MS will select one neighboring AP as its backup AP. From the above description, the selection of the backup AP is mainly based on the vicinity consideration. The status of the selected backup AP is not considered. In addition, the selected backup AP may be also fail simultaneously. It is possible that a failure-affected MS selects an overloading AP or faulty AP as its backup AP. The second problem is to enhance the backup AP selection with the status consideration. As for the third problem, it is to make each failure-affected MS follow the given direction to move itself to the coverage range of its selected backup AP. If each failure-affected MS randomly moves without following the direction, it is possible that many failure-affected MSs move to a common hot-spot coverage range, as shown in Fig. 3. (Usually, the hop-spot coverage range is close to the coverage range of the faulty AP since all failure-affected MSs want to move as short as possible). Due to serving too many MSs, the AP corresponding to the hop-spot coverage range will incur significant performance degradation. The main goal of the third problem is to avoid the performance degradation.

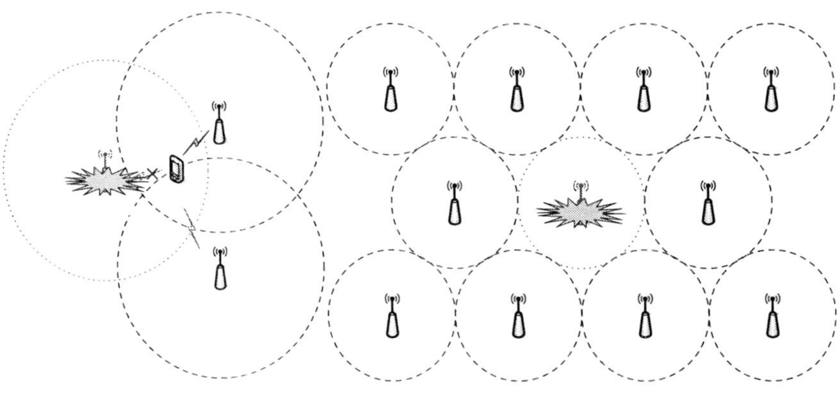

(a) (b)

Figure 2. The selection of the backup AP. (a) Overlapping coverage range. (b) Non-overlapping coverage range.

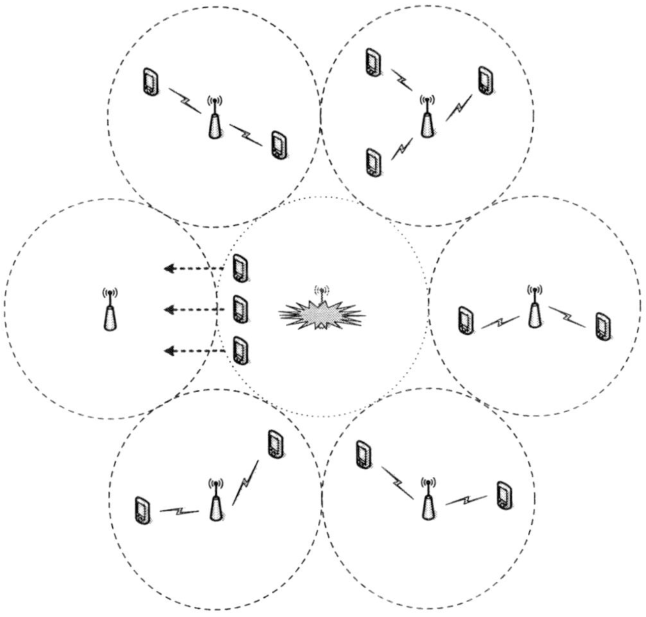

Figure 3. Movement to a hop-spot coverage range.

3.2. Problem Solution

For solving the above problems, the following data structures and procedures are required to put in the SNMP server, AP, and MS, respectively.

- Data Structures:
  SNMP Server: The AP deployment map and AP location table with loading information
  AP: The overloading record
  MS: The AP deployment map and AP location table

- Procedures:
  SNMP Server: The loading inquiry routine and backup AP recommendation procedure
  AP: The loading control procedure
  MS: The fault-tolerant procedure and map direction procedure

Initially, each MS is pre-installed a fault-tolerant procedure. The AP deployment map and AP location table for a public area are also pre-stored in the SNMP server of the public WLAN. For the AP location table, it stores the location information of each AP. Each MS downloads the map and table from the SNMP server when it first visits the public area. Once detecting a failure in an AP, the failure-affected MS executes the internal fault-tolerant procedure, which is divided into two stages.

```
The first stage fault-tolerant procedure
Add the identity of the faulty AP to Faulty_AP list
If (the failure-affected MS has received the beacon signal from other APs)
/* The failure-affected MS is now located within the overlapping area of several APs*/
    backup AP ← The AP with the strongest beacon signal sent to the failure-affected MS
Else
    backup AP ← Call the routine of Find_backup_AP
    If (backup AP is not null)
        Switch the scanning mode to the active scanning
        Change the scanning channel to the operating channel of the backup AP
        Execute the map direction procedure
    Else
        Pop up message "No available APs in the public area"
    End
End

Routine Find_backup_AP
    neighbor_APs ← Ø
    hop_dist ← one hop distance
    entry ← find the entry number of the faulty AP in the AP location table
    up_entry ← entry – 1 ; down_entry ← entry + 1
    Repeat
        While (up_entry ≥ 1)
            If (The distance between AP_up_entry and AP_entry) ≤ hop_dist)
                If (AP_up_entry is not in faulty_AP list)
                    neighbor_APs ← neighbor_APs ∪ AP_up_entry
                End
                up_entry ← up_entry – 1
            Else
                Exit While
            End
        End While
        While (down_entry ≤ The number of entries in the AP location table)
            If (The distance between AP_up_entry and AP_entry) ≤ hop_dist)
                If (AP_down_entry is not in faulty_AP list)
                    neighbor_APs ← neighbor_APs ∪ AP_down_entry
                End
                down_entry ← down_entry + 1
            Else
                Exit While
            End
        End While
        IF (neighbor_APs is not empty)
            Exit Repeat
        Else
            hop_dist ← hop_dist + one hop distance
            /*extend the search range to two hops, three hops, or more*/
        End
    Until (All the APs in the AP location table have been searched)
    If (neighbor_APs is not Ø)
        Use the AID of the failure-affected MS to randomly select an AP from neighbor_APs
        as the backup AP
        Return the selected backup AP
    Else
        Return Null
    End
End
```

(a)

Figure 4. Continued on next page.

The second stage fault-tolerant procedure

```
When (the Probe Response message is received from the backup AP)
    If (the Probe Response message includes the new backup AP information)
            change the scanning channel to the operating channel of the new backup AP
            call the map direction procedure
    Else
            perform the authentication and association to connect with the backup AP
    End
End
```

(b)

The map direction procedure

```
Pop-up the AP deployment map
scale ← the coordinate scale in the AP deployment map
coord_faulty_AP ← the coordinate of the current faulty AP in the AP location table × scale
coord_backup_AP ← the coordinate of the backup AP in the AP location table × scale
Draw an arrow line in the AP deployment map based on the given coord_faulty_AP and coord_backup_AP
```

(c)

The loading control procedure

```
If (the load status of the backup AP ≤ threshold)
    Send the normal Probe Response message to the failure-affected MS
Else /* backup AP is overloading*/
    If (the AP recommendation information recorded in the overload record is invalid)
            Ask the SNMP server to find a new backup AP
            new_BU_AP ← the identity of the found backup AP
            Update the overload record with the new_BU_AP and reset the lifetime of the overload record
    Else
            new_BU_AP ← the value of the recommending-AP field in the overload record
    End
    Send an extended Probe_Response message attached with new_BU_AP to the failure-affected MS
End
```

(d)

The backup AP recommendation procedure

```
found_APs ← Ø
entry ← find the entry number of the backup AP in the AP location table
up_entry ← entry – 1 ; down_entry ← entry + 1
While (up_entry ≥ 1)
    If (the load status of the AP_up_entry ≤ threshold AND AP_up_entry is not in faulty_AP list)
            found_APs ← AP_up_entry
            Exit While
    End
    up_entry ← up_entry – 1
End While
While (down_entry ≤ The number of entries in the AP location table)
    If (the load status of the AP_down_entry ≤ threshold AND AP_down_entry is not in faulty_AP list)
            found_APs ← found_APs ∪ AP_down_entry
            Exit While
    End
    down_entry ← down_entry + 1
End While
IF (found_APs is not Ø)
    Among found_APs, select an AP which distance from the AP_entry is the minimum, as the new backup AP
    Return the selected new backup AP
Else
    Return Null
End
```

(e)

Figure 4. Procedures for achieving fault tolerance. (a) The first stage fault-tolerant procedure. (b) The second stage fault-tolerant procedure. (c) The map direction procedure. (d) The loading control procedure. (e) The backup AP recommendation procedure.

In the first stage, the identity of the detecting faulty AP is first recorded. Then, a backup AP is selected from survival APs. To quickly resume the wireless connectivity, the backup AP is selected based on the vicinity consideration (see lines of 19-44 of Fig.4.(a)) by using the faulty AP as the center to search one-hop neighboring APs. If more than one one-hop neighboring AP exist, the failure-affected MS randomly selects one of neighboring APs. Conversely, if there is no one-hop neighboring APs, the search range is extended to two-hop, three-hop and so on. After finding a suitable survival AP as the backup AP, the map direction procedure is executed to generate a map for directing the failure-affected MS to move to the coverage range of the backup AP. In the meantime, the failure-affected MS sets the scanning mode to the active mode, and switches its scanning channel to the operating channel of the backup AP (see line 7-8 of Fig.4. (a)). Due to switching the scanning channel, if the failure-affected MS arrives at the directed range (the coverage range of the backup AP), it will detect a strong signal with large SNR value. Otherwise, the failure-affected MS will not receive any signal or detect a weak signal with small SNR value. Based on the SNR strength of the received signal, the failure-affected MS can easily determine whether it has arrived at the directed range or not. However, there is an exception in the above scenario. If the backup AP also fails, the failure-affected MS cannot receive a strong signal from the AP when it arrives at the directed range. In such case, the failure-affected MS must be aware of this event by itself, and then it uses the backup AP as the new faulty AP to re-initiate the first-stage fault tolerant procedure. If the exceptional event does not occur, the second stage fault-tolerant procedure will be initiated. Since the failure-affected MS has set its scanning mode to the active mode (see the above description in this section), the backup AP will receive a *probe_request* message from the failure-affected MS [14]. The backup AP calls its loading control procedure to verify the loading status. If its loading status is below a threshold, it can serve the failure-affected MS and reply a normal *probe_response* message to the MS. Otherwise, the backup AP is inappropriate to serve the failure-affected MS and it will recommend a new backup AP. The recommendation of the new backup AP can be assisted by the SNMP server or by increasing an *overloading record* in each AP. The overloading record has two fields (*recommending AP* and *lifetime)*. When the backup AP cannot serve a failure-affected MS due to overloading, it first checks the validity of the recommending AP field in the overloading record according on the lifetime field.

If the recommending AP field is null or expired, the overloading backup AP asks the SNMP server to find a new backup AP. The SNMP server has the loading information of all APs since it periodically inquires each AP to acquire its loading information. Note that the loading inquiry is an intrinsic routine task of the SNMP server [15], not an additional task introduced by our proposed approach. Then, the SNMP server initiates the backup AP recommendation procedure to find an appropriate AP as the new backup AP by adopting the similar method used in the first-stage fault-tolerant procedure. The new backup AP is close to the old backup AP (the overloading backup AP) and its loading status is below the threshold. Then, the SNMP server sends the identity of the new backup AP to the old backup AP. The old backup AP records the identity information in the recommending AP field of its overloading record. The lifetime field is set to be the loading inquiry interval. From the above description, the overloading record is mainly used to reduce the overhead of finding the new backup AP without frequently retrieving this information from the SNMP server.

If the identity recorded in the recommending AP field is valid, the new backup AP information is directly retrieved from the filed.

After the overloading backup AP acquires the new backup AP information from the SNMP server or its overloading record, the information is attached on the probe_response message to be sent to the failure-affected MS. The failure-affected MS calls the map direction procedure to generate a new map for directing it to the coverage range of the new backup AP.

4. Evaluation

This section evaluates the failure-free and fault-tolerant overheads of the proposed approach.

4.1. Failure-free Overhead

During the failure-free period, each MS needs to download the AP deployment map and AP location table from the SNMP server when it first visits the area of the public WLAN. The cost for downloading the map and table can be represented as:

$$T_{data\_load} = T_{wiredline} + T_{wireless} = \left(\frac{S_{map} + S_{table}}{B_{wiredline}} \right) + \left(\frac{S_{map} + S_{table}}{B_{wireless}} \right) \quad (1)$$

where the first term $T_{wiredline}$ is the time to download the two desired data from the SNMP server to an AP, and the second term $T_{wireless}$ is time for transmitting the two data from an AP to a MS. S_{map} and S_{table} are the sizes of the AP deployment map and the AP location table, respectively. $B_{wiredline}$ and $B_{wireless}$ are the transmission bandwidth from the SNMP server to an AP and the transmission bandwidth from the AP to a MS, respectively. Here, the failure-free overhead metric T_{data_load} will be further quantified in section 6.

In addition, the SNMP server is also required to periodically inquire the loading status of each AP [4]. This task is an intrinsic routine of the SNMP server, which is not additionally introduced by the proposed fault-tolerant approach. The proposed approach only uses the given information of the routine to select possible backup APs. If the execution cost of the routine is counted into the failure-free overhead of the proposed approach, this cost can be expressed as

$$T_{loading} = T_{inquiry} + T_{response} \quad (2)$$

where the first term is the average time for broadcasting the loading inquiry message from SNMP server to APs, and the second term is average time for sending the response message from an AP to SNMP server.

The execution cost of the loading inquiry routine ($T_{loading}$) is trivial, which is explained as follows. The loading inquiry message is broadcasted to all the APs via an Ethernet network. The Ethernet network is usually used as the distribution system of a public WLAN for connecting all APs and the SNMP server [1, 14]. After receiving the loading inquiry message, each AP sends its loading status to the SNMP server via Ethernet network. Due to the Ethernet network, the transmission bandwidth can be up to 100Mbps or above. Furthermore, the sizes of the loading inquiry and loading status messages are also very small. Therefore, $T_{loading}$ is trivial. This will be further validated by simulations in next section 6.

4.2. Fault-tolerant Overhead

With the fault-tolerant overhead, we concern the execution cost of the two stage fault-tolerant procedure. The proposed approach utilizes survival APs to take over the workload of the faulty AP. The performance affection on a survival AP is also concerned. To analyze the execution cost and the performance affection, the following parameters are made.

- The arrivals of MSs to an AP follow a Poisson distribution with the mean arrival rate λ_a.
- The association time of an MS with an AP is not dependent on any specific distribution. The mean association time is $\frac{1}{\mu_r}$.
- The maximum number of MSs associated with an AP is c.
- The failure rate of an AP is λ, which is the average number of failures occurring in an AP within a given time period.
- The recovery rate of an AP is μ, which is the average number of recoveries performed within a given time period.
- The probability that an AP is on the overloading status is $P_{overload}$.
- The probability that an AP is on the faulty status is $P_{failure}$.
- The average movements of a failure-affected MS to associate with a qualified backup AP is MS_{moves}.

Based on the above first three parameters, the behavior of MSs served by an AP can be modeled as the *M/G/c/c* queuing model [16]. According to the *M/G/c/c* queuing model, $P_{overload}$ can be derived as:

$$P_{overload} = \sum_{n=k}^{c} P_{M_n} = \sum_{n=k}^{c} \frac{\frac{\left(\frac{\lambda_a}{\mu_r}\right)^n}{n!}}{\sum_{i=0}^{c} \frac{\left(\frac{\lambda_a}{\mu_r}\right)^i}{i!}}, \tag{3}$$

where P_{M_n} is the probability that there are n MSs within the coverage range of an AP. To conveniently derive $P_{overload}$, the determination of an AP being on the overloading state is based on the number of MSs associated with the AP, not the throughput of the AP. If the number of MS associated with the AP is greater than or equal to k, the AP is regarded to be in the overloading status. The value of k can be inferred from the throughput, as follows:

$$k = \frac{thru_{\max}}{thru_{MS}} \tag{4}$$

where $thru_{\max}$ is the threshold of the AP throughput, and $thru_{MS}$ is the average throughput of an MS.

For $P_{failure}$, it can be gotten based on the fourth and fifth parameter [16], as follow:

$$\begin{aligned} P_{failure} &= 1 - \frac{1}{1 + \frac{\lambda}{\mu}} \\ &= \frac{\lambda}{\lambda + \mu} \end{aligned} \tag{5}$$

where the term $\frac{1}{1+\frac{\lambda}{\mu}}$ is the availability of AP at an instant of time

With MS_{moves}, its value is dependent on the number of times for finding a qualified backup AP. In section 3.2, the first stage fault-tolerant procedure first finds one neighbor of the faulty AP to be the possible backup AP. If the status of the possible backup AP is faulty or overloading, the second stage fault-tolerant procedure will recommend another survival AP to be the possible backup AP until a qualified backup AP is assigned to the failure-affected MS. The probability that a failure-affected MS will move i times for acquiring a qualified backup AP to serve it is $\left(P_{failure} + P_{overload}\right)^{i-1}\left(1 - \left(P_{failure} + P_{overload}\right)\right)$. The number of movements of a failure-affected MS is a geometric distribution, and the expected number is $\frac{1}{1-\left(P_{failure} + P_{overload}\right)}$. The MS_{moves} can be represented as:

$$MS_{moves} = \frac{1}{1 - \left(P_{failure} + P_{overload}\right)} \tag{6}$$

Substituting $P_{overload}$ and $P_{failure}$ into (6), MS_{moves} can be rewritten as:

$$MS_{moves} = \frac{1}{1-\left(\frac{\lambda}{\lambda+\mu}+\sum_{n=\frac{thru_{\max}}{thru_{MS}}}^{c}\frac{\frac{\left(\frac{\lambda_a}{\mu_r}\right)^n}{n!}}{\sum_{i=0}^{c}\frac{\left(\frac{\lambda_a}{\mu_r}\right)^i}{i!}}\right)} \quad (7)$$

Based on (7), the execution cost C_{exec} of the two stage fault-tolerant procedure can be obtained as:

$$C_{exec} = \frac{1}{1-\left(\frac{\lambda}{\lambda+\mu}+\sum_{n=\frac{thru_{\max}}{thru_{MS}}}^{c}\frac{\frac{\left(\frac{\lambda_a}{\mu_r}\right)^n}{n!}}{\sum_{i=0}^{c}\frac{\left(\frac{\lambda_a}{\mu_r}\right)^i}{i!}}\right)} \times e \quad (8)$$

where e is the execution time for performing the two stage fault-tolerant procedure, which time complexity is donated by O(N_{AP}), where N_{AP} is the number of APs in a public WLAN.

For the other fault-tolerant overhead metric: the performance affection, it is represented as the affection on a normal MS in a survival AP. In the proposed approach, the failure-affected MSs are directed to the coverage ranges of survival APs. For a survival AP, its radio channel is contended by the MSs originally located within its coverage range and the failure-affected MSs newly moving to its coverage range. Compared to pre-failure, the performance of an MS in a survival AP is affected by some failure-affected MSs. Since there are two radio access modes in an AP: distributed coordination function (DCF) and point coordination function (PCF), the performance affection is respectively evaluated under these two access modes.

In the DCF mode, the radio access is based on contention. Before an AP failure, an MS only contended the radio channel with other MSs in the same APs. Based on the traffic model of a working AP: *M/G/c/c*, the average number N_{MS} of MSs in the coverage range of a working AP is

$$N_{MS} = \sum_{n=0}^{c} n \times P_{M_n}$$

$$= \sum_{n=0}^{c} n \times \frac{\dfrac{\left(\dfrac{\lambda_a}{\mu_r}\right)^n}{n!}}{\sum_{i=0}^{c} \dfrac{\left(\dfrac{\lambda_a}{\mu_r}\right)^i}{i!}} \tag{9}$$

where P_{M_n} is the probability that there are n MSs in the coverage range of a working AP.

Before an AP failure, the collision probability P_{b_c} of a MS in a working AP is

$$P_{b\_c} = 1 - (1 - P_s)^{(N_{MS}-1)}$$

$$= 1 - (1 - P_s)^{\left(\sum_{n=0}^{c} n \times \frac{\frac{\left(\frac{\lambda_a}{\mu_r}\right)^n}{n!}}{\sum_{i=0}^{c} \frac{\left(\frac{\lambda_a}{\mu_r}\right)^i}{i!}} - 1 \right)} \tag{10}$$

where P_s is the probability that a MS would like to send data at an instant of time. The term $(1 - P_s)^{(N_{MS}-1)}$ is the probability that other MSs in the same AP do not send data.

After an AP failure, the failure-affected MSs in the faulty AP are directed to move to the coverage ranges of survival APs. In the proposed approach, the overloading situation is also considered. Therefore, the maximum number $N_{Affected\_MS}$ of failure-affected MSs moving to a survival AP is limited as

$$N_{Affected\_MS} = \frac{thru_{\max}}{thru_{MS}} - N_{MS}$$

$$= \frac{thru_{\max}}{thru_{MS}} - \sum_{n=0}^{c} n \times \frac{\dfrac{\left(\dfrac{\lambda_a}{\mu_r}\right)^n}{n!}}{\sum_{i=0}^{c} \dfrac{\left(\dfrac{\lambda_a}{\mu_r}\right)^i}{i!}} \tag{11}$$

From (9), (10), and (11), the collision probability P_{a_c} of a MS in a survival AP after an AP failure can be derived as:

$$P_{a\_c} = 1 - (1 - P_s)^{(N_{MS} + N_{Affecte\_MS} - 1)}$$

$$=1-\left(1-P_s\right)^{\left(\frac{thru_{\max}}{thru_{MS}}-1\right)} \tag{12}$$

From (10) and (12), the performance affection on a normal MS based on the DCF mode can be represented as

$$P_{a\_c}-P_{b\_c}=1-\left(1-P_s\right)^{\left(\frac{thru_{\max}}{thru_{MS}}-1\right)}-1+\left(1-P_s\right)^{\left(\sum_{n=0}^{c} n\times\frac{\frac{\left(\frac{\lambda_a}{\mu_r}\right)^n}{n!}}{\sum_{i=0}^{c}\frac{\left(\frac{\lambda_a}{\mu_r}\right)^i}{i!}}-1\right)}$$
$$=\left(1-P_s\right)^{\left(\sum_{n=0}^{c} n\times\frac{\frac{\left(\frac{\lambda_a}{\mu_r}\right)^n}{n!}}{\sum_{i=0}^{c}\frac{\left(\frac{\lambda_a}{\mu_r}\right)^i}{i!}}-1\right)}-\left(1-P_s\right)^{\left(\frac{thru_{\max}}{thru_{MS}}-1\right)} \tag{13}$$

As for the performance affection on a normal MS based on the PCF mode, it concerns the increase of the transmission waiting interval for a MS. In the PCF model, the AP uses round robin to inquire each MS to send its data frames. In the worst case, if a MS has data frames to send, the transmission waiting interval for the MS is dependent on the number of MSs in the coverage range of a working AP. After an AP failure, failure-affected MSs are directed to move to the coverage ranges of survival APs. The number of MSs in a working AP becomes larger, and it is up to $N_{MS}+N_{Affected\_MS}$. Therefore, the increase of the transmission waiting interval for a MS is dependent on $N_{Affected\_MS}$. The performance affection on a normal MS based on the PCF mode is represented as

$$N_{Affected\_MS}\times t=\left(\frac{thru_{\max}}{thru_{MS}}-\sum_{n=0}^{c} n\times\frac{\frac{\left(\frac{\lambda_a}{\mu_r}\right)^n}{n!}}{\sum_{i=0}^{c}\frac{\left(\frac{\lambda_a}{\mu_r}\right)^i}{i!}}\right)\times t \tag{14}$$

where t is the average time for sending a data frame between an MS and an AP.

5. Comparison

This section makes the comparisons between the proposed approach and the previous approaches. The comparisons are in terms of the hardware support, fault-tolerant capability, failure-free overhead, and fault-tolerant overhead as shown in Table 1. The previous approaches are divided into access-point replication, overlapping coverage, and link multiplexing (see section 2.3) to compare with our proposed approach.

Table 1. Comparisons between the previous approaches and the proposed approach.

Comparing Metrics	Access-point replication	Overlapping coverage	Link multiplexing	Proposed approach
Hardware support	Yes (Redundant AP)	No	Yes (Multiple wireless network cards)	No
Fault-tolerant capability	Dependent on the equipped redundant AP	Dependent on the overlapping APs	Dependent on the overlapping APs	Dependent on the working APs in the system
Failure-free overhead	Monitor the state of the primary AP	No	Multiplex and demultiplex packets	Download map and inquire loading
Fault-tolerant overhead	• Activate the redundant AP • Associate with the redundant AP	• Associate with a working AP • Degrade the performance of existing working APs	• Reduce one communication path	• Execute the proposed fault-tolerant process • Degrade the performance of existing working APs, but not overloading

- Hardware support: In the access-point replication approach, each working AP is equipped with one redundant AP as its backup AP. For the link multiplexing approach, each MS uses multiple wireless network cards to maintain multiple communication paths to different APs. For the overlapping coverage approach and proposed approach, they do not need additional hardware mechanisms. Although the proposed approach uses the SNMP server, it is a basic component of a public WLAN.
- Fault-tolerant capability: For the access-point replication approach, if a failure also occurs in the redundant AP, the access-point replication approach is not workable. For the overlapping coverage and link multiplexing approaches, their fault-tolerant capabilities are based on a suitable overlapping AP. The fault-tolerant idea in these two approaches is to let one or more overlapping APs take over the workload of the faulty AP. However, as mentioned in [1, 2], the overlapping coverage approach cannot ensure that the overlapping coverage is available in each place of a public WLAN. If an MS is not located in an overlapping coverage range and its serving AP fails, the overlapping coverage approach cannot assist the MS to connect another AP. To conquer this problem, the link multiplexing approach can provide the overlapping coverage everywhere based on multiple wireless network cards [2]. However, if the overlapping APs of the faulty AP all fail or their statuses are overloading, the link multiplexing approach cannot handle such failure situation. As for the proposed approach, if an AP fails, the proposed approach can direct the failure-affected MSs to survival APs based on the vicinity and loading considerations (see section 3). The proposed approach can allow multiple working APs to fail simultaneously. In theory, if there are n working APs in a public WLAN, the proposed approach can tolerate up to $n-1$ faulty APs.
- Failure-free overhead: For the access-point replication approach, each equipped redundant AP needs to periodically monitor the status of its corresponding primary AP. If a failure occurs in the primary AP, the equipped redundant AP is activated to take over the workload of the primary AP. For the overlapping coverage approach, it does not take

any actions against a failure during the normal time (the failure-free period). For the link multiplexing approach, multiple communication paths are provided for an MS to make it immune from the failure affection. However, packets to/from an MS are required to be multiplexed over the communication paths. Compared to other fault-tolerant approaches, the packet transmission in the link multiplexing approach incurs the additional delay due to packet multiplexing and demultiplexing. As for the proposed approach, its failure-free overhead is determined on the operation of downloading a map and a location table and the operation of inquiring the AP loading information (see section 4.1). The two operations include few simple instructions, which costs will be quantified in section 6.

- Fault-tolerant overhead: For the access-point replication approach, if an AP fails, its corresponding redundant AP is activated and all the failure-affected MSs are re-associated with the redundant AP. For the overlapping coverage approach, the failure-affected MS actively detects the AP failure. Once detecting an AP failure, the failure-affected MS select one AP from its AP list as its new serving AP (Note that the AP list of an MS records which APs signal to the MS). For the link multiplexing approach, if an AP fails, the failure-affected MS does not care this failure event since its connectivity is still sustained by the remaining already-established communication paths. The failure affection on the failure-affected MS is that one of communication paths cannot be used to transmit packets. For the proposed approach, two stage fault-tolerant procedures are executed to make each failure-affected MS move to a survival AP. Compared to previous approaches, the proposed approach may take higher fault-tolerant overhead. However, the fault-tolerant overhead is still small, which will be validated in next section.

6. Simulation

Compared to previous approaches, the proposed approach has the best fault-tolerant capability (see Table 1). The best fault-tolerant capability is based on the software support, not the hardware support. For quantifying the overhead of the software support in the proposed approach, section 4 has given the numerical analysis. To validate the numerical analysis, we extend the given wireless LAN module of the Network Simulator version 2 (NS-2) [17] to perform simulations. In the simulations, the used public WLAN model refers to the simulation model of [18], as shown in Fig. 5. There are 25 APs and 10 application servers in the public WLAN. A 100Mbps Ethernet is set among the APs and application servers. The wireless bandwidth between an AP and an MS is set to 11 Mbps. The arrivals of MSs to an AP follow a Poisson distribution. The association time of an MS with an AP is random. The MS intensity (the ratio of MS arrival rate over the MS association rate) is controlled to be 10, 30, 60, and 90, respectively. The maximum number of MSs associated with an AP is set to 100. If the number of MSs associated with an AP is over 90, the AP is regarded in the overloading status. Each MS in an AP randomly issues a data service to an application server, and the service time is also random. For the failure rate and recovery rate of an AP, the ratio between them refers to [1] and is set to 0.0033. The AP deployment map and location table of the public LAN, their sizes are set to 32KB and 4KB, respectively.

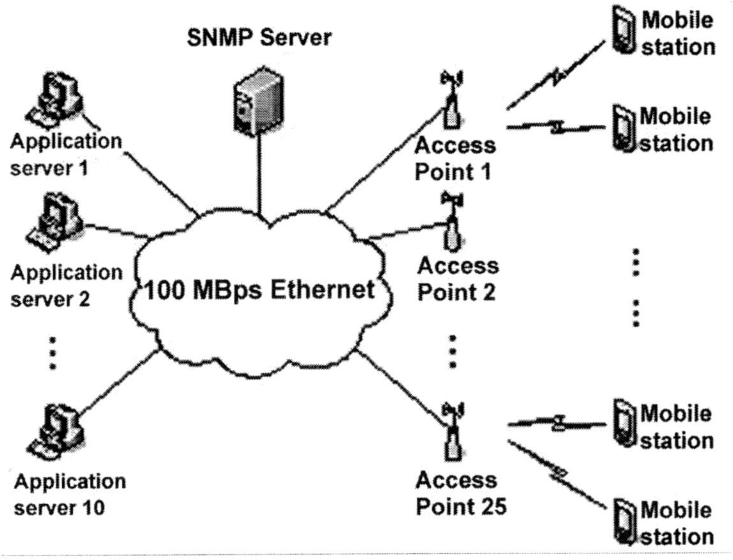

Figure 5. The public WLAN model used for simulations.

Based on the simulations, the mentioned two failure-free overheads of the proposed approach (see section 4.1) are trivial. The average time for an MS to download the AP deployment map and location information is 1.17 second. For the time for the SNMP server to inquire the AP loading, it is far small, which is only 0.027 second.

With the two fault-tolerant overheads of the proposed approach (see section 4.2), their simulation results and analytical results are illustrated in Fig. 6 and Fig. 7. The simulation result of each fault-tolerant overhead metric is derived from the average result of 500 simulation runs. Fig. 6 shows that the simulation results match closely the analytical results of equation (8). From Fig. 6, we can also see that if the number of AP failures is less than 8, the execution time (cost) for the two stage fault-tolerant procedure does not obviously increase as the number of AP failures increases. The execution cost has a large increase only when the number of AP failures is 8 and the MS intensity ($\frac{\lambda_a}{\mu_r}$) is 90. In this case, each failure-affected MS needs to move several times to find its backup AP. In addition, in Fig. 6, it also shows that the execution cost is almost invariant except $\frac{\lambda_a}{\mu_r} = 90$. When $\frac{\lambda_a}{\mu_r} \leq 60$, the average number of MSs associated with an AP is far below the overloading threshold (90). If an AP fails, each survival AP has enough available resources to serve some failure-affected MSs. By simulations, the number of average movements for a failure-affected MS to find its qualified backup AP is not larger than 2. The number of average movements is 1.14. In each movement, the execution cost is around 84 μs. Therefore, when $\frac{\lambda_a}{\mu_r} \leq 60$, the average execution cost is 95.76 (1.14*84) μs. In contrast, when $\frac{\lambda_a}{\mu_r} = 90$, there are already many MSs associated with an AP and the average number of MSs in the coverage range of an AP is

88. In such case, if an AP fails, each failure-affected MS may need to move more than 2 times to find a non-overloading AP as its backup AP. The number of average movements is 2.4, which are 2 times of the average movements (1.14) of $\frac{\lambda_a}{\mu_r} \leq 60$. However, the corresponding execution cost is still very small, which is only 202 μs.

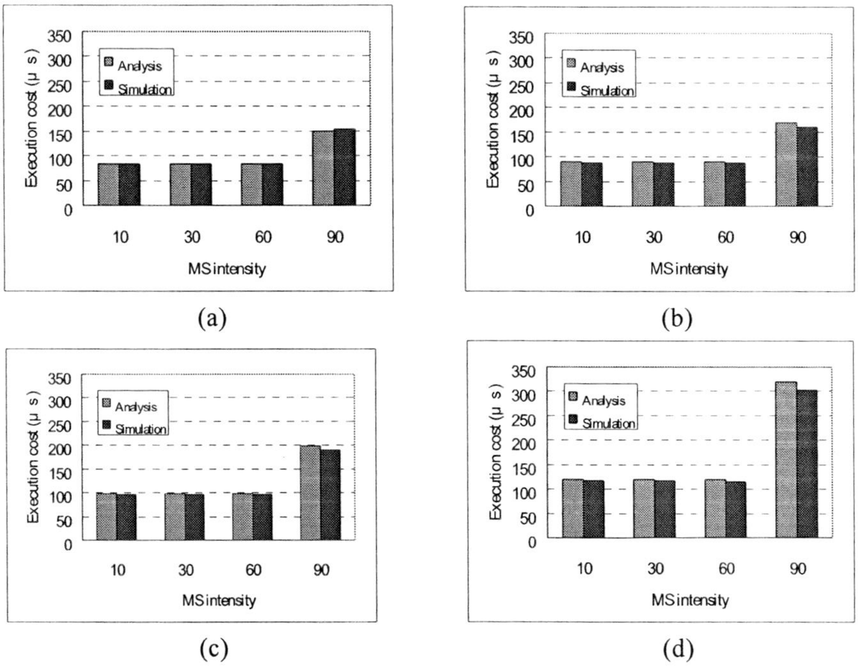

Figure 6. The time for executing the proposed fault-tolerant process. (a) Number of simultaneous AP failures = 1. (b) Number of simultaneous AP failures = 2. (c) Number of simultaneous AP failures = 4. (d) Number of simultaneous AP failures = 8.

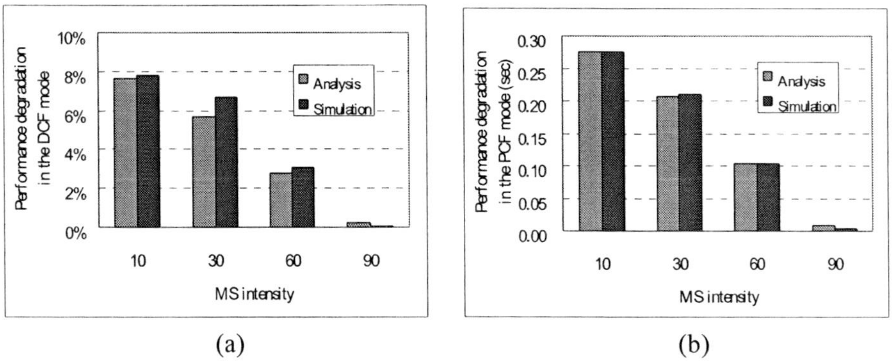

Figure 7. Performance affection on a survival AP. (a) DCF mode. (b) PCF mode.

Fig. 7 illustrates the simulation results and analytics result of the other fault-tolerant metric. Both results are also very close. Note that the given performance affection in Fig. 7 is independent of the number of AP failures. Regardless of how many AP fail simultaneously, there is an overloading threshold in each survival AP to limit the number of failure-affected MSs served by it. Also as shown in Fig. 7, the performance affection on a survival AP decreases as $\frac{\lambda_a}{\mu_r}$ increases. The reason is explained as follows. When $\frac{\lambda_a}{\mu_r}$ is not large, the average number of MSs associated with an AP is below the overloading threshold. If an AP fails, each survival AP can associate some failure-affected MSs to serve them. When $\frac{\lambda_a}{\mu_r}$ becomes larger, the available resources in each AP become few. Compared to smaller $\frac{\lambda_a}{\mu_r}$, if an AP fails, each survival AP can serve fewer failure-affected MSs. Without regard to adopting the DCF or PCF as the radio access mechanism, the performance affection at smaller $\frac{\lambda_a}{\mu_r}$ is larger than that at larger $\frac{\lambda_a}{\mu_r}$. Especially, when $\frac{\lambda_a}{\mu_r} = 90$, the performance affection is almost close to 0. Before an AP failure, each AP has already served a large number of MSs. After an AP fails, the number of failure-affected MSs being able to serve by a survival AP is very small. From the simulations, this average number is 2. Therefore, when $\frac{\lambda_a}{\mu_r} = 90$, the performance of a survival AP incurs a very small degradation. In the DCF mode, the performance affection is in terms of the increase in the contention probability, which is 0.05%. As to the PCF mode, the performance degradation is in terms of the increase in the transmission waiting interval, which is 0.004 second.

7. Conclusion

This paper has presented an efficient approach to tolerating AP failures in a public WLAN. Whenever an AP failure is detected, the proposed approach utilizes the survival APs to dynamically constitute a backup AP set. After the failure-affected MSs move to the coverage ranges of the backup AP set, their wireless connectivity can be reconnections. To reduce the fault-tolerant overhead of the proposed approach, the movements of fault-affected MSs are first according to the vicinity consideration. Each failure-affected MS is served by the survivable AP close to it. To avoid a survival AP incurring the significant performance affection, the proposed approach also concerns the overloading situation. Compared to the previous approaches, the proposed approach has the following advantages:

- Not requiring the hardware support.
- Avoiding overloading situation.
- Having the best fault-tolerant capability.

Finally, extensive numerical analysis and simulation experiments were performed to evaluate the failure-free and fault-tolerant overheads of the proposed approach. The results show that the two overheads are small.

References

[1] D. Chen, C. Kintala, S. Garg, K. S. Trivedi, "Dependability Enhancement for IEEE 802.11 Wireless LAN with Redundancy Techniques," *Proceedings of the International Conference on Dependable Systems and Networks*, pp. 521-528, June 2003.

[2] Gandhi R., "Tolerance to Access-Point Failures in Dependable Wireless Local-Area Networks," *The Ninth IEEE International Workshop on Object-Oriented Real-Time Dependable Systems* , pp. 136-143, Oct. 2003.

[3] IEEE, "Part 11: Wireless LAN Medium Access Control (MAC) and Physical Layer (PHY) Specifications," *IEEE Standard 802.11*, September 1999.

[4] J. Case, M. Fedor, M. Schoffstall, J. Davin, "A Simple Network Management Protocol (SNMP)," *Technical Report IETF RFC 1157*, May 1990.

[5] Cisco Systems Inc., "Data Sheet for Cisco Aironet 350 Series Access Points," June 2001.

[6] Proxim Corp., "User's Guide for ORiNOCO AP-2500," June 2003.

[7] Anand Balachandran, Geoffrey M. Voelker, Paramvir Bahl, P. Venkat Rangan, "Characterizing User Behavior and Network Performance in a Public Wireless LAN," *In Proceedings of the ACM Sigmetrics Conference on Measurement and Modeling of Computer Systems*, pp. 195-205, June 2002.

[8] Carlos Oliveira, Jaime Bae Kim, Tatsuya Suda, "Long-Range Dependence in IEEE 802.11b Wireless LAN Traffic: An Empirical Study," *Computer Communications, 2003. CCW 2003. Proceedings. 2003 IEEE 18th Annual Workshop on*, pp. 17-23, Oct. 2003.

[9] F. K. Al-Bin-Ali, P. Boddupalli, N. Davies, "An Inter-Access Point Handoff Mechanism for Wireless Network Management: The Sabino System," *In Proceedings of The 2003 International Conference on Wireless Networks*, pp. 225-230, June 2003.

[10] Anand Balachandran, Paramvir Bahl, Geoffrey M. Voelker, "Hot-Spot Congestion Relief in Public-Area Wireless Networks," *Mobile Computing Systems and Applications, 2002. Proceedings Fourth IEEE Workshop on*, pp. 70-80, June 2002.

[11] Shiann-Tsong Sheu, Chih-Chiang Wu, "Dynamic Load Balance Algorithm (DLBA) for IEEE 802.11 Wireless LAN," *Tamkang Journal of Science and Engineering*, pp. 45-52, June 1999.

[12] I. Papanikos, M. Logothetis, "A Study on Dynamic Load Balance for IEEE 802.11b Wireless LAN," *In Proceedings of the 8th International Conference on Advances in Communication & Control, COMCON 8, Rethymna,Crete/Greece*, June 2001.

[13] Hector Velayos, Victor Aleo, Gunnar Karlsson, "Load Balancing in Overlapping Wireless LAN Cells," *ICC 2004 - IEEE International Conference on Communications*, pp. 3833-3836, June 2004.

[14] Gast, Matthew S., "802.11 Wireless Networks: The Definitive Guide," *O'Reilly & Associates*, April 2002.

[15] Colubris Networks Inc., "Data Sheet for Colubris Networks Management System (CNMS)," 2004.

[16] D. Gross, C. M. Harris, "Fundamentals of Queuing Theory," *John Wiley & Sons*, 1985.

[17] NS-2 Network Simulator, Available: http://www.isi.edu/nsnam/

[18] Huan-Yun Wei, Ching-Chuang Chiang, Ying-Dar Lin, "Co-DRR: An Integrated Uplink and Downlink Scheduler for Bandwidth Management over Wireless LANs," *IEEE Symposium on Computers and Communications*, pp. 1415-1420, June 2003.

In: Wireless Communications Research Trends
Editor: Tong S. Lee, pp. 349-370
ISBN: 978-1-60021-674-9

Chapter 12

AN EFFICIENT MOBILE-AGENT-BASED PLATFORM FOR DYNAMIC SERVICE PROVISIONING IN 3G/UMTS

Yuan-Lin Ko**[*]**, ***Kuochen Wang*****[**]**
and Hung-Cheng Shih**[#]**
Department of Computer Science
National Chiao Tung University, Hsinchu, 300, Taiwan

Abstract

An important key concept of the Virtual Home Environment (VHE) is dynamic service provisioning. In 3G/B3G, the mobile network will have such a capability. The users can dynamically subscribe new services anytime, and the system operator or service provider can dynamically provide services to subscribed users immediately. Based on the UMTS CAMEL (Customized Applications for Mobile Enhanced Logic) architecture, we propose an efficient mobile-agent–based platform to provide services dynamically, which can greatly reduce signaling traffic. To demonstrate the efficiency of our platform, we used the operations of incoming and outgoing calls to illustrate the operation of mobile agents. In an existing approach, a CORBA agent-based platform was deployed in a distributed processing environment, and it requires a standard, OMG Mobile Agent System Interoperability Facility (MASIF), to be interoperable between agent environments of different vendors or operators. However, there are some problems in this approach, such as problems in the aspects of security and performance. Analysis results have shown that the signaling traffic in our CAMEL mobile-agent-based platform can be reduced 40% compared to that in the CORBA agent-based platform. Our platform can provide efficient mobility management, and enhance network performance, security and interoperability.

---

[1] This work was supported by the NCTU EECS-MediaTek Research Center under Grant Q583 and National Science Council under Grant NSC 94-2213-E-009-043.
[*] E-mail address: kolin.cis89g@nctu.edu.tw. +886-3-5712121 Ext. 59274
[**] E-mail address: kwang@cs.nctu.edu.tw. +886-35131363. FAX: +886-3-5721490 (Corresponding author)
[#] E-mail address: hcshih@cs.nctu.edu.tw. +886-3-5712121 Ext. 59274.

Keywords: Customized Applications for Mobile Enhanced Logic (CAMEL), Dynamic Service Provisioning, Mobile Agent, Universal Mobile Telecommunication System (UMTS), Virtual Home Environment.

I. Introduction

In recent years, GSM grows rapidly. Although the subscribers of GSM increased, the users still didn't feel satisfied with the services, due to low data bit rates and insufficient service types, etc. In order to overcome these problems, the Universal Telecommunication Mobile System (UMTS) has been developed to provide advanced capabilities and services. The 3rd generation mobile communication system has been deployed in some countries. In the GSM, service providers developed all kinds of services, but the services could not be used by different system operators. This shortcoming restricts the development of services, service portability, and personal mobility. In order to deal with this incapability, the concept of Virtual Home Environment (VHE) was introduced. The VHE is an importance concept in the UMTS for personalized service portability between networks boundaries and between terminals. It enables users to get access to their personalized services in remote networks, just like *feel and look* in home networks [1][2][3][4].

In the second generation mobile communication systems, the trend of services evolution was driven by integrating the *intelligent network* (IN) concept into these systems [14]. In the 3rd generation mobile communication systems, service control and service management are also based on the IN concept. The specification of the IN concept in the UMTS, which was defined by 3GPP, is the *Customized Applications for Mobile Enhanced Logic* (CAMEL) [5][6][15].

Mobile Agent Technology (MAT) is a technology that has gained momentum in telecommunications [1][2]. Mobile agents (MAs) are autonomously, asynchronously intelligent software entities, which in order to fulfill their tasks can migrate to other nodes in the network [8][27]. Flexibility and scalability are two important motivations for deploying MAs in telecommunications systems. In such systems, the MAs platform can provide an alternative aspect to realize the dynamic service provisioning toward the VHE. So the MAs in telecommunications will draw more and more attention in the near future. In Europe, several projects have been conducted in this area, like ACTS CLIMATE and CAMELEON projects [20][21]. Besides, the MAT that enables *on-demand* provision of customized services offers interesting aspects for realizing the concept of VHE [8]. Because MAs can migrate as close as possible to a resource node, the traffic of signaling and service downloading can greatly be reduced. The characteristics of MAs can be used to enhance telecommunication network features. Owing to these benefits, the MAT has been applied to the UMTS.

II. Existing Approaches

In [8], an MA platform using CORBA in UMTS was studied. Figure 1 shows this MA platform. The MAs were deployed in the Distributed Agent Environment (*DAE*) that was built on top of the Distributed Processing Environment (*DPE*) [8]. All involved system nodes, including end user systems (*UE: User Equipment*), provider systems (*Core Network*) and 3rd party service provider (*SP*) systems, must provide a corresponding agent environment that enables MAs to download and migrate in the DPE. The end user system is a UE that includes the UMTS Subscriber Identity Module (*USIM)*, and the USIM must provide an agent execution environment (*Agency*). The core network is a provider system that contains switches (*Visited MSC*, *MSC*, and *Gateway MSC*). All system nodes must also provide agencies. A user can use a UE to communicate with an MSC through the UMTS Terrestrial Radio Access Network (*UTRAN*). MAs that execute on the agency can fit into various kinds of application areas, such as mobility management, service control, and even end user applications.

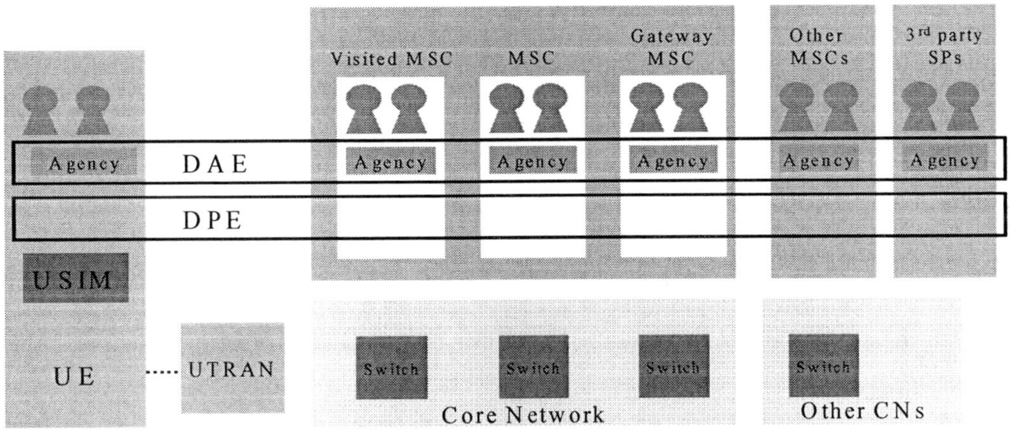

Figure 1. A mobile agent platform using CORBA in UMTS [8].

Hence, setting up MA standards is necessary, like the Object Management Group (OMG) Mobile Agent System Interoperability Facility (MASIF), or the Foundation of Intelligent Physical Agents (FIPA), which are to enable basic interoperability between agent environments of different venders or operators [12][13]. The MASIF standard has two interfaces: the *MAFAgentSystem* interface and *MAFFinder* interface. The MAFAgentSystem interface provides agents operations with the managements and transformations of agents, including of create, suspend, resume, terminate agent, etc.[9] [10]. The MAFFinder interface supports the localization of agents, agencies, and places in the scope of a region, including of register, de-register, look up agent/place/system, etc. [9][10]. Therefore, a mobile user may roam between different networks, and he/she must register temporarily at different UMTS system providers. On the other hand, a USIM agency in UE must register at the MAFFinder of the region that belongs to the system operator [9][10]. In this platform, the security of agents communication in the DPE is weak and its mobility management can be improved.

In [7], dynamic service provisioning was deployed on a CORBA agent-based platform in UMTS. In this platform, the GMSC (Gateway Mobile Switch Center) must have the MAFFinder component that manages the life cycles of agents and the registrations of agent services. The VHE-agent and PCS-agent are important personal agents. The VHE-agent realizes the VHE concept and it is responsible for outgoing calls [8]. The PCS-agent realizes the personal communication support (PCS) concept and it handles incoming calls [8]. All the visiting user-related agents (including VHE and PCS agents) have to register at the MAFFinder of the region belonging to the provider system [7]. The Provider Agent (PA) manages Service Agents (SAs) that must register in the MAFFinder. If the PA does not find the required SA in the MAFFinder, then the PA can find service in the home environment or third party service providers. In this platform, its communication steps (23 steps) are too complex to have good performance. To improve the performance, we propose a CAMEL agent-based agent platform to provide services dynamically in 3G/UMTS [22][23][28].

III. Preliminary

A. Introducing CAMEL Concept

Taking the increase of GSM operators into account, a standard of services needs to be created for cooperation and communication between different operators. The ETSI started with the specification of IN functionality of GSM in 1994, named *Customized Applications for Mobile Network Enhance Logic* (CAMEL) [17]. The CAMEL provides GSM operators with the ability to offer the IN based services to its own subscribers, operator specific services when the subscriber is roaming within or outside the home GSM network [16]. The CAMEL has four phases. Phase 1 describes the basic concept about CAMEL and provides a set of basic services. Phase 2 enhances the basic services with the addition of many supplementary services for voice calls [16]. Phase 3 describes how CAMEL interworks with GPRS, and mobile originated SMS and services they offer. Phase 4 introduces the increased level of functionality as compared to phase 3 [5][6]. The CAMEL Application Protocol (CAP), which can support CAMEL phase 3 and phase 4, is based on ETSI core Intelligent Network Application Protocol (INAP) Capability Set-2 (CS-2) [17]. Figure 2 shows the CAMEL phase 1 architecture and explains the CAMEL basic concept. In CAMEL phase 1, it identifies two functions: *GSM Service Control Function (gsmSCF)* in the *Service Control Point (SCP)* and *GSM Service Switch Function (gsmSSF)* in the *Service Switch Point (SSP)*. The gsmSCF acts as an entity where the execution of an operator specific service takes place [17]. The gsmSSF acts as an interface from GMSC or VMSC towards the gsmSCF [17]. In the home network, gsmSCF is along with HLR that communicates with each other by using the MAP protocol. HLR communicates with GMSC and VLR using the MAP protocol. The Interrogating Network performs the interrogation of the home network for information on the treatment of terminating CAMEL calls, and it contains GMSC and gsmSSF [17]. The visiting network contains VLR, VMSC and gsmSSF. The gsmSCF communicates with gsmSSF using the CAP protocol.

The CAMEL divides call processing into different phases in GMSC or MSC. And call states and events are modeled by using an abstract representation called a Basic Call State Model (BCSM) [17]. The BCSM has two forms, one for originating events (O-BCSM) and the other for terminating events (T-BCSM) [17]. Both the O-BCSM and T-BCSM of CAMEL phase 1 are the simplification of the BCSM for CS-1 (Capability Set 1). CAMEL phase 2 is based on CAMEL phase 1. CAMEL phase 2 adds a new component, named *gsmSRF*. The specialized resource functions provide resources for access by other network entities [17]. CAMEL phase 3 indicates the functional entities interworking with GPRS. The function entities involve in a GPRS session requiring the CAMEL support. The new functional entity is introduced as *gprsSSF* that is along with SGSN [6] (see Figure 3). The gprsSSF also communicates with gsmSCF using the CAP protocol. In the home network, the gsmSCF is still along with the HLR, and the HLR communicates the SGSN with the MAP protocol. The CAMEL can offer more flexible and scalable services even when users roam in different networks that also provide a bridge to future evolution toward UMTS. The specification of the UMTS in 3GPP has adopted the CAMEL architecture to achieve the VHE.

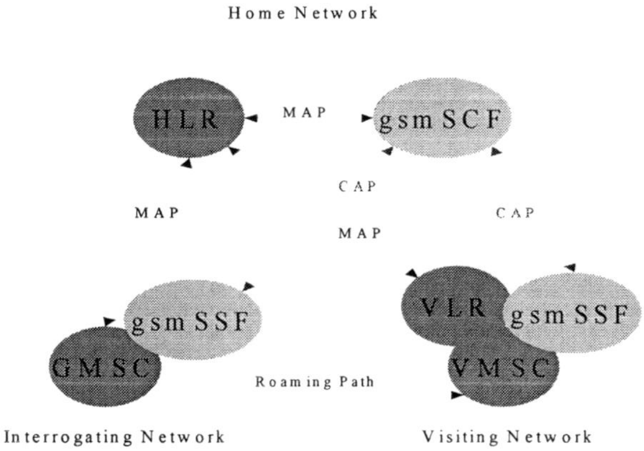

Figure 2. CAMEL phase 1 architecture [17].

The CAMEL architecture in UMTS can support authentication and mobility management. The CAMEL defines many interfaces that can help create service development environments or facilitate cooperation in different networks, etc. Therefore, it can provide some benefits, such as service implementation independence, multi-vendor interworking, multi-network interworking, and rapid service delivery and deployment, etc. [16]. These benefits help the service providers or system operators develop services or communicate with each other more easily and more efficiently. Furthermore, it also brings more kinds of services to satisfy user's need.

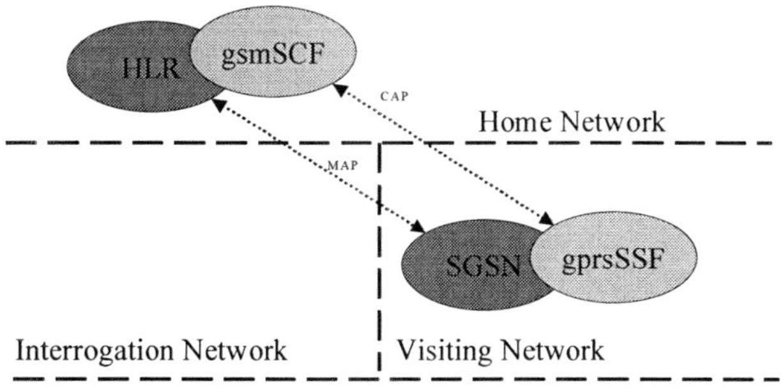

Figure 3. GPRS functional architecture for supporting CAMEL phase 3 [4].

B. UMTS CAMEL Architecture Integrated with Mobile Agent Technology

In the CAMEL agent-based service environment with MAT, MAs can be deployed in end-user systems, Service Switch Points, or distributed Service Control Points. MAs can bring intelligent services as close as possible to the end user or any resource node. Thereby, it can reduce the signaling traffic and resource usages in the network. The intelligent services can be regarded as service personalization or user environment customization. Thus, the integration of CAMEL and MAT will provide great flexibility. Furthermore, the MAT will be deployed in the distributed environment (e.g., CORBA) in CAMEL instead of the centralized environment in CAMEL. The MAT deployed in CORBA can solve the bottleneck problem of the traditional centralized CAMEL mode to improve performance.

1.) Integration of an Agent System into CAMEL

The integration of MAT into CAMEL in distributed environments needs some specific interfaces and functions. That is, the Physical Entities (PEs) and Function Entities (FEs) of CAMEL will be changed. Comparing with the GSM in CAMEL, the UMTS in CAMEL will adopt new FEs and PEs. According to the specification of CAMEL phase 3, *gprsSSF* has been defined. In the future CAMEL version, *umtsSSF* or *umtsSCF* will be used. So we used prefix *umts* with original CAMEL entities to deploy new functional entities, such as *umtsSSF*, *umtsSCF*, and *umtsSDF*, which reside within UMTS network components (USMP: UMTS Service Management Point, USCP: UMTS Service Control Point, UMSC: UMTS Mobile Switch Center, and UE: User Equipment) [11].

The FEs offer places for the installation and maintenance of agent systems to execute or control. Figure 4 shows the integration of an agent system into UMTS CAMEL architecture. The following FEs are defined [11]:

- UMTS Service Creation Environment (*umtsSCE*)
- Combined UMTS Service Control Function (*umtsSCF*) and UMTS Service Data Function (*umtsSDF*).
- Combined UMTS Service Control Function (*umtsSCF*) and UMTS Service Switch Function (*umtsSSF*)
- UMTS Mobile Control Function (*umtsMCF*)
- USIM (UMTS Subscriber Identity Module) Agent System (*UAS*)

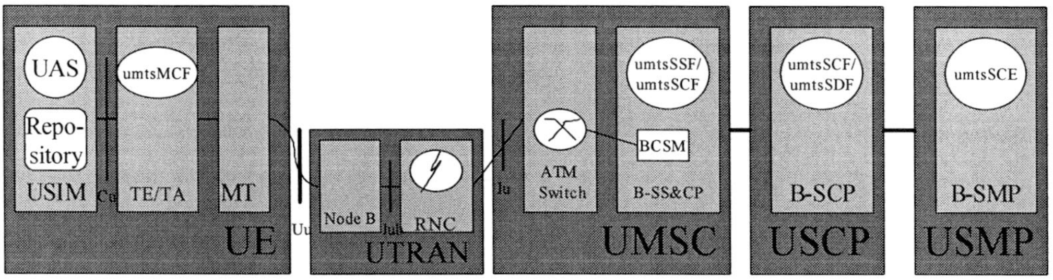

Figure 4. The integration of an agent system into UMTS CAMEL architecture.

USMP is a Broadband SMP (B-SMP) that provides a *umtsSCE* agent environment [19]. The *umtsSCE* is the design and test environment. Service providers develop their distinct service agents in the *umtsSCE* place. USCP is also a Broadband SCP (B-SCP) that provides a *umtsSCF* agent environment. The *umtsSCF* is a centralized FE that integrates agent systems that have access to a secure database that stores the user profiles and service profile [11]. The *umtsSDF* is a centralized database (just like HLR in GSM) and is attached to the *umtsSCF*. In UMSC, it contains an ATM switch and a Broadband Service Switch & Control Point (B-SS&CP). The *umtsSCF* in UMSC is a decentralized FE that offers a place for agent-based mobility management [11]. If a Location Update (LU) has taken place in the agent place, the Detection Points (DPs) trigger the *umtsSSF*. Basically, the *umtsSSF* is responsible for establishing a bearer connection to enable communication [11]. In UTRAN, it contains a Radio Network Control (RNC) and a Node B. A UE communicates with UMSC through UTRAN. It contains a Mobile Terminal (MT), Terminal Agency (TA) and USIM. The *umtsMCF* integrates a permanent agent system that provides a place to execute an MA or download an agent from a UE to another FE in the network [11]. The *umtsMCF* also offers the interface that is used to request bearer connection for communication. The UAS, which can be removed or exchanged by another UAS, resides in a USIM [11].

2.) Agent Entities in UMTS CAMEL Architecture

In [11], some employed agents that are delegated special functionalities for the VHE have been introduced:

- Adaptive Profile Manager (APM) is a static agent that maintains a centralized register for storing subscriber information, such as HLR in GSM. The APM stores user profiles, including the user execution environment profile, security profile, and application profile.
- Personal Communication Manager (PCM) is a static intelligent agent that provides the user with a configurable communication environment. This includes the automatic handling of incoming calls [24].
- User Profile Agent (UPA) is identified as a personalized mobile profile agent and exists for each user. It can follow a user to roam or stay in the home network.
- Terminal Profile Agent (TPA) that resides locally in every terminal provides a start-up interface to the user, which allows the user to register at an APM in a Serving Network (SN).
- Service Logic Manager (SLM) is a static agent that implements the functionalities of a registry server in a Registry System. The SLMs store the location information and the service description for each service represented by the SPA. A roamed UPA can request a specific SPA of a subscribed service from the SLM.
- Service Profile Agent (SPA) is an MA, which must be implemented by every service provider who wants his/her service to be offered by a service provider.

IV. Proposed Mobile Agent-based Platform for Dynamic Service Provisioning in UMTS CAMEL Architecture

Figure 5 shows the proposed agent-based platform in UMTS CAMEL architecture. In USMP, the *umtsSCE* agency can develop all kinds of agent services. In USCP, the *umtsSCF* agency can deploy service agents, such as the APM, SLM, and PCM, in it, and *umtsSDF* can store service agents in it, In UMSC, the *umtsSSF* agency also has the SLM, which stores service agents that users had ever used. In UE, the *umtsMCF* agency contains the TPA that has terminal information and capabilities. In this platform, system operators and service providers can develop all kinds of services easily. And users also can access services dynamically.

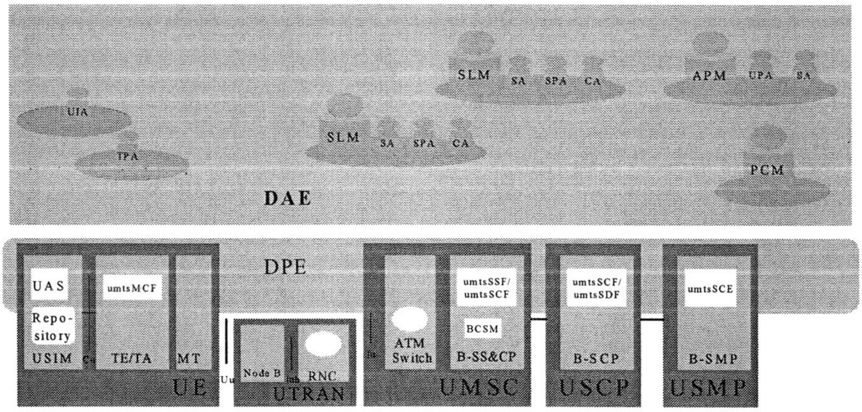

Figure 5. Proposed agent-based platform for dynamic service provision in UMTS CAMEL architecture.

A. Agent-based CAMEL Call Processing in UMSC

In the CORBA environment, service logic and data will have to be implemented in an object-oriented manner within the DPE. MAs can be regarded as a plain CORBA object, because MAs are not only object-oriented but also have special characteristics, such as being autonomous and intelligent. MAs can help finish the work more easily. So the service objects can be implemented as MAs residing in an agency.

Figure 6 shows the proposed agent-based CAMEL call processing in UMSC. It illustrates how to process the agent-based CAMEL incoming call in UMSC. It also shows the integration of MAs with the CAMEL service. In UMSC, an enhanced service, *CAMEL Adapter Service* is used to control the resources. It represents a bridge between the agent environment and basic switching capabilities [18]. The trigger table that has formerly been maintained by the switch itself is now provided by the stationary trigger agent [18]. This trigger agent collects responsive services information and maps the trigger of a DP and finds a correct service agent with the Service Login Manager (SLM). The SLM has all information about the service that registers in local or other places. A service agent can be downloaded dynamically (on-demand) to the switch or executed by remote control where they are currently required. Then the service agent communicates with the trigger agent and delivers the necessary trigger information. During the call processing, an incoming call is going through the UTRAN. In UMSC, the B-SS&CP processes this incoming call. And if it knows the incoming call has CAMEL services, it is based on the BCSM to process the call. When a DP in BCSM detects service events in UMSC (steps 1-2), the processing is suspended. Then the trigger agent contacts with the corresponding service and determines which service shall be chosen (step 3). The trigger agent sends the service information to the SLM and finds a correct service in the SLM. If the service is found, it will be accessed and executed (steps 4-5). After returning the result, the switch resumes the basic call processing (steps 6-7) and delivers the call to the destination. With this agent-based CAMEL call processing, the UMTS system operators and service providers will provide services dynamically to their subscribed users.

The concept of VHE allows users to get access to their personal services in different networks. It provides service portability and personal mobility between terminals and networks. To achieve the goal, the user must have his/her own profile that stores his/her own information or subscribed services. The user profile, which is implemented by the User Profile Agent (UPA), will always follow with the user. The benefits from the UPA that follows the user are reduction of signaling response time and service execution time [11]. According to the user profile, the dynamic provision services will provide to the subscribers. To understand the dynamic service provisioning, we discuss the registration and location update in the following.

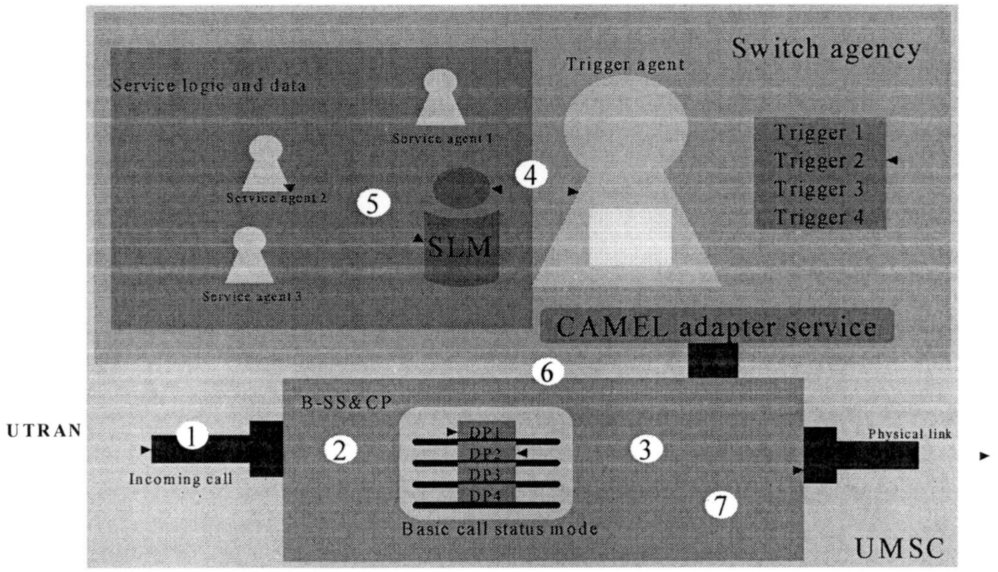

Figure 6. Agent–based CAMEL call processing in UMSC.

B. Registration and Location Update

Before using their subscribed services, the users must register in the communication system. The registration procedure in the agent-based platform is shown in Figure 7. The user turns on his mobile terminal (step 1) and registers his/her identification and terminal information to the APM in the service environment by the TPA interface (step 2). The TPA tells the APM about the capability of the user terminal. When the APM receives the user registration information, it stores the user information in its database (like VLR in GSM) and finds usable services for the TPA. Then the APM in the SE (service environment) forwards the user information to the APM in the HE (home environment) (step 3). When the APM in the HE receives the data, it updates the information of user location and finds the corresponding User Profile Agent (UPA) (step 4). The UPA has all user information, which includes user preference, subscribed services, execution environment, and security parameters, etc. The UPA copies itself and migrates to the serving UMSC, where the user is (step 5). The UPA in UMSC can communicate with the APM and also can store information in the APM of the SE (step 6). When the APM knows the existence of the UPA, it sends a message to the TPA and tells the user that the registration procedure has completed (step 7). Furthermore, the APM in the SE will communicate with the SLM in the SE and finds the required services that the user often uses in the past according to the UPA (steps 8-9). Next, the required services agent will migrate to the serving UMSC (step 10). When the user wants to make a call or use some service, he/she can use the service immediately and the signaling traffic and response time can be reduced. Location updating happens at the first time during the registration procedure, and that happens again when the user moves to another UMSC. The location updating procedure is almost the same as the registration procedure, because the UPA will follow the user closely.

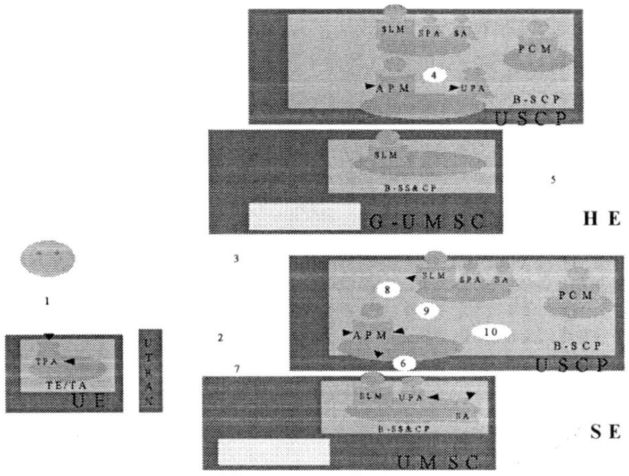

Figure 7. Registration procedure in the agent-based platform.

C. Dynamic Service Provision

To illustrate the proposed agent-based architecture, we use incoming and outgoing call scenarios to explain the operation of MAs in UMTS. The first example is based on incoming call scenarios, and the second one is based on outgoing call scenarios.

1.) Incoming Calls Scenarios

Figure 8 shows an incoming call scenario that a user has subscribed a required service to make an incoming call but the SE cannot provide it. When the G-UMSC receives an incoming call, the Call Control Unit (CCU) will process this call (step 1). Then the CCU contacts the APM in the HE and finds the correct user UPA (step 2). When the UPA is founded, it will be executed (step 3). The UPA communicates with the PCM in order to know the operation of the call from the user call configuration (steps 4-5). The PCM allows the user to flexibly process the incoming call. The incoming call can be a voice call, e-mail, fax, SMS, etc., and can be converted into different types, e.g., voice call to e-mail, e-mail to SMS, or fax to SMS. The UPA in the HE communicates with the UPA in the SE to inform the type of this call (step 6). Later the UPA in the SE communicates with the TPA, and knows the capabilities of the user terminal (steps 7-8). Afterwards, the UPA will check the SLM in UMSC. Here, we assume the SE cannot provide the service in this scenario. If the SLM does not have the service that the user wants, the UPA will check the master SLM in USCP (steps 9-10). After checking the SLM in USCP, the UPA knows that the subscribed service can not be provided by the SE and returns the result to the UPA in the HE (step 11). Then the UPA contacts the SLM in the HE and searches for the correct service (steps 12-13). If the SLM is found, the service agent will migrate to the UMSC in the SE (step 14). Otherwise, the UPA will contact a third party service provider (steps 12a-14a). When the UPA has found the call service, it will inform the CCU in the HE to deliver the call to the CCU in the SE (steps 15-16). Within the UMSC in the SE, the CCU triggers the Service Agent (SA) to execute (step 17) and then the SA sends the Service Profile Agent (SPA) to the user mobile terminal (step 18). The SPA is an interface for users to communicate with the SA. Finally, this call can be established with the SPA (step 19).

2.) Outgoing Calls Scenarios

Figure 9 shows an outgoing call scenario that a user has subscribed a required service to make an outgoing call but the SE cannot provide it. First, the user uses the TPA interface to inform the UPA that he/she wants to make an outgoing call. At the same time, the TPA also sends the terminal capabilities to the UPA in the SE (steps 0-1). Because the UMSC in the SE cannot provide the required service, the UPA requests the service to the master SLM in USCP. If the USCP also cannot provide the service, it will return the result to the UPA (steps 2-3). Then the UPA searches for the SA according to the configuration of the user profile. The SA may be in the USCP of the HE (steps 4-6), or in a third party service provider (steps 4a-6a). When the UPA in the SE receives the result and knows that a required SA is in the UMSC, the UPA will inform the CCU to execute the SA and establish the physical link (steps 7-10). That completes the total outgoing call procedure.

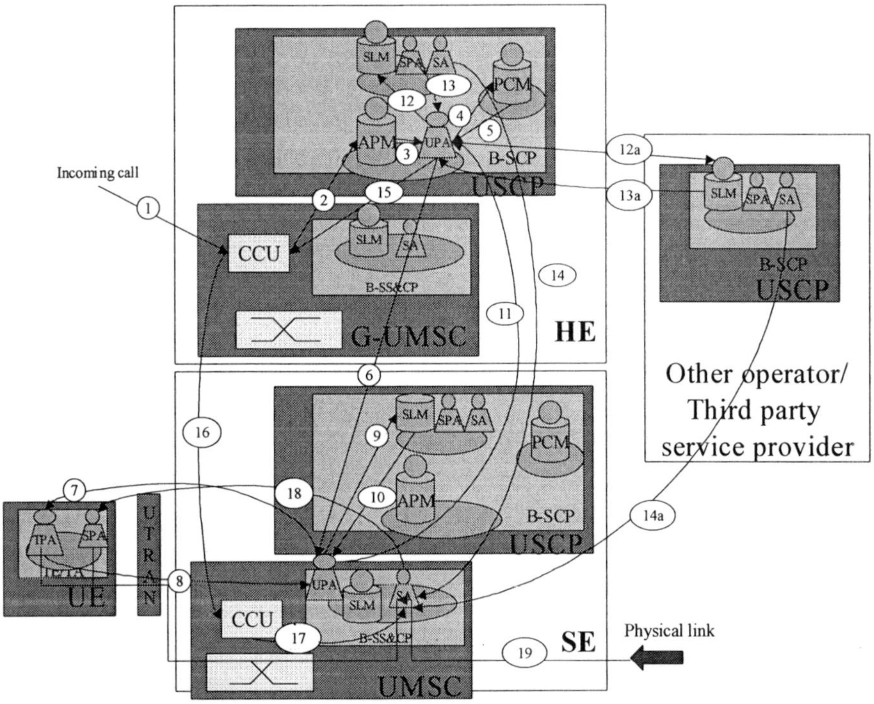

Figure 8. An incoming call scenario: a user has subscribed a required service to make an incoming call but the SE cannot provide it.

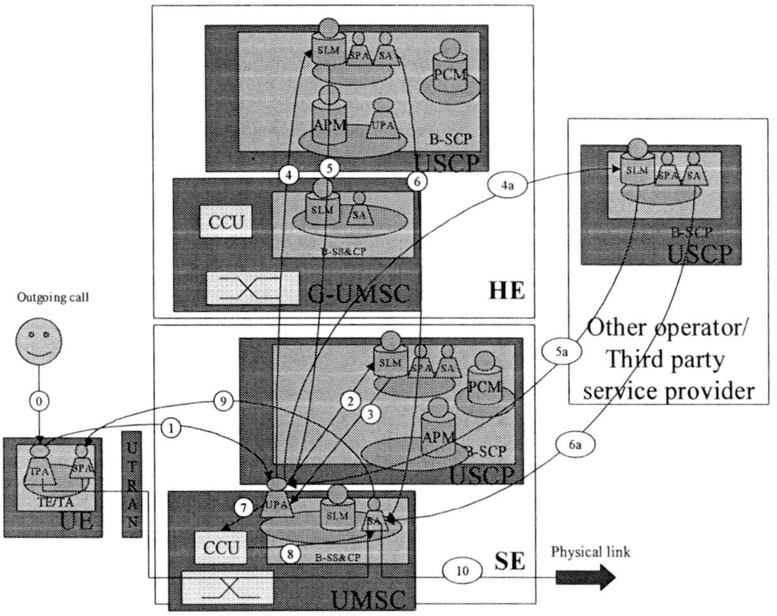

Figure 9. An outgoing call scenario: a user has subscribed a required service to make an outgoing call but the SE cannot provide it.

V. Analysis and Discussion

To analyze performance of the proposed UMTS CAMEL agent-based platform, we will compute the expected traffic volumes of signaling and service downloading for dynamic service provisioning based on the above two different service scenarios [7]. For evaluation, three probabilities related to the user, the UMSC, and the USCP are defined as follows [7]:

- P_U: the probability that the user has subscribed the required service to handle a call.
- P_S: the probability that the UMSC of the current serving environment provides the required service.
- P_C: the probability that the USCP of the current serving environment provides the required service.

The following three factors are taken into account to form six incoming/outgoing call scenarios [7]:

1. If the user has subscribed the required service or not;
2. If the UMSC of the current serving environment provides the required service or not;
3. If the USCP of the current serving environment provides the required service or not.

The above three factors are independent, because the UMSC can provide the required service independently, regardless whether the USCP provides the service or not. Therefore, when the UMSC has provided the required service, the service provided by the USCP will not be taken into account. When the UMSC cannot provide the required service, the service provided by the USCP will be considered. If a user does not have subscribed the required services, the user will be asked for subscribing the service immediately; otherwise the system operator will reject the incoming call/outgoing call.

Table 1 shows six call scenarios according to the above three factors. The character *v* indicates that the statement is true; otherwise, false. The character *x* indicates that the statement is *don't care*. According to Table 1, the occurrence probability of each incoming/outgoing call scenario can be computed as follows:

$$P_{S1} = P_U \times P_S \tag{1}$$

$$P_{S2} = P_U \times (1 - P_S) \times P_C \tag{2}$$

$$P_{S3} = P_U \times (1 - P_S) \times (1 - P_C) \tag{3}$$

$$P_{S4} = (1 - P_U) \times P_S \tag{4}$$

$$P_{S5} = (1 - P_U) \times (1 - P_S) \times P_C \tag{5}$$

$$P_{S6} = (1 - P_U) \times (1 - P_S) \times (1 - P_C) \tag{6}$$

Without loss of generality, we assume the probability that the user subscribes the required service equals to 0.9. And the probability that the USCP has the required service to provide to the user also equals to 0.9. Then according to Equations (1) - (6), we can compute the probabilities of P_{S1} - P_{S6} as P_S varies from 0 to 1.

Table 1. Incoming/outgoing call scenarios.

Scenario	User has subscribed the required service (U)	UMSC provides required service (S)	USCP provides required service (C)
S1	*v*	*v*	*x*
S2	*v*		*v*
S3	*v*		
S4		*v*	*x*
S5			*v*
S6			

1.) Incoming Call Analysis

According to Figure 8, which belongs to scenario S3, we can compute the number of signaling and times of service downloading. The figure shows that the number of signaling between G-UMSC and UMSC is 3. The number of signaling between UMSC and USCP is 0. And the number of signaling between UMSC and UE is 2, etc. Table 2 shows the numbers of signaling and times of service downloading under different incoming call scenarios.

The expected number of signaling between G-UMSC and UMSC is denoted by *sig(GUMSC-UMSC)* and is computed as follows:

$$sig(GUMSC - UMSC) = 3 \times (P_{S1} + P_{S2} + P_{S3} + P_{S4} + P_{S5} + P_{S6}) = 3 \quad (7)$$

Similarly, the others can be computed as follows:

$$sig(GUMSC - UMSC(3rd)) = 2 \times (P_{S3} + P_{S6})$$
$$= 2 \times (1 - P_S)(1 - P_C) \quad (8)$$

$$sig(UMSC - USCP) = 2 \times (P_{S2} + P_{S5} + P_{S6})$$
$$= 2 \times (1 - P_S)(P_U \times P_C + 1 - P_U) \quad (9)$$

$$sig(UMSC - UE) = 2 \times (P_{S1} + P_{S2} + P_{S3} + P_{S4} + P_{S5} + P_{S6}) = 2 \quad (10)$$

The expected times of service downloading between G-UMSC and UMSC is denoted by *serv(GUMSC-UMSC)* and is computed as follows:

$$serv(GUMSC - UMSC) = P_{S3} = P_U \times (1 - P_S) \times (1 - P_C) \quad (11)$$

Similarly, the others can be written as follows:

$$serv(GUMSC - UMSC(3^{rd})) = P_{S3} + P_{S6} = (1 - P_S) \times (1 - P_C) \quad (12)$$

$$serv(UMSC - USCP) = P_{S2} + P_{S5} = (1 - P_S) \times P_C \quad (13)$$

$$serv(UMSC - UE) = (P_{S1} + P_{S2} + P_{S3} + P_{S4} + P_{S5} + P_{S6}) = 1 \quad (14)$$

Table 3 shows the result under the conditions of $P_U = 0.9$ and $P_C = 0.9$. We can see that the increase of P_S will result in the decrease of the number of signaling and times of service downloading between G-UMSC and USCP (3rd), and between UMSC and USCP. That means when P_S is larger, and the traffic volumes of signaling and service downloading will be less. In addition, the signaling between GUMSC and UMSC, and between UMSC and UE cannot be reduced and will not be affected by the variation of P_S.

Table 2. The number of signaling and times of service downloading under different incoming call scenarios.

Incoming Call Scenario	Signaling				Service download			
	G-UMSC – UMSC	G-UMSC – USCP(3rd)	UMSC – USCP	UMSC – MS	G-UMSC – UMSC	UMSC – USCP(3rd)	UMSC – USCP	UMSC – MS
S1	3	0	0	2	0	0	0	1
S2	3	0	2	2	0	0	1	1
S3	3	2	0	2	1	1	0	1
S4	3	0	0	2	0	0	0	1
S5	3	0	2	2	0	0	1	1
S6	3	2	2	2	0	1	0	1

Then, we compare the number of signaling of incoming call scenarios between the CORBA agent-based [7] and our CAMEL agent-based platforms, as shown in Table 4. Figure 10 compares the total signaling numbers between the CORBA agent-based and our CAMEL agent-based platforms under different incoming call scenarios. For incoming call scenarios, the performance of our CAMEL agent-based platform is almost equal to the CORBA agent-based platform. Note that the CORBA agent-based platform assumes that the service of MSC must be provided by the GMSC. Thus, if the GMSC does not have the required service for a user, the MSC will not provide the required service to the user. When an incoming call arrives, it just decides if the required service is in the GMSC or not. Therefore, the signaling traffic between GMSC and VMSC in the CORBA agent-based platform is slightly less than that in our CAMEL agent-based platform under incoming call scenarios 1 - 3.

Table 3. Expected traffic volumes of signaling and service downloading under incoming call scenarios (given $P_U = 0.9$, $P_C = 0.9$).

P_S	Signaling				Service download			
	G-UMSC – UMSC	G-UMSC – USCP(3rd)	UMSC – USCP	UMSC – MS	G-UMSC – UMSC	UMSC – USCP(3rd)	UMSC – USCP	UMSC – MS
0	3	0.2	1.82	2	0.09	0.1	0.9	1
0.1	3	0.18	1.638	2	0.081	0.09	0.81	1
0.2	3	0.16	1.456	2	0.072	0.08	0.72	1
0.3	3	0.14	1.274	2	0.063	0.07	0.63	1
0.4	3	0.12	1.092	2	0.054	0.06	0.54	1
0.5	3	0.1	0.91	2	0.045	0.05	0.45	1
0.6	3	0.08	0.728	2	0.036	0.04	0.36	1
0.7	3	0.06	0.546	2	0.027	0.03	0.27	1
0.8	3	0.04	0.364	2	0.018	0.02	0.18	1
0.9	3	0.02	0.182	2	0.009	0.01	0.09	1
1	3	0	0	2	0	0	0	1

Table 4.Comparing signaling numbers between CORBA agent-based and CAMEL agent-based platforms under different incoming calls scenarios.

Incoming call Scenarios	CORBA agent platform					UMTS CAMEL agent platform				
	Signaling number					Signaling number				
	GMSC – VMSC	GMSC – GMSC (3rd)	GMSC – GMSC	VMSC – MS	Total	G-UMSC – UMSC	G-UMSC – USCP (3rd)	UMSC – USCP	UMSC – UE	Total
S1	1	0	0	2	3	3	0	0	2	5
S2	1	0	0	2	3	3	0	2	2	7
S3	1	2	1	2	6	3	2	0	2	7
S4	3	0	0	4	7	3	0	0	2	5
S5	3	0	0	4	7	3	0	2	2	7
S6	3	2	1	4	10	3	2	2	2	9

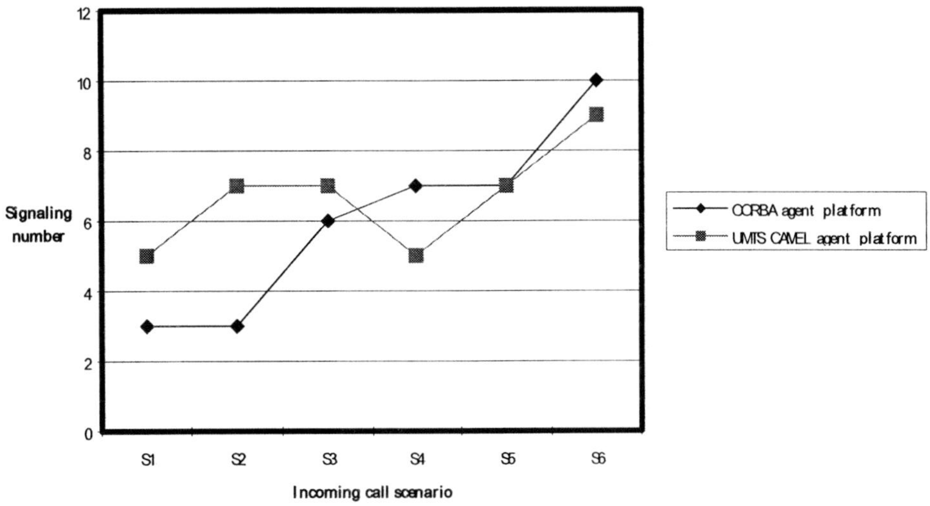

Figure 10. Comparing signaling numbers between CORBA agent-based and CAMEL agent-based platforms under different incoming call scenarios.

2.) Outgoing Call Analysis

From Figure 9, we can compute the number of signaling and times of service downloading under outgoing call scenarios. It shows that the number of signaling between G-UMSC and UMSC is 2. The number of signaling between UMSC and USCP is 2. And the number of signaling between UMSC and UE is 1, etc. Table 5 shows the numbers of signaling and times of service downloading under six outgoing call scenarios (as specified in Table 1).

The expected number of signaling between G-UMSC and UMSC is denoted by *sig(GUMSC-UMSC)*, and is computed as follows:

$$sig(GUMSC - UMSC) = 2 \times (P_{S3} + P_{S6})$$
$$= 2 \times (1 - P_S) \times (1 - P_C) \quad (15)$$

Table 5. The number of signaling and times of service downloading under different outgoing call scenarios.

Outgoing Call Scenario	Signaling				Service download			
	G-UMSC | UMSC	G-UMSC | USCP(3rd)	UMSC | USCP	UMSC | MS	G-UMSC | UMSC	UMSC | USCP(3rd)	UMSC | USCP	UMSC | MS
S1	0	0	0	1	0	0	0	1
S2	0	0	2	1	0	0	1	1
S3	2	2	2	1	1	1	0	1
S4	0	0	0	3	0	0	0	1
S5	0	0	2	3	0	0	1	1
S6	2	2	2	3	1	1	0	1

Similarly, the others can be computed as follows:

$$sig(GUMSC - UMSC(3rd)) = 2 \times (P_{S3} + P_{S6})$$
$$= 2 \times (1 - P_S) \times (1 - P_C) \quad (16)$$

$$sig(UMSC - USCP) = 2 \times (P_{S2} + P_{S3} + P_{S5} + P_{S6})$$
$$= 2 \times (1 - P_S) \quad (17)$$

$$sig(UMSC - UE) = 1 \times (P_{S1} + P_{S2} + P_{S3}) + 3 \times (P_{S4} + P_{S5} + P_{S6})$$
$$= 3 - 2 \times P_U \quad (18)$$

The expected times of service downloading between GUMSC and UMSC is denoted by *serv(GUMSC-UMSC)*, and is computed as follows:

$$serv(GUMSC - UMSC) = P_{S3} + P_{S6} = (1 - P_S) \times (1 - P_C) \quad (19)$$

$$serv(GUMSC - UMSC(3rd)) = P_{S3} + P_{S6} = (1 - P_S) \times (1 - P_C) \quad (20)$$

$$serv(UMSC - USCP) = P_{S2} + P_{S5} = (1 - P_S) \times P_C \quad (21)$$

$$serv(UMSC - UE) = (P_{S1} + P_{S2} + P_{S3} + P_{S4} + P_{S5} + P_{S6}) = 1 \quad (22)$$

Table 6 shows the result under the condition of $P_U = 0.9$ and $P_C = 0.9$. We can see that the increase of P_S will result in the decrease of the number of signaling and times of service downloading between G-UMSC and UMSC, between G-UMSC and USCP(3rd), and between UMSC and USCP. That means when P_S is larger, the traffic volumes of signaling and service downloading will be less. In addition, the signaling between UMSC and UE cannot be reduced and will not be affected by the variation of P_S.

Signaling numbers between CORBA agent-based and CAMEL agent-based platforms under different outgoing call scenarios are compared in Table 7. The signaling traffic can be reduced in our platform, particularly between GUMSC and UMSCP and between UMSC and UE. In addition, the total signaling number can be also reduced greatly in our platform. Therefore, our UMTS CAMEL agent platform has better performance than the CORBA agent-based platform. Figure 11 compares the total signaling numbers between CORBA agent-based and CAMEL agent-based platforms under different outgoing call scenarios. It shows that our UMTS CAMEL agent-based platform has better performance and is more efficient than the CORBA agent-based platform. The signaling number in the proposed UMTS CAMEL agent-based platform can be reduced 40% compared to that in the CORBA agent-based platform.

Table 6. Expected traffic of signaling and service downloading under outgoing call scenarios (given $P_U = 0.9$ and $P_C = 0.9$).

P_S	Signaling				Service download			
	G-UMSC | UMSC	G-UMSC | USCP(3rd)	UMSC | USCP	UMSC | MS	G-UMSC | UMSC	UMSC | USCP(3rd)	UMSC | USCP	UMSC | MS
0	0.2	0.2	2	1	0.1	0.1	0.9	1
0.1	0.18	0.18	1.8	1	0.09	0.09	0.81	1
0.2	0.16	0.16	1.6	1	0.08	0.08	0.72	1
0.3	0.14	0.14	1.4	1	0.07	0.07	0.63	1
0.4	0.12	0.12	1.2	1	0.06	0.06	0.54	1
0.5	0.10	0.1	1	1	0.05	0.05	0.45	1
0.6	0.08	0.08	0.8	1	0.04	0.04	0.36	1
0.7	0.06	0.06	0.6	1	0.03	0.03	0.27	1
0.8	0.04	0.04	0.4	1	0.02	0.02	0.18	1
0.9	0.02	0.02	0.2	1	0.01	0.01	0.09	1
1	0	0	0	1	0	0	0	1

Table 7. Comparing signaling numbers between CORBA agent-based and CAMEL agent-based platforms under different outgoing call scenarios.

Outgoing call scenarios	CORBA agent platform					UMTS CAMEL agent platform				
	Signaling number					Signaling number				
	GMSC | VMSC	GMSC | GMSC (3rd)	GMSC | GMSC	VMSC | MS	Total	G-UMSC | UMSC	G-UMSC | USCP (3rd)	UMSC | USCP	UMSC | UE	Total
S1	1	0	3	3	7	0	0	0	1	1
S2	2	0	3	3	8	0	0	2	1	3
S3	2	2	4	3	11	2	2	2	1	7
S4	1	0	3	5	9	0	0	0	3	3
S5	2	0	3	5	11	0	0	2	3	5
S6	2	2	4	5	13	2	2	2	3	9

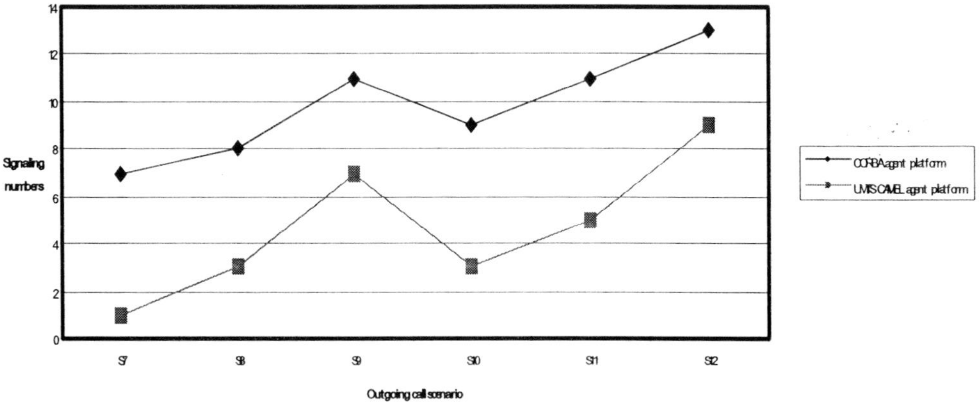

Figure 11. Comparing signaling numbers between CORBA agent-based and CAMEL agent-based platforms under different outgoing call scenarios.

VI. Conclusions

In this paper, we have described the concept of CAMEL and Virtual Home Environment (VHE), and have also shown how UMTS CAMEL can be integrated with mobile agents (MAs) to provide dynamic services. The CAMEL is an application architecture that combines the intelligent network concept and mobile communication systems to provide useful and powerful services to users. The evolution trend of mobile communications moves toward the VHE. The mobile agent technology provides flexibility and scalability that enables integrated terminal mobility, personal mobility, and service mobility, as envisaged by the VHE [25][26]. In addition, we have presented a CAMEL mobile-agent-based platform to provide dynamic services in UMTS. We have used the operations of incoming calls and outgoing calls to evaluate our platform. The analysis results have shown that our platform is indeed more efficient than the CORBA agent-based platform in terms of generating 40% less signaling traffic.

References

[1] V. D. Modak, D. D. Langan, and T. F. Hain, “A Pattern-based development tool for mobile agents,” in *Proc. ACM SIGCSE*, 2005, vol. 37, pp. 72-75.

[2] C. Panayiotou and G. Samaras, “mPERSONA: Personalized portals for the wireless user: an agent approach,” *Mobile Networks and Applications*, vol. 9, pp. 663-677, Dec. 2004.

[3] 3GPP TS 23.955, “Virtual home environment (VHE) concepts,” V0.1.0, Jan. 2001.

[4] 3G TS 23.127, “Virtual home environment,” V4.1.0, April 2001.

[5] 3G TS 22.078, “Customized Applications for Mobile network Enhanced Logic (CAMEL),” Service Description, Stage 1, V5.2.0, Dec. 2001.

[6] 3G TS 23.078, “Customized Applications for Mobile network Enhanced Logic (CAMEL),” Service Description, Stage 2, V5.2.0, Dec. 2001.

[7] C. Dou, Y. P. Chen, and H. K. Chen, "An agent-based platform for dynamic service provisioning in 3G mobile systems: scenarios and performance analyses," in *Proc. of the 15th International Conference on Information Networking*, Jan.- Feb. 2001, pp. 883-888.

[8] L. Hagen, M. Breugst, and T. Magedanz, "Impacts of mobile agent technology on mobile communication system evolution," *IEEE Personal Communications*, vol. 5, pp. 56-69, Aug. 1998.

[9] ACTS Project AC341 D06, "Communication Agents for Mobility Enhancements in a Logical Environment of Open Network (CAMELEON): Concept Description," April 1999.

[10]ACTS Project AC341 D08, "Communication Agents for Mobility Enhancements in a Logical Environment of Open Network (CAMELEON): Specification of Prototype," July 1998.

[11]ACTS Project AC341 D09, "Communication Agents for Mobility Enhancements in a Logical Environment of Open Network (CAMELEON): Performance Assessment Results," Nov. 1999.

[12]OMG Mobile Agent System Interoperability Facility (MASIF), http://www.fokus.gmd.de/research/cc/ecco/masif/.

[13]Foundation of Intelligent Physical Agents (FIPA), http://www.fipa.org/.

[14]M. Finkelstein, J. Garrahan, D. Shrader, and G. Weber, "The future of the intelligent network," *IEEE Communication,* vol. 38, pp. 100-106, June 2000.

[15]M.L.F. Grech, "Providing seamless services for VoIP mobile data networks using CAMEL/IN concepts," in *Proc. of the 1st International Conference on 3G Mobile Communication Technologies*, March 2000, pp.133-137.

[16]J. D. Humphrey, "Interworking and the IN platform: detailing the development of the GSM CAMEL standard for interworking IN," in *Proc. of the 6th IEE Conference Telecommunications*, March 1998, pp. 250 –257.

[17]P. Meskauskas, "Customized applications for mobile enhanced logic (CAMEL)," in *Proc. of the Research Seminar on Nomadic Computing*, Department of Computer Science, University of Helsinki, Aug. 2001.

[18]M. Breugst, T. Magedanz, "Mobile agents - enabling technology for active intelligent network implementation," *IEEE Network*, vol. 12, pp. 53-60, May-June 1998.

[19]F.G. Chatzipapadopoulos, M.K. Perdikeas, and I.S. Venleris, "Mobile agent and CORBA technologies in the broadband intelligent network," *IEEE Communications*, vol. 38, pp. 116-124, June 2000.

[20]ACTS Cluster for Intelligent Mobile Agents in Telecommunication Environments (CLIMATE), http://www.fokus.gmd.de/research/cc/ecco/climate/.

[21]ACTS CAMELEON project, http://www.comnets.rwth-aachen.de/~cameleon.

[22]IKV++ Grasshopper2 agent platform, http://www.grasshopper.de/.

[23]IBM Aglet agent platform, http://www.trl.ibm.com/aglets/index_e.htm.

[24]R. P. Zeletin, and T. Magedanz, "Towards intelligence on demand - on the impact of intelligent agent in IN," in *Proc. of the International Conference on Intelligent Networks*, Nov. 1996.

[25]S. Lipperts and A. S. B. Park, "An agent-based middleware: a solution for terminal and user mobility," *Computer Networks*, vol. 31, no.19, pp. 2053-2062, Aug. 1999.

[26]L. Hagen, J. Mauersberger, and C. Weckerle, "Mobile agent based service subscription and customization using the UMTS virtual home environment," *Computer Networks*, vol. 31, no19, pp. 2063-2078, Aug. 1999.

[27]C. S. Hong, Y. Koga, and Y. Matsushita, "A network architecture for mobility services using mobile agent approach," in *Proc. of TINA on Telecommunications and Distributed Object Computing*, Nov. 1997, pp. 297-307.

[28]J. Bigham, L. Yokarchuk, D.J. Cuthbert, J. Lisalina, M. Dinis and S. Robles, "Agent-based resource management for 3G networks", in *Proc. of the 2nd International Conference on 3G Mobile Communication Technologies*, March 2001, pp. 236-240.

INDEX

A

A - Arcs, 193
access, vii, viii, ix, xi, 76, 83, 84, 85, 86, 87, 88, 89, 90, 91, 96, 97, 98, 99, 100, 101, 102, 103, 104, 105, 107, 116, 118, 120, 123, 126, 127, 128, 129, 133, 139, 140, 143, 144, 145, 146, 148, 149, 150, 151, 152, 153, 155, 160, 165, 170, 183, 185, 231, 301, 325, 327, 328, 338, 340, 341, 342, 345, 350, 352, 354, 355, 356
accounting, 86, 104, 196
accuracy, 10, 106, 115, 194, 203, 207, 212, 213, 238, 255
achievement, 2
activation, 100, 141
actuation, 259, 260
actuators, ix, 229, 259, 260
ad hoc network(ing), vii, 49, 50, 51, 55, 56, 61, 70, 71, 74, 76, 80, 81, 82, 85, 174, 188, 189, 226
adaptability, 130
adaptation, 139, 144
adjustment, 258
advertisements, 88, 156, 157
advertising, 3
age, 136, 168, 182
agent, xii, 90, 349, 351, 353, 354, 355, 356, 357, 358, 359, 360, 362, 363, 364, 365, 366, 367, 368, 369
aggregates, 178
aggregation, xi, 167, 168, 169, 170, 177, 301
aging, 231
agriculture, 166
airports, 325
algorithm, 9, 12, 16, 19, 21, 24, 28, 31, 32, 36, 39, 42, 45, 52, 64, 71, 72, 73, 76, 86, 100, 115, 126, 129, 130, 132, 133, 135, 136, 137, 138, 139, 147, 148, 149, 150, 151, 160, 171, 172, 173, 175, 176, 182, 183, 194, 269, 271, 272, 274, 275, 279, 295, 296
alternative, 173, 350
aluminium, 238
aluminum, 244, 246
analytical framework, 119
anatomy, 232
animals, 166
annealing, 260
architecture design, 118
argument, 178
ARs, 96, 98, 112
Asia, 284, 303
aspect ratio, 252, 254
assignment, 126, 138, 150, 151, 160, 172, 173, 187, 222, 296
assumptions, 2, 3, 22, 57, 105, 207, 239, 241
attachment, 85, 87, 90, 101, 108, 115
attacker, 63, 64, 65, 66, 67, 70, 74, 283
attacks, viii, 6, 55, 56, 57, 58, 60, 61, 70, 79, 268, 274, 283
attention, 98, 100, 128, 138, 239, 287, 350
authentication, 58, 59, 62, 63, 64, 71, 76, 86, 93, 104, 105, 108, 110, 268, 269, 271, 272, 273, 274, 275, 283, 284, 353
authenticity, 282
authority, 59, 70, 71, 89, 274
automata, 194
automation, 166
availability, ix, x, 7, 89, 100, 101, 110, 129, 191, 211, 230, 231, 235, 252, 298, 337
averaging, 210, 218, 291
avoidance, 143, 158
awareness, 92

B

background noise, 290
bandwidth, xi, 4, 5, 7, 61, 62, 63, 84, 85, 87, 88, 89, 96, 97, 100, 101, 102, 103, 104, 112, 197, 217, 222, 230, 232, 233, 245, 246, 247, 250, 253, 254,

258, 262, 268, 274, 278, 297, 301, 302, 309, 310, 335, 336, 342
basic services, 352
batteries, viii, x, 123, 126, 133, 138, 139, 144, 229, 232, 233, 234, 260
beams, 171
behavior, 16, 70, 101, 116, 129, 133, 140, 143, 150, 151, 156, 159, 160, 170, 186, 210, 243, 253, 336
benchmarks, 81
bile, 84
binding, 88, 90, 93, 94, 95, 107, 108, 109, 110, 115, 116
biomedical applications, 230, 232, 247
Bit Error Rate (BER), 195
bladder, 231, 234
bleeding, 231
blocks, 268
Bluetooth, x, 2, 196, 267
bonding, 230, 248, 257, 262
border crossing, 106
bounds, 194
bowel, 231
broadband, 368
browser, 281
buffer, 14, 96, 101, 111, 116, 142, 146
buildings, 171

C

C - Network Configurations, 198
C++, 37, 40, 53
calibration, 245
CAMEL, xii, 349, 350, 352, 353, 354, 355, 356, 357, 360, 362, 363, 364, 365, 366, 367, 368
Canada, 81, 83
candidates, 19, 27, 42, 251
CAP, 352
capsule, ix, 229, 231
carrier, xi, 18, 124, 126, 143, 153, 255, 302, 303
catheters, 231
cation, 76, 81
CE, 119
cell, x, 88, 106, 170, 181, 182, 195, 236, 237, 240, 287, 288, 289, 290, 291, 292, 293, 294, 297, 298
ceramic(s), 235
certificate, 270, 272, 274, 275, 281, 283
certification, 274
channel interference, 289
channels, x, 105, 131, 132, 142, 197, 267, 282, 288, 289, 292, 297, 302, 303, 328
chemical composition, 230
children, 20, 27, 28, 30, 32, 146, 155, 156, 157, 158, 173, 182
China, 187
chopping, 21
cipher, viii, 55, 268, 269, 270, 271, 274, 275, 276, 279, 283
classes, 101
classification, 102, 125, 171, 190
clients, 4, 5, 270, 277, 280
clustering, 3, 56, 65, 73, 222
clusters, viii, 55, 57, 58, 61, 63, 64, 67, 68, 69, 70, 71, 72, 170, 222
coding, 170
coffee, 329
collaboration, 98, 152
collisions, 125, 126, 127, 128, 129, 131, 137, 139, 140, 143, 145, 147, 151, 160, 161, 288, 290
colon, 231
colonoscopy, 231
commodity, 196
communication, vii, ix, x, xi, 4, 7, 11, 12, 14, 31, 32, 56, 59, 60, 74, 76, 102, 105, 123, 124, 125, 127, 128, 131, 132, 139, 141, 142, 143, 146, 147, 148, 149, 153, 166, 169, 171, 172, 177, 178, 181, 182, 187, 193, 195, 199, 229, 230, 233, 267, 270, 287, 294, 297, 302, 304, 325, 326, 329, 341, 342, 350, 351, 352, 355, 357
communication overhead, 12
communication systems, vii, x, 229, 350
community, vii, 1, 12, 18, 23, 98
compatibility, 263
complementarity, 85
complex interactions, 215
complexity, viii, 26, 27, 28, 29, 31, 32, 45, 47, 48, 55, 56, 61, 79, 102, 106, 110, 172, 175, 194, 249, 250, 251, 253, 302, 338
complications, 232
components, 16, 22, 24, 32, 44, 86, 103, 213, 215, 235, 245, 269, 327, 354
composition, 44
computation, 9, 31, 64, 87, 112, 167, 168, 171, 172, 185, 194
computers, vii, 52
computing, vii, 4, 51, 133, 325, 327
concentration, 232
conception, 146
concreteness, 289
concurrency, 3
conductivity, 244, 246
conductor, 238, 240
confidence, 194
configuration, ix, 2, 15, 24, 49, 88, 90, 94, 112, 128, 166, 191, 192, 195, 196, 198, 199, 200, 201, 202, 203, 208, 223, 238, 243, 290, 291, 295, 358, 359
conflict, 150

confusion, 21
connectivity, 16, 22, 45, 89, 96, 104, 139, 175, 192, 195, 202, 208, 209, 210, 216, 217, 218, 329, 330, 342
conservation, 52, 138
constraints, 196, 207, 232, 293
construction, vii, 1, 3, 4, 21, 23, 25, 26, 32, 45, 46, 48, 172
consumers, 4, 5
consumption, viii, 84, 89, 99, 100, 123, 124, 125, 130, 131, 139, 140, 144, 149, 153, 154, 155, 158, 159, 160, 230, 234, 260
consumption rates, 139
continuity, 85, 90, 91, 104, 116
continuous data, 275
control, vii, ix, 1, 28, 47, 86, 87, 91, 93, 100, 101, 104, 108, 109, 110, 119, 120, 123, 124, 127, 131, 132, 133, 134, 135, 136, 138, 139, 140, 142, 145, 146, 149, 154, 158, 165, 167, 169, 184, 231, 234, 235, 271, 277, 283, 288, 299, 331, 333, 334, 350, 351, 354, 356
convergence, 12, 21, 40, 84, 85, 118, 295
copper, 244
CORBA, xii, 225, 349, 350, 351, 354, 356, 362, 363, 364, 365, 366, 367, 368
correlation, 167, 168, 219
cost saving, ix, 115, 229
Costa Rica, 188
costs, 3, 101, 109, 110, 122, 178, 258, 288, 342
coupling, 85, 235, 247, 251
coverage, xi, 52, 85, 87, 100, 102, 106, 107, 110, 135, 143, 145, 195, 207, 208, 209, 210, 211, 214, 215, 217, 218, 287, 325, 326, 327, 328, 329, 330, 331, 334, 335, 337, 338, 339, 340, 341, 342, 343, 345
covering, 327
CPU, 4, 5, 201, 203, 204, 280
cryptography, 76, 189, 269, 272, 274, 281
cycles, 4, 5, 140, 160

D

data analysis, ix
data availability, 180, 185
data base, ix, 165
data collection, 166, 185
data communication, 303
data distribution, 185
data gathering, 153
data processing, 167, 170
data structure, 5, 182, 331
data transfer, 147, 170, 269, 280, 302
database, vii, 4, 6, 167, 168, 170, 176, 177, 178, 180, 185, 186, 187, 354, 357
decay, 149
decentralization, vii, 1, 2, 48
decisions, 207, 215, 222
decoding, 124
decomposition, 180
defects, 4
definition, 15, 20, 21, 22, 208, 233, 243
degradation, xi, 102, 197, 250, 325, 330, 345
delivery, viii, 10, 55, 57, 73, 75, 88, 96, 101, 108, 110, 115, 116, 117, 139, 140, 169, 353
demand, viii, 3, 52, 74, 76, 79, 82, 83, 101, 189, 216, 217, 218, 220, 350, 356, 368
democracy, 5
denial, 6
density, 181, 195, 232, 259, 313, 316, 318, 320, 322
Department of Defense, 196, 224
deregulation, 233
designers, 193
detection, 88, 90, 93, 94, 119, 141, 166, 231, 258, 260, 327
deviation, 220, 224, 308
dielectric constant, 234, 236, 239, 241, 242, 248, 254
dielectric permittivity, 244
diffusion, 167, 168, 169, 189
digestive tract, 231
directional antennas, 171
directionality, 189
directives, ix, 123
discrete random variable, 66, 68
disorder, 86, 231
displacement, 259, 260, 261
disseminate, 169
distributed applications, 12
distributed computing, 5, 11, 16, 18
distribution, 5, 8, 9, 10, 13, 20, 48, 57, 67, 68, 73, 106, 107, 126, 140, 147, 156, 184, 193, 202, 209, 211, 217, 223, 271, 272, 274, 275, 281, 290, 327, 335, 336, 337, 342
division, xi, 126, 156, 157, 158, 184, 301, 303
doctors, 231
dominance, 4, 23
downlink, xi, 143, 299, 302, 303, 309, 310, 311, 312, 313, 314, 315, 316, 318, 320, 322
draft, 49, 80, 119, 120, 121
drug delivery, 231
duodenum, 231
duration, 106, 130, 133, 135, 141, 154, 170, 177, 209, 210, 217, 218, 294, 295
dynamic systems, 12

E

eavesdropping, 56, 70, 283
education, 166
election, 58, 138
electrical properties, x, 229, 235, 236, 239, 241, 255
electrodes, 259
electromagnetic, 236, 260, 261, 262, 263
electronic communications, 267
email, 32
emergence, 4
emergency response, 188
emission, 142, 233
emitters, 125
encryption, 86, 268, 269, 271, 272, 273, 275, 277, 279, 281, 282, 283, 284
endoscopy, ix, 229, 231
energy, viii, 99, 123, 124, 125, 126, 128, 129, 130, 131, 132, 133, 134, 135, 137, 138, 139, 140, 141, 143, 146, 147, 148, 149, 155, 158, 160, 161, 168, 169, 170, 171, 172, 174, 178, 233, 260, 302
energy consumption, 99, 123, 124, 129, 137, 138, 147, 155, 158, 160
energy efficiency, 128, 158, 160, 170, 178
English Language, 50
enhanced service, 356
enthusiasm, vii, 1
entropy, 66, 275, 283
environment, xii, 27, 102, 127, 144, 150, 166, 232, 273, 274, 277, 283, 302, 309, 315, 320, 329, 349, 351, 352, 353, 354, 355, 356, 357, 360, 367, 369
equality, 4, 292
equilibrium, 293
equipment, 167, 263
estimating, 159, 194
etching, 255
Europe, 303, 350
evolution, vii, viii, 12, 83, 84, 129, 130, 167, 181, 194, 288, 298, 350, 353, 367, 368
execution, 18, 24, 42, 87, 88, 89, 90, 104, 105, 155, 177, 204, 335, 336, 338, 343, 351, 352, 355, 357
extraction, 238

F

fabrication, x, 57, 230, 235, 242, 243, 246, 251, 253, 254, 255, 256, 258, 261, 262
failure, xi, 4, 5, 6, 22, 32, 58, 88, 94, 105, 112, 137, 180, 195, 196, 208, 211, 223, 325, 326, 327, 328, 329, 330, 331, 334, 335, 336, 337, 338, 339, 340, 341, 342, 343, 345
fairness, 137
family, 4, 48, 172
fault tolerance, 4, 5, 326, 328, 329, 333
Federal Communications Commission (FCC), xi, 301, 302, 319, 322, 323
feedback, 259, 283
FEM, 242, 243, 252, 254
filament, 188
filters, 76
Finland, 162
flexibility, ix, 12, 23, 191, 192, 209, 210, 234, 354, 367
flight, 113
flood(ing), 6, 12, 70, 74, 166, 172
flow value, 138
fluctuations, 292
fluid, 106
forecasting, 190
fragmentation, 272
France, 51, 165, 166, 168, 170, 172, 174, 176, 178, 180, 182, 184, 186, 188, 190, 263, 287
freedom, 4, 196, 232
freedom of choice, 232
functional architecture, 353

G

G - Network, 193
gastrointestinal tract, 231
generation, viii, xi, 15, 83, 84, 103, 106, 118, 119, 120, 184, 272, 273, 274, 281, 287, 301, 302, 350
GERAN, x, 287, 288, 290, 294, 295, 296
Germany, 50, 52
glass, 235, 239, 240, 248, 250, 251, 252, 253, 254, 255, 256, 262
global mobility, 91
goals, 24, 101, 158, 215
gold, 49
gossip, 11, 12, 14, 16
government, 7
GPS, 171, 187, 303, 323
graph, 25, 169, 173, 174, 175, 181, 194
Greece, 267, 346
grid computing, 5
grouping, 289
groups, 137, 180
growth, 31

H

habitat, 190
handoff, vii, viii, 83, 85, 86, 87, 88, 89, 90, 91, 92, 93, 94, 95, 96, 97, 98, 99, 100, 101, 102, 103,

104, 105, 108, 109, 110, 111, 112, 113, 114, 115, 116, 117, 118, 119, 120, 195
harm, 67
harvesting, 233, 260
Hawaii, 50
HE, 357, 358, 359
health care, ix, 229, 231
heterogeneity, viii, 83, 84, 85
homogeneity, 2
Hong Kong, 80, 187, 188, 190
Hops, 10
host, 4, 7, 185, 210, 215, 247, 267
hotels, 325
housing, 235
humidity, 230
hybrid, x, 6, 76, 104, 118, 120, 160, 267, 268, 283
hypothesis, 214

I

identification, 90, 273, 275, 357
identity, 56, 64, 66, 67, 74, 79, 92, 98, 119, 275, 334
imaging, xi, 301
imaging systems, xi, 301
immunity, xi, 302
implants, 230, 233, 258
implementation, viii, x, 7, 55, 57, 58, 86, 88, 93, 95, 98, 135, 143, 148, 177, 195, 210, 211, 219, 222, 248, 251, 267, 269, 282, 284, 353, 368
incompatibility, 98
independence, 290, 353
indexing, viii, 57, 61, 64, 65, 79
indication, 232
industry, vii
inefficiency, 100
infinite, 66
information exchange, 132, 150, 160
information sharing, 51
infrastructure, ix, 2, 48, 56, 74, 76, 93, 129, 143, 166, 182, 189, 191, 192, 193, 195, 196
initiation, 87, 97, 104, 119
input, 13, 18, 19, 33, 36, 38, 42, 148, 174, 178, 198, 201, 205, 207, 209, 214, 215, 217, 218, 243, 245, 275
insight, 196, 222
instability, 100
insulation, 244
integrated circuits, viii, x, 123, 129, 230, 235
integration, viii, ix, x, 83, 85, 92, 98, 99, 229, 230, 232, 233, 234, 235, 247, 248, 250, 251, 254, 258, 261, 262, 354, 356
integrity, 59, 60, 61, 63, 70, 188, 268, 269, 276, 282, 283
intelligence, 368
intensity, 342, 343
interaction(s), 98, 99, 100, 214, 288
interface, 7, 8, 88, 99, 100, 102, 120, 123, 124, 129, 131, 132, 149, 160, 168, 177, 233, 235, 257, 260, 327, 351, 352, 355, 357, 359
interference, x, xi, 98, 124, 222, 235, 245, 287, 288, 289, 290, 291, 292, 297, 298, 301, 302, 303, 304, 305, 306, 307, 308, 309, 310, 311, 312, 313, 314, 315, 316, 317, 323, 328
Internet, vii, x, 1, 4, 5, 6, 21, 22, 49, 50, 52, 74, 80, 82, 85, 90, 101, 105, 118, 121, 196, 224, 267, 268, 269, 284, 285
interoperability, xii, 349, 351
interval, 14, 86, 88, 90, 132, 154, 155, 156, 157, 158, 334, 340, 345
intestine, 231
IP address, 4, 6, 10, 22, 24, 87, 88, 92, 93, 94, 95, 97, 105, 110
IP networks, 2, 4, 22, 84, 85, 93, 105, 109, 121
iris, 51
Israel, 7
Italy, 49, 52, 165, 187, 190, 265

J

Japan, 302
Java, 37, 40
joint admission, 288
JRRM, x, 287, 288, 294, 298

K

kindergarten, 190

L

language, 177, 178, 281
laptop, 129, 201
laser ablation, 257
latency, viii, 15, 83, 84, 88, 90, 91, 92, 93, 94, 96, 97, 102, 104, 105, 110, 111, 112, 113, 115, 116, 118, 128, 129, 130, 131, 133, 134, 135, 136, 144, 145, 151, 155, 158, 159, 160, 161, 204, 282
Latin America, 188
learning, 194
lesions, 231
life cycle, 351
lifestyle, 263
lifetime, x, 101, 107, 137, 169, 184, 230, 274, 334
limitation, 177
linkage, 195

links, 12, 59, 65, 92, 111, 173, 176, 192, 193, 194, 195, 196, 199, 202, 209, 212, 214, 215, 217, 223
listening, 124, 133, 134, 135, 136, 140, 141, 142, 143, 145, 147, 149, 153, 155, 156, 159
load balance, 9, 184, 186
local mobility, 91
localization, 151, 170, 171, 187, 190, 351
location, ix, 12, 14, 52, 56, 57, 63, 64, 76, 85, 86, 87, 90, 92, 93, 94, 101, 108, 109, 122, 130, 153, 166, 170, 171, 179, 181, 182, 189, 191, 207, 315, 316, 328, 331, 335, 342, 343, 355, 357
location information, 87, 101, 189, 331, 343, 355
longevity, 99, 195
Louisiana, 50
low temperatures, 265

M

MAC protocols, ix, 22, 123, 124, 125, 129, 130, 131, 133, 144, 148, 149, 155, 158, 161
machinery, 166
magnetic field, 259, 265
magnetic sensor, 260, 262
magnetization, 258
management, vii, viii, xii, 55, 83, 84, 85, 86, 87, 88, 89, 90, 91, 92, 93, 94, 98, 100, 101, 103, 104, 105, 108, 110, 111, 115, 116, 118, 119, 122, 129, 132, 133, 138, 146, 158, 166, 167, 168, 169, 185, 187, 327, 349, 350, 351, 353, 355
MANETs, vii, 1, 2, 3, 4, 16, 21, 22, 23, 24, 27, 30, 32, 45, 48, 53, 55, 56
mapping, 23, 60, 65, 66, 109, 110, 115, 184
market, x, 2, 230
Markov chain, 295
matrix, 172, 193, 202, 204, 205, 208, 209, 216, 223, 295
maturation, 196
measurement, 100, 137, 236, 239, 256
measures, x, 13, 26, 56, 70, 79, 137, 167, 198, 259, 287, 288, 295, 296
media, 282, 283
median, 149, 183
medium access control, ix, 134, 158, 160
membership, 15, 57
memory, 5, 23, 73, 133, 172, 179, 269
men, 231
message passing, 16, 24
messages, 9, 12, 14, 16, 18, 26-33, 36, 38, 39, 42, 45, 47, 61, 70, 90-92, 94-95, 97-98, 100, 103, 105, 110, 115, 125-126, 131-132, 135, 142, 144-147, 154-155, 158, 169, 171, 174, 176, 180-182, 184, 269, 270-271, 273-274, 276-278, 283, 336
Miami, 80, 187
Microsoft, 7
microstructures, 265
microwaves, 234
migration, 118
military, ix, 56, 166, 191, 192, 196
miniaturization, ix, x, 229, 232, 233, 235, 248
mixing, 70
mobile communication, x, xi, 213, 267, 301, 304, 350, 367, 368
mobile device, 52, 87, 103, 268
Mobile Node (MN), 195
mobile phone, 268, 302, 303
mobility, viii, ix, xii, 32, 44, 55, 57, 59, 72, 73, 76, 79, 83, 84, 85, 86, 87, 88, 89, 90, 91, 92, 93, 94, 97, 98, 100, 101, 102, 104, 105, 106, 108, 110, 114, 115, 116, 118, 119, 120, 130, 144, 181, 191, 192, 193, 195, 196, 207, 208, 210, 211, 213, 215, 216, 217, 287, 299, 327, 349, 350, 351, 353, 355, 356, 367, 368, 369
modeling, 122, 192, 195, 243, 252, 265
models, ix, x, 21, 85, 102, 106, 165, 167, 168, 170, 179, 185, 186, 194, 195, 196, 197, 242, 251, 280, 287, 288, 294, 298
modules, 124, 230, 231
modulus, 20
momentum, 350
money, 171
monopoly, 23
Monte Carlo method, 194, 213
Moon, 121
motion, 195, 259
movement, vii, 1, 22, 86, 87, 88, 89, 90, 94, 95, 105, 106, 107, 108, 111, 144, 155, 196, 259, 343
MTS, 145, 146
multimedia, vii, viii, 83, 84, 91, 101, 121, 122
multimedia services, 84
music, 4, 5, 6

N

naming, 7
nation, 61, 67, 176, 179
natural disasters, 166
natural evolution, 167
neglect, 149
negotiating, 272
negotiation, 103, 104, 121, 122
network(ing), vii-ix, xi, 1-5, 7-18, 20-28, 31, 44-45, 47, 55-58, 61, 66-67, 70-74, 76, 79, 84-94, 96-107, 109, 115, 118, 121, 124-141, 143-145, 147-148, 150-155, 157-161, 165-174, 176-186, 189, 191-200, 203-211, 214-215, 217-219, 222-224, 236, 239, 245, 257, 267, 273-274, 277-278, 280-

281, 283, 287-289, 294-295, 298, 303, 325-329, 335, 341, 349-355, 367-369
network elements, viii, 55, 56, 61, 70, 76
next generation, viii, 83, 118, 121
nl - Number of linked pairs, 199
nodes, viii, 2-3, 5-10, 12-16, 18, 20-33, 36-39, 42, 45, 47, 55-70, 72-74, 76, 79, 86, 90, 101-102, 104-105, 109, 123-161, 166-183, 185, 192-196, 199, 201-206, 208-219, 223-224, 350
noise, xi, 66, 101, 149, 197, 217, 235, 297, 301, 302, 305, 306, 309, 310, 312, 313, 314, 315, 316, 318, 320
normal distribution, 217, 220, 308
numerical analysis, 342, 345
nutrition, 230

O

observations, 72
one dimension, 13
operating system, 280
operator, xi, 87, 298, 349, 351, 352, 361
optical communications, 233
optimization, 15, 22, 90, 116, 144, 176, 178, 192, 197
organ, 185
organization, 2, 124, 125, 126, 128, 148, 319
orientation, 259
originality, 192
orthogonality, 297
output, 13, 124, 178, 182, 209
oxide thickness, 246

P

Pacific, 284, 303
packaging, x, 230, 247, 250, 262
packet forwarding, 76, 110, 112
pairing, 76
parameter, 6, 46, 93, 132, 138, 182, 197, 223, 236, 239, 242, 253, 272, 273, 274, 275, 276, 279, 283, 337
parents, 27, 155, 156, 157
Pareto, 106
partition, 3, 183, 288, 289
passive, 14, 56, 70, 89, 102, 147, 148, 235, 247, 250
patterning, 258
pause time, 73
PCM, 355, 358
peer review, 207
peers, 4, 5, 6, 16, 18, 19, 36, 269, 270
PEP, 100
permeability, 236, 238
permittivity, 236, 237, 238, 239, 240, 241, 243, 251
personal communication, vii, 351
PET, 81
PHB, 101
physical properties, 197
planning, xi, 155, 325
plasma, 255
poison, 166
polarization, 245
polling, 153
pollution, 166
polyimide, 248
poor, 4, 10, 27, 48, 166, 168, 169, 185, 272
portability, 350, 356
Portugal, 187, 229, 263
power, viii, x, xi, 2, 5, 22, 28, 52, 84, 88, 89, 99, 100, 103, 123, 124, 125, 129, 130, 131, 132, 133, 136, 137, 138, 139, 140, 143, 144, 147, 149, 151, 153, 154, 155, 158, 159, 160, 171, 177, 197, 215, 217, 223, 224, 230, 232, 233, 234, 245, 260, 261, 269, 290, 292, 293, 295, 297, 301, 302, 303, 304, 305, 307, 308, 310, 313, 316, 318, 320, 322
predicate, 178
prediction, 115, 116, 117, 146
preference, 12, 357
pressure, 260
privacy, 4, 56, 57, 64, 76, 78, 93, 268, 269, 282
probability, 2, 10, 66, 68, 105, 106, 107, 108, 109, 110, 112, 115, 116, 117, 145, 146, 147, 151, 154, 161, 179, 184, 193, 194, 195, 196, 197, 198, 199, 200, 201, 205, 206, 207, 211, 216, 223, 288, 289, 290, 296, 297, 298, 336, 337, 339, 345, 360, 361
probability density function, 106
probability distribution, 107
probe, 132, 236, 239, 334, 335
probe station, 236, 239
producers, 4
production, x, 166, 230
profit, 160
program, 12, 20, 21, 167, 168, 169, 176, 185, 186, 201
programming, 44, 167, 168, 177
proliferation, ix, 191, 196
propagation, 16, 169, 171, 197, 237, 258, 288, 304, 305, 307, 308, 309, 310, 312, 318, 322, 323
prosthesis, ix, 229, 234
protocol, vii-viii, x, 1, 3-4, 11-12, 14-21, 24, 26, 32, 44, 48-49, 55-57, 59, 72-74, 76, 85, 91, 93, 95, 97, 98, 101-104, 108, 110, 118-125, 128-132, 136-140, 142-144, 148-151, 153-155, 158-159, 169-170, 174-175, 179, 181-182, 185, 188, 196, 267-279, 281, 283-285, 327, 352

prototype, 244, 256, 257
pulse, xi, 301, 303

Q

QoS, viii, 56, 83, 84, 85, 86, 88, 90, 96, 100, 101, 102, 103, 104, 105, 108, 116, 118, 121, 185, 186, 294, 297
quality of life, ix, 229
quality of service, viii, 52, 83, 121, 129, 130, 274, 277
quartz, 235
query, 6, 7, 167, 168, 169, 170, 171, 177, 178, 180, 182, 187, 189

R

radiation, 235, 238, 240, 242, 245, 251, 253, 261, 262, 263, 265, 322
radio, viii, xi, 21, 22, 45, 49, 83, 86, 88, 89, 98, 99, 100, 103, 104, 121, 123, 124, 125, 129, 131, 132, 138, 140, 146, 148, 149, 160, 166, 171, 207, 210, 218, 230, 233, 247, 264, 267, 287, 301, 302, 303, 327, 328, 338, 345
radius, 248, 297
random numbers, 273
range, viii, xi, 12, 13, 21, 22, 44, 45, 49, 55, 57, 86, 143, 166, 171, 172, 173, 176, 181, 182, 184, 207, 208, 209, 210, 213, 214, 215, 218, 223, 230, 231, 232, 234, 236, 242, 244, 245, 259, 260, 262, 263, 269, 301, 303, 304, 305, 306, 307, 308, 309, 310, 311, 312, 313, 314, 315, 316, 317, 318, 319, 320, 321, 322, 325, 326, 327, 328, 329, 330, 331, 334, 335, 337, 338, 339, 340, 341, 343
ratings, 177
reading, 259
real time, x, 268, 274, 275, 277, 287, 297
reality, 129
reasoning, 20
reception, 95, 105, 139, 141, 145, 169
recovery, 3, 12, 268, 336, 342
reduction, 118, 124, 129, 130, 131, 136, 154, 155, 158, 194, 235, 248, 250, 251, 253, 254, 261, 262, 310, 311, 312, 313, 314, 315, 316, 319, 320, 322, 357
redundancy, xi, 12, 92, 179, 185, 186, 325
refining, 12
reflection, 244
reforms, 192
Registry, 355
regulations, 302
reinforcement, 169
relationship(s), 2, 4, 5, 14, 155, 156, 192, 197, 206, 214, 220
relatives, 188
reliability, ix, 87, 101, 166, 167, 179, 185, 191, 192, 193, 194, 195, 196, 198, 199, 200, 201, 205, 206, 207, 208, 209, 210, 211, 213, 214, 215, 216, 217, 218, 219, 220, 222, 223, 225
repair, 169
replacement, 207
replication, 180, 185, 186, 326, 328, 340, 341, 342
reputation, 15
reservation protocol, 121
residential buildings, 106
resistance, 60
resolution, 215, 302
resource management, 104, 369
resources, 4, 5, 56, 85, 87, 89, 97, 99, 100, 101, 102, 103, 104, 125, 131, 133, 138, 148, 160, 168, 186, 288, 298, 343, 345, 352, 356
response time, 104, 269, 357, 358
retention, 97
returns, 7, 14, 17, 20, 36, 39, 42, 48, 132, 146, 171, 174, 179, 181, 359
reusability, 99
rings, vii, 1, 31, 45, 48, 176, 297
robustness, 3, 12, 23, 87, 101, 120
routing, vii, viii, 1, 2, 3, 5, 7, 8, 9, 22, 23, 24, 36, 39, 42, 48, 51, 52, 55, 56, 57, 59, 60, 61, 62, 63, 64, 65, 66, 67, 69, 70, 71, 72, 73, 74, 76, 77, 79, 81, 82, 86, 94, 96, 99, 101, 105, 109, 110, 137, 144, 147, 151, 161, 168, 169, 170, 172, 173, 174, 175, 176, 177, 178, 181, 182, 184, 186, 187, 188, 189
RTS, 131, 132, 133, 134, 135, 141, 142, 145, 149, 152, 158

S

safety, 192, 196
sample, 177, 203, 238, 265
sampling, ix, 142, 143, 165, 168, 177, 178, 195
satellite, 267, 303
satellite service, 303
satisfaction, viii, 83, 101
savings, 129, 230
scalability, vii, 3, 4, 6, 9, 55, 56, 73, 74, 76, 84, 86, 87, 91, 96, 98, 101, 131, 350, 367
scaling, 5, 172, 180, 188
scarcity, 85, 102
scattering, 236
scheduling, 101, 126
schema, 147, 148, 214
SCP, 153, 154, 352, 354

search(es), 5, 6, 7, 9, 10, 16, 17, 18, 21, 26, 27, 30, 31, 32, 33, 34, 35, 40, 48, 49, 72, 115, 132, 133, 187, 202, 204, 209, 217, 334, 359
searching, 20, 23, 25, 26, 27, 30, 31, 32, 33, 35, 36, 39, 40, 42, 48, 49, 133
second generation, 7, 303, 350
secure communication, 76
security, vii, x, xii, 55, 56, 57, 59, 61, 63, 71, 74, 76, 79, 85, 86, 89, 96, 99, 105, 267, 268, 269, 270, 271, 272, 273, 281, 282, 283, 284, 285, 349, 351, 355, 357
seed, 3
segmentation, 222
selecting, 175, 294, 330
selectivity, 178
self-organization, 12
semantics, 64
semiconductor, x, 229, 258
sensing, x, 126, 152, 166, 180, 229, 230, 232, 259, 260
sensitivity, 197, 205, 207, 258, 259, 272, 305
sensor nodes, 143, 165, 167, 168, 181
sensors, ix, x, 128, 137, 144, 147, 160, 165, 166, 167, 168, 169, 170, 171, 172, 174, 176, 177, 178, 180, 182, 183, 185, 186, 190, 229, 245, 258, 265
separation, 2, 208, 215, 216, 304, 307, 308, 310, 311, 312, 313, 314, 315, 316, 317
Serbia, 187
series, ix, 18, 37, 40, 42, 123, 128, 195, 259
service provider, xi, 84, 89, 96, 100, 101, 349, 350, 351, 352, 353, 355, 356, 359
severity, 276, 283
shape, 106, 170, 217, 236, 255
shaping, 255
shares, 76, 152
sharing, 3, 4, 5, 6, 7, 8, 70, 129, 149, 172, 289
sign, 84, 274
signalling, 131, 132, 133
signals, xi, 67, 172, 301, 329
silicon, x, 229, 234, 242, 244, 248, 251, 258, 262
similarity, 177
simulation, ix, 12, 15, 21, 31, 32, 44, 45, 47, 48, 49, 71, 72, 73, 74, 76, 79, 120, 191, 193, 194, 195, 198, 202, 203, 205, 206, 209, 210, 211, 212, 216, 217, 218, 219, 220, 223, 243, 246, 251, 252, 258, 325, 326, 342, 343, 344, 345
sites, 303
slaves, 4
small intestine, 231
SMS, 352, 358
social costs, 231
software, 329, 342, 350
soil, 230
SPA, 355, 359
Spain, 51, 123, 301
spectrum, vii, xi, 263, 301, 302, 303, 322
spectrum allocation, vii, 302
speech, 303
speed, 12, 15, 17, 18, 40, 73, 75, 89, 106, 144, 193, 195, 207, 267, 302
spinal cord, 231, 232, 258
sputtering, 254, 256
stabilization, vii, 1, 3, 22, 24
stages, 222, 331
standard deviation, 217, 220, 308
standardization, 100, 104, 233
standards, 171, 264, 274, 351
statistics, 212
stimulus information, 234
stomach, 231
storage, ix, 4, 5, 7, 14, 23, 26, 31, 32, 96, 101, 165, 167, 168, 169, 170, 173, 179, 182, 185, 186, 190
strategies, x, 123, 167, 170, 178, 185, 186, 269, 287, 288, 294, 296, 297, 298
streams, 177, 178, 185
strength, 87, 89, 99, 101, 144, 171, 233, 327, 334
stress, 192
subgroups, 5
subscribers, 303, 350, 352, 357
substitution, 2, 166
substrates, x, 7, 229, 235, 236, 240, 248, 251, 252, 256
Sun, 52
supply, 151, 230
surgical intervention, 232
survival, 84, 126, 329, 330, 334, 336, 337, 338, 339, 340, 341, 342, 343, 344, 345
susceptibility, 198
Sweden, 263
switching, 105, 124, 149, 328, 334, 356
Switzerland, 7, 53
symmetry, 26
symptoms, 231
synchronization, 14, 15, 97, 98, 124, 129, 134, 143, 146, 149, 152, 153, 154, 155, 157, 159, 160, 268, 271, 279, 283
systems, vii, ix, x, xi, 1, 2, 3, 4, 5, 7, 21, 24, 48, 52, 53, 70, 85, 118, 166, 167, 170, 171, 172, 179, 185, 186, 187, 190, 191, 192, 198, 205, 207, 211, 213, 217, 222, 229, 230, 231, 233, 234, 258, 267, 287, 288, 294, 296, 297, 298, 299, 301, 302, 303, 309, 318, 322, 323, 350, 353, 354, 367, 368

T

Taiwan, 187, 325, 349

targets, 11, 214
TCP, 51, 89, 92, 93, 98, 103, 196, 224, 281
TCP/IP - Transmission Control Protocol/Internet Protocol, 196
technology, vii, viii, ix, x, 2, 83, 85, 87, 88, 89, 165, 166, 188, 191, 192, 196, 213, 222, 229, 230, 231, 232, 234, 235, 266, 325, 327, 350, 367, 368
telecommunications, 350
TEM, 236
temperature, 168, 177, 178, 184, 230
terminals, 4, 84, 86, 88, 194, 217, 218, 219, 269, 350, 356
theory, 12, 48, 111, 341
therapy, ix, x, 229, 230
thinking, 141
threat, 5
threshold, 59, 160, 298, 334, 337, 343, 345
TIA, 303
time, viii, 2-7, 12, 14, 15, 18, 20-22, 26-29, 31-32, 37, 39-40, 42, 45, 48-49, 55, 57, 60, 70, 72-75, 83-84, 86-93, 95-96, 100-108, 111-112, 115, 117, 121-126, 128-133, 135-136, 138-141, 143-145, 147-161, 166, 169-171, 177, 180, 185-186, 192-195, 197-198, 203-204, 207-213, 216-218, 230, 249, 251, 253, 258-259, 284, 288-289, 295, 297-298, 302, 328, 335-340, 342-344, 357-359
time increment, 208, 209, 210, 217, 218
time use, 45
time variables, 106
timing, 107, 135, 137, 145, 146
tissue, x, 229, 232, 233
topology, vii, 1, 2, 3, 5, 6, 8, 11, 12, 14, 15, 16, 17, 20, 21, 23, 24, 32, 40, 48, 50, 53, 73, 74, 111, 126, 128, 129, 144, 145, 146, 147, 148, 151, 155, 157, 158, 160, 170, 171, 173, 180, 196, 198
topology management, 50, 144, 155, 157, 158
torus, 8, 13
TPA, 355, 357, 358, 359
tracking, vii, 9, 188
trade, 26, 89, 100, 101, 154
trade-off, 26
traffic, vii, xii, 1, 56, 57, 59, 66, 70, 73, 79, 86, 87, 90, 91, 94, 98, 101, 105, 106, 108, 112, 116, 122, 130, 137, 138, 139, 140, 141, 143, 146, 148, 150, 151, 154, 155, 158, 159, 161, 206, 269, 273, 275, 288, 289, 338, 349, 350, 353, 358, 360, 362, 363, 365, 366, 367
trajectory, 144
transactions, 74
transducer, 178
transformation(s), 13, 351
transition(s), 104, 131, 146, 195, 295
transition rate, 295
translation, 282
transmission, xi, 65, 92, 101, 105, 109, 110, 119, 126, 128, 132, 133, 134, 135, 136, 139, 140, 141, 142, 143, 144, 145, 147, 150, 151, 152, 153, 155, 156, 173, 178, 197, 207, 209, 210, 214, 215, 217, 218, 223, 234, 236, 242, 247, 268, 269, 271, 272, 274, 277, 279, 281, 289, 292, 293, 295, 297, 301, 335, 336, 340, 342, 345
transmits, 76, 141, 269
transplantation, vii, 1, 2
transport, x, 93, 119, 267, 268, 269, 281, 283, 284, 285
transportation, 12
trees, 173, 177, 187
trend, viii, 2, 83, 123, 220, 350, 367
triangulation, 173, 175
triggers, 94, 95, 120, 359
tumours, 231
tunneling, 97, 118
Two-Terminal Reliability (2TR), 194

U

UK, 7
uncertainty, 66, 69, 194, 203
uniform, 9, 13, 25, 27, 45, 48, 156, 179, 182, 184, 185, 186, 193, 202, 211
unions, 178
universities, 325
updating, 358
uplink, 302, 303
urinary bladder, 263
urinary tract, 231
urine, 231
user data, 47
users, vii, viii, ix, xi, 1, 5, 6, 55, 56, 65, 83, 84, 85, 86, 88, 89, 104, 105, 110, 151, 165, 192, 222, 288, 292, 293, 294, 297, 298, 299, 302, 325, 326, 349, 350, 353, 355, 356, 357, 359, 367
UTRAN, x, 287, 288, 294, 295, 296, 351, 355, 356
UV, 254

V

validation, 246
validity, 60, 274, 334
values, 61, 63, 101, 111, 116, 159, 173, 176, 178, 184, 201, 203, 207, 214, 235, 239, 240, 242, 245, 246, 251, 253, 257, 258, 270, 283, 295, 297, 310, 312, 313, 315, 322
variability, 185

variable(s), 18, 32, 41, 67, 85, 102, 106, 108, 111, 130, 150, 157, 158, 194, 215, 223, 277, 308
variance, 106, 194, 203, 213
variation, 21, 146, 196, 203, 220, 239, 241, 271, 362, 365
vector, 13, 36, 37, 39, 40, 42, 52, 172, 189, 202, 223, 236, 239, 245, 257, 275, 290, 295
velocity, 208, 209, 210, 211, 215, 217
versatility, 12
Virtual Home Environment (VHE), xi, 349, 350, 367
viscosity, 258, 265
vision, 102, 167
voice, 84, 94, 192, 267, 289, 293, 294, 295, 296, 297, 310, 352, 358
volatility, 209, 212, 215, 220
voting, 138

W

waking, 149, 153
wave number, 238
weakness, 49
web, 74, 268, 280, 281, 282
Weibull distribution, 211
well-being, 231
windows, 154
wireless connectivity, 302, 328, 329, 330, 334, 345
wireless devices, x, 165, 230, 232
wireless LANs, 325
wireless networking, 195
wireless networks, viii, xi, 49, 51, 52, 83, 84, 85, 86, 87, 88, 89, 91, 98, 103, 105, 106, 107, 116, 118, 122, 125, 127, 129, 131, 132, 192, 195, 281, 301
wireless sensor networks, 124, 125, 130, 131, 133, 141, 143, 148, 160, 166, 167, 170, 174, 186, 187, 189, 230
wireless systems, 87, 118
wireless technology, 84
wires, 257
WLANs, xi, 120, 325
women, 231
workload, 280, 328, 336, 341
workstation, 177
World Wide Web, 4, 269
WWW, 51, 269

Y

yield, 49